Biophysical Techniques in Biosciences

This book details the latest advancements in spectroscopic, analytical, and imaging techniques, emphasizing their crucial roles in both research and biomedical diagnostics. The initial chapters introduce the fundamental principles of the techniques, highlighting the use of optical spectroscopies for disease diagnosis, such as oral cancer. The book also explores their innovative applications, such as quantitative optical phase imaging, and the examination of biopolymers like starch through spectroscopy and microscopy. Further, the book discusses cutting-edge developments in biomaterials essential for understanding tissue engineering and the innovative use of synthesized bioactive glasses. The chapters also examine revolutionary methods such as HPLC and HPTLC techniques for detailed analysis at unprecedented scales and for observing various processes in health and disease. Importantly, the book reviews the impact of machine learning in enhancing the accuracy of disease diagnoses through nonlinear optical microscopy. The book also presents technological breakthroughs in the transformative impact of these techniques in developing diagnostic and therapeutic solutions. This book is intended for students, researchers, and professionals in biophysics, medical imaging, and biomedical engineering.

Key Features:

- Highlights innovative applications such as quantitative optical phase imaging and the use of spectroscopy in disease diagnosis
- Explores the fundamental principles of advanced spectroscopic and imaging techniques
- Demonstrates the role of new technologies like synthesized biomaterials and applications of HPLC techniques
- Discusses the integration of machine learning with nonlinear optical microscopy to enhance the accuracy of disease diagnoses
- Presents the latest developments in biomaterials that are revolutionizing tissue engineering

Biophysical Techniques in Biosciences

From Fundamentals to Advanced Applications

Edited by
Nirmal Mazumder, Rajib Biswas, Guan-Yu Zhuo

CRC Press
Taylor & Francis Group
Boca Raton London New York

CRC Press is an imprint of the
Taylor & Francis Group, an informa business

Designed cover image: Shutterstock

First edition published 2025
by CRC Press
2385 NW Executive Center Drive, Suite 320, Boca Raton FL 33431

and by CRC Press
4 Park Square, Milton Park, Abingdon, Oxon, OX14 4RN

CRC Press is an imprint of Taylor & Francis Group, LLC

ISBN: 9781032899831 (hbk)
ISBN: 9781041007654 (pbk)
ISBN: 9781003545712 (ebk)

DOI: 10.1201/9781003545712

Typeset in Times
by Deanta Global Publishing Services, Chennai, India

Contents

Preface

The field of biosciences is evolving at a rapid pace, driven by the continuous advancement of biophysical and other bioanalytical techniques. These techniques have revolutionized our understanding of biological systems, providing unprecedented insights into the molecular and cellular processes that underpin life. Furthermore, understanding the physicochemical properties of biomolecules and biomaterials is crucial for expanding their applications and uncovering their potential in the biomedical field. Characteristics such as viscosity, solubility, and mechanical properties are influenced by factors like composition, chain structure (e.g., linear or branched), linkage patterns, and molecular weight. These properties can be investigated using advanced biophysical techniques, including scanning electron microscopy (SEM), Fourier transform infrared (FT-IR) spectroscopy, X-ray diffraction (XRD), Raman scattering, and optical microscopy.

In *Biophysical Techniques in Biosciences: From Fundamentals to Advanced Applications*, we aim to bridge the gap between fundamental principles, biophysical techniques, and modern biomedical applications. Each chapter is meticulously crafted to provide a balance between theory and practice, with real-world examples and case studies illustrating the application of these techniques in contemporary research. In the first chapter, titled "Cancer Metabolism: Advanced Biophysical Techniques for Diagnosis and Research," authored by Babu et al., the authors discuss the various biophysical techniques employed for the diagnosis and research of cancer metabolism, thus illuminating the critical role of biophysical techniques in cancer biology. Spectroscopy has revolutionized the diagnosis and understanding of biological and medical disorders by providing detailed insights into the composition, structure, and dynamics of biomolecules. Spectroscopic techniques such as X-ray fluorescence, magnetic resonance imaging (MRI) spectroscopy, Raman spectroscopy, FT-IR spectroscopy, reflectance spectroscopy, fluorescence spectroscopy, infrared (IR) spectroscopy, photoacoustic spectroscopy, and laser-induced breakdown spectroscopy (LIBS) are frequently used to study biological macromolecules through molecular fingerprinting. The second chapter, titled "Applications of Spectroscopy in Biology and Medicine," written by Nidhi et al., highlights the applications of these spectroscopy techniques in the assessment of breast cancer, along with a detailed discussion on the currently available spectroscopy-based commercial devices for breast cancer diagnosis. Apart from breast cancer, oral squamous cell carcinomas and other malignant oral mucosa abnormalities are highly prevalent and can be fatal. Early diagnosis and intervention are crucial for improving survival rates; however, conventional methods have limitations. In the third chapter, titled "Advances and Challenges with Oral Cancer Diagnosis," the authors Nanda et al. discuss the recent advancements in omics and technology that demonstrate a precise diagnosis of oral cancer. Furthermore, with the increasing incidence of breast cancer, advanced screening methods like photoacoustic assessments are gaining popularity to address the limitations of standard techniques. Photoacoustic signals, induced by photons, provide biochemical information along with high-resolution, and high-contrast images of the tissue samples. Devika et al., in the fourth chapter, titled "Current Updates on Photoacoustic-Based Techniques for Breast Cancer Diagnosis," review the design, functions, and applications of photoacoustic-based instruments, highlighting their potential for early breast cancer screening and monitoring. Moving on to the study of biomaterials, polysaccharides, composed of multiple monosaccharides, play essential roles in bacterial functions. Research in applied microbiology focuses on identifying novel bacterial strains and characterizing their exopolysaccharides for biological activities like anticancer, antibacterial, and immunomodulatory effects. The fifth chapter, titled "Characterization and Identification of Novel Bacterial Exopolysaccharides" authored by Prabhune et al., discusses the techniques employed for the characterization of bacterial strains and the study on the exopolysaccharides. Furthermore, the authors elaborate on the challenges and the need for standardized protocols in such research. Recent research trends highlight the shift from traditional methods to tissue engineering techniques for solutions to various problem statements in biosciences. This shift is highlighted in the sixth and seventh chapters. The sixth chapter, titled "Bioactive Glasses: Synthesis, Characterization, and Bioactivity," authored by Goel et al., discusses the dissolution behavior of bioactive glasses and the techniques to control their bioactivity to develop effective bone graft materials. The seventh chapter, titled "Eggshell Membrane as Biomaterial for Tissue Engineering" by Dugar et al., explores eggshell membrane extraction techniques and their applications in designing functional scaffolds for tissue regeneration. Imaging techniques are an integral part of biosciences, offering detailed insights into biological structures and functions. These techniques enable precise analysis and visualization, advancing research and diagnostic capabilities in the field. Quantitative optical phase imaging techniques provide 3D structural and functional insights into transparent specimens by measuring optical path length changes. The eighth and ninth chapters, authored by Pillai et al., focus on interference-based techniques like digital holographic microscopy and spatial light interference microscopy, which use interference to convert phase information into intensity variations. The chapters also discuss the principles, applications, and advantages of these techniques in biomedical imaging. Interference-based techniques are discussed in the eighth chapter, while non-interference-based imaging is highlighted in the ninth chapter. Nonlinear optical microscopy (NLO) technologies like stimulated Raman scattering (SRS), coherent anti-Stokes Raman scattering (CARS), two photon excitation fluorescence (TPEF), second harmonic generation (SHG), and third harmonic generation (THG) enable label-free, deep optical sectioning. These techniques offer advantages like depth penetration, ultrahigh resolution, and high sensitivity, making NLO imaging valuable for applications in biomedical research. The tenth chapter, titled "Machine Learning Approach in Nonlinear Optical Microscopy," by Amin et al., discusses the various NLO microscopy techniques along with detailed elucidation of the analysis of images obtained from NLO microscopy using a machine learning approach. Obsessive-compulsive disorder (OCD) is a neuropsychological disease characterized by intrusive obsessions and compulsions, causing functional impairment and neural disruption. Biosciences play a crucial role in understanding OCD's underlying mechanisms, developing effective treatments, and creating animal models to study its pathology. Jijimon et al. in the 11th chapter, titled "Animal Models of Obsessive-Compulsive Disorder," examine the popular animal models used in OCD research, their associated phenotypes, along with a critical review of the transgenic, pharmacological, behavioral, and neurodevelopmental models of OCD. Chromatography is a vital analytical technique in biosciences, enabling the separation, identification, and quantification of biomolecules. Widely used in research and clinical diagnostics, chromatography and its variations play a crucial role in studying proteins, nucleic acids, metabolites, and other biological compounds. Its precision and versatility make it indispensable for advancing our understanding of biological processes and developing new therapeutic strategies. The next two chapters discuss the instrumentation and applications of advanced chromatographic techniques. Chapter 12, titled "Multidimensional HPTLC: Instrumentation, Techniques and Applications" by Pai et al., discusses various aspects of high-performance thin-layer chromatography (HPTLC), which is an advanced analytical technique based on traditional thin-layer chromatography. Chapter 13, titled "High-Performance Liquid Chromatography (HPLC): Analytical Technique for Component Identification and Purification Across Diverse Industries," by Dey et al., critically analyzes the HPLC methodologies developed and implemented for the identification and purification of compounds fields such as pharmaceuticals, food industry, drug delivery research, and textile industries. Current research trends emphasize green technology across various applications, with starch – a versatile homopolysaccharide from plants – playing a key role due to its remarkable chemical and physical properties. In the 14th chapter, titled "Starch-Based Aanalytical Schemes for Various Applications," the authors, Daimari et al., explore the origin, properties, and modifications of starch to enhance its applications, particularly in sensing technologies. The field shows promising potential for future exploration.

One of the unique features of this book is its interdisciplinary approach. Biophysical techniques are inherently multidisciplinary, intersecting with fields such as physics, chemistry, engineering, and computational biology. We have endeavored to highlight these intersections, demonstrating how the integration of diverse disciplines can lead to innovative solutions and groundbreaking discoveries. This book is intended for a diverse audience, including undergraduate and graduate students, researchers, and professionals in the biosciences. Whether you are a novice seeking to understand the basics or an experienced scientist looking to stay abreast of the latest developments, this book offers valuable knowledge and practical guidance. We are fortunate to have contributions from distinguished experts who bring a wealth of knowledge and experience to this book. Their insights and expertise ensure that each chapter is not only informative but also reflective of the current trends and future directions in the field. We are grateful to the reviewers for devoting their time to provide insightful recommendations and comments on the chapters. We also acknowledge our colleagues who helped us with this volume's suggestions. Bhavik Sawhney (CRC Press, Taylor and Francis Group) deserves special thanks for his contributions to the book. May this book be a source of inspiration, a catalyst for scientific inquiry, and a cornerstone for the ongoing advancement of biophysical research. More importantly, we wish our readers a pleasant and productive browsing experience.

Dr. Nirmal Mazumder, India
Dr. Rajib Biswas, India
Dr. Guan-Yu Zhuo, Taiwan

Editors

Nirmal Mazumder is an associate professor at the Department of Biophysics, Manipal School of Life Sciences at Manipal Academy of Higher Education (MAHE) in India. He earned his Ph.D. in 2013 from the National Yang Ming Chiao Tung University in Taipei, Taiwan. Following his doctoral studies, he served as a postdoctoral fellow at the University of Virginia in the USA and the Italian Institute of Technology in Genoa, Italy, from 2013 to 2016. Dr. Mazumder specializes in the development of nonlinear optical microscopes, including technologies such as two-photon fluorescence, second harmonic generation, and coherent anti-Stokes Raman scattering, all aimed at biomedical applications. His research extends into developing microscopy-spectroscopy techniques, biomedical devices, biophotonics, smart materials, and microfluidics, with particular attention to their biomedical uses. He has authored more than 100 peer-reviewed research articles, 11 books, and 40 book chapters and has been an active participant in numerous national and international conferences through posters, oral presentations, and invited lectures. Dr. Mazumder leads, as the principal investigator, several national and bilateral research projects funded by the Government of India. He also contributes as a reviewer, guest editor, and associate editor for various prestigious journals. He is an active member of various national and international scientific societies and organizations, such as the Optical Society (OPTICA), where he is a senior member, and the International Society for Optical Engineering (SPIE). Additionally, he holds lifetime memberships in the Society of Biological Chemists (India), the Environmental Mutagen Society of India, the Asian Polymer Association, the Indian Laser Association (ILA), and the Research Society for the Study of Diabetes in India (RSDI).

Rajib Biswas is an associate professor in the Department of Physics at the School of Sciences, Tezpur University, India. He completed his Ph.D. at CSIR-NEIST in India and has more than 13 years of teaching experience and 15 years of research experience. Dr. Biswas has an impressive publication record, with 144 articles in peer-reviewed international journals, 23 book chapters, and three patents. He has also edited two books. His research primarily focuses on applied photonics, quantum dots, nanomaterials, and material photonics. Dr. Biswas has secured research funding from various agencies, including the Biotechnology Industry Research Assistance Council (BIRAC) and the Ministry of Earth Sciences (MoES), Government of India, India. He is a Life Member of the Earthquake Society of India and an Emeritus Member of the Optical Society of America. Currently, he holds an associate editor position at *Frontiers in Physics* in the photonics category and serves on the editorial boards of several other peer-reviewed journals.

Guan-Yu Zhuo is an associate professor at the Institute of New Drug Development at China Medical University in Taiwan. He earned his B.S. and M.S. degrees in Physics from National Chung Cheng University, Taiwan, in 2005 and 2007, respectively, and completed his Ph.D. in Physics at National Taiwan University in 2012. Dr. Zhuo's research primarily focuses on developing advanced optical imaging techniques for applications in biophotonics. His long-term goal is to translate these imaging techniques into clinical settings, enhancing diagnostic and therapeutic methods, particularly using endoscopes and non-invasive treatment of collagenous tissues with short-pulse lasers. His work aims to address and overcome existing challenges in diagnosis and therapy with promising demonstrations and tests in clinical environments.

Contributors

Manvitha Amin
Department of Biophysics
Manipal School of Life Sciences, Manipal Academy of Higher Education
Manipal, Karnataka, India

Anand K Asundi
d'Optron Pte Ltd
Singapore

Mani Shankar Babu
Department of Botany
University College
Thiruvananthapuram, Kerala, India

Vidhu Sankar Babu
Department of Plant Sciences
Manipal School of Life Sciences, Manipal Academy of Higher Education
Manipal, Karnataka, India

Shaik Basha
Department of Biophysics
Manipal School of Life Sciences, Manipal Academy of Higher Education
Manipal, Karnataka, India

Shimul Biswas
Department of Biophysics
Manipal School of Life Sciences, Manipal Academy of Higher Education
Manipal, Karnataka, India

Rajib Biswas
Department of Physics
Applied Optics and Photonics Research Laboratory
Tezpur University
Assam, India

Pohar Borah
Department of Physics
Applied Optics and Photonics Research Laboratory
Tezpur University
Assam, India

Nikhil Kumar Daimari
Department of Physics
Applied Optics and Photonics Research Laboratory
Tezpur University
Assam, India

Upama Das
Department of Physics
Applied Optics and Photonics Research Laboratory
Tezpur University
Assam, India

Devika
Department of Biophysics
Manipal School of Life Sciences, Manipal Academy of Higher Education
Manipal, Karnataka, India

Bipasa Dey
Department of Biophysics
Manipal School of Life Sciences, Manipal Academy of Higher Education
Manipal, Karnataka, India

Aman Dugar
Department of Biotechnology
Manipal School of Life Sciences, Manipal Academy of Higher Education
Manipal, Karnataka, India

Gireesh Gangadharan
Department of Ageing Research
Manipal School of Life Sciences, Manipal Academy of Higher Education
Manipal, Karnataka, India

Yash Goel
Department of Biotechnology
Manipal School of Life Sciences, Manipal Academy of Higher Education
Manipal, Karnataka, India

Ameera Hanif
Department of Biophysics
Manipal School of Life Sciences, Manipal Academy of Higher Education
Manipal, Karnataka, India

Feyba Jijimon
Department of Ageing Research
Manipal School of Life Sciences, Manipal Academy of Higher Education
Manipal, Karnataka, India

Manikanth Karnati
Department of Biophysics
Manipal School of Life Sciences, Manipal Academy of Higher Education
Manipal, Karnataka, India

Yury V Kistenev
Laboratory of Laser Molecular Imaging and Machine Learning
Tomsk State University
Tomsk, Russia

Krishna Kishore Mahato
Department of Biophysics
Manipal School of Life Sciences, Manipal Academy of Higher Education
Manipal, Karnataka, India

Nirmal Mazumder
Department of Biophysics
Manipal School of Life Sciences, Manipal Academy of Higher Education
Manipal, Karnataka, India

Sonam Fathima Mehak
Department of Ageing Research
Manipal School of Life Sciences, Manipal Academy of Higher Education
Manipal, Karnataka, India

Parichita Mishra
Department of Biotechnology
Manipal School of Life Sciences, Manipal Academy of Higher Education
Manipal, Karnataka, India

Darshan Chikkanayakanahalli Mukunda
Department of Biophysics
Manipal School of Life Sciences, Manipal Academy of Higher Education
Manipal, Karnataka, India

Arathi Narayanan
Department of Medicine
Government Medical College
Thiruvananthapuram, Kerala, India

Padmini Pai
Department of Biophysics
Manipal School of Life Sciences, Manipal Academy of Higher Education
Manipal, Karnataka, India

Ajeetkumar Patil
Department of Atomic and Molecular Physics
Manipal Academy of Higher Education (MAHE)
Manipal, Karnataka, India

Anusha Pillai
Department of Atomic and Molecular Physics
Manipal Academy of Higher Education (MAHE)
Manipal, Karnataka, India

Sudharshan Prabhu
Department of Cell and Molecular Biology
Manipal, Karnataka, India

Nupura Manish Prabhune
Department of Cell and Molecular Biology
Manipal School of Life Sciences, Manipal Academy of Higher Education
Manipal, Karnataka, India

Deva Nanda A R
Department of Biophysics
Manipal School of Life Sciences, Manipal Academy of Higher Education
Manipal, Karnataka, India

Yashaswini Reddy
Department of Biophysics
Manipal School of Life Sciences, Manipal Academy of Higher Education
Manipal, Karnataka, India

Jackson Rodrigues
Department of Biophysics
Manipal School of Life Sciences, Manipal Academy of Higher Education
Manipal, Karnataka, India

Nidhi S
Department of Biophysics
Manipal School of Life Sciences, Manipal Academy of Higher Education
Manipal, Karnataka, India

Swathi Sadashiva
Department of Biotechnology
Manipal School of Life Sciences, Manipal Academy of Higher Education
Manipal, Karnataka, India

Manasa Gangadhar Shetty
Department of Biophysics
Manipal School of Life Sciences, Manipal Academy of Higher Education
Manipal, Karnataka, India

Bhisham Narayan Singh
Department of Biotechnology
Manipal School of Life Sciences, Manipal Academy of Higher Education
Manipal, Karnataka, India

Babitha Kampa Sundara
Department of Biophysics
Manipal School of Life Sciences, Manipal Academy of Higher Education
Manipal, Karnataka, India

Guan-Yu Zhuo
Institute of New Drug Development
China Medical University
Taichung, Taiwan

1 Cancer Metabolism
Advanced Biophysical Techniques for Diagnosis and Research

Mani Shankar Babu[1*], *Arathi Narayanan*[2], *and Vidhu Sankar Babu*[3]

[1]Department of Botany, University College, Thiruvananthapuram, Kerala, India
[2]Department of Medicine, Government Medical College, Thiruvananthapuram, Kerala, India
[3]Department of Plant Sciences, Manipal School Life Sciences, Manipal Academy of Higher Education, Manipal, Karnataka, India
[*]Corresponding Author

INTRODUCTION

Cancer cells exhibit distinct metabolic adaptations that fuel their rapid proliferation. Unlike normal cells that mainly utilize oxidative phosphorylation (OXPHOS) for energy production, cancer cells frequently prefer glycolysis even when oxygen is available, a phenomenon referred to as the Warburg effect. This metabolic shift provides advantages such as rapid Adenosine triphosphate (ATP) generation and redirection of glycolytic intermediates for biosynthesis. Proliferating tumour cells require ample building blocks for deoxyribonucleic acid (DNA), ribonucleic acid (RNA), proteins, and lipids. Glycolysis not only provides ATP but also contributes to nucleotide synthesis through the pentose phosphate pathway (PPP). Cancer cells exhibit metabolic flexibility, adapting to nutrient availability and microenvironmental cues. Glutamine fuels the tricarboxylic acid (TCA) cycle, while lipid metabolism supports membrane formation and signalling. Understanding cancer metabolism helps identify therapeutic targets, and inhibiting glycolysis or targeting specific metabolic enzymes shows promise (Ganapathy-Kanniappan and Geschwind 2013). Moreover, metabolic reprogramming influences immune responses and tumour-stroma interactions, offering avenues for combination therapies and improved patient outcomes.

Cancer Metabolism: A Historical and Molecular Perspective

The investigation of cancer metabolism traces back to the early 20th century when Otto Warburg initially noted that cancer cells exhibit a high rate of glycolysis followed by lactic acid fermentation in the cytosol, rather than pyruvate oxidation in the mitochondria. Warburg's work led to the formulation of the Warburg hypothesis, suggesting that cancer arises due to irreversible damage to cellular energy production. Over time, our understanding has evolved. Oncogenes (such as c-Myc) and tumour suppressors (like p53) have emerged as key players in metabolic reprogramming. These genetic alterations influence metabolic pathways, driving aberrant energy production and biosynthesis in cancer cells. In recent years, the development of high-throughput technologies

DOI: 10.1201/9781003545712-1

and metabolomics has allowed for an in-depth analysis of metabolic networks in cancer, offering a detailed understanding of the metabolic landscape of tumours (Yu et al. 2024).

The Significance of Investigating Cancer Metabolism

Studying cancer metabolism is crucial for several reasons. First, it offers insights into the fundamental mechanisms of cancer biology, revealing how metabolic alterations contribute to tumour initiation, growth, and metastasis. By understanding these processes, researchers can identify novel biomarkers for early detection and prognosis, improving patient clinical outcomes. Second, cancer metabolism presents unique therapeutic opportunities. Since cancer cells depend on specific metabolic pathways for survival and proliferation, targeting these pathways can selectively kill cancer cells while sparing normal cells. Several metabolic inhibitors, such as glycolysis inhibitors and glutaminase inhibitors, are currently being investigated in clinical trials, highlighting the potential of metabolic-targeted therapies. Third, the study of cancer metabolism has implications beyond cancer. Metabolic reprogramming is a hallmark of several diseases, including diabetes and neurodegenerative disorders. Insights gained from cancer metabolism can thus inform our understanding of other pathological conditions, leading to broader therapeutic applications.

AIM AND SCOPE OF THE CHAPTER

This chapter aims to provide a comprehensive overview of cancer metabolism, detailing the key metabolic pathways involved in tumour growth and survival. It will explore the field's historical context, highlighting the seminal discoveries shaping our current understanding. The chapter will discuss the molecular mechanisms driving metabolic reprogramming in cancer, focusing on the role of oncogenes, tumour suppressors, and the tumour microenvironment. Additionally, the chapter will examine the therapeutic implications of targeting cancer metabolism, reviewing the latest advances in metabolic inhibitors and their clinical potential. This chapter will offer a holistic view of cancer metabolism by integrating insights from basic research and clinical studies, underscoring its importance in cancer biology and therapy. In summary, this chapter will serve as a valuable resource for researchers, clinicians, and students, providing a detailed and up-to-date account of cancer metabolism and its implications for cancer treatment and beyond.

BASICS OF CANCER METABOLISM

Metabolic reprogramming in cancer cells is a hallmark of cancer and is crucial for supporting the rapid proliferation of a cell's energetic and biosynthetic demands. Cancer cells undergo significant alterations in their metabolism to sustain their growth, survival, and metastasis. This reprogramming involves the upregulation of glycolysis, even in the presence of oxygen (a phenomenon known as aerobic glycolysis or the Warburg effect), alterations in the tricarboxylic acid (TCA) cycle, and modifications in lipid and amino acid metabolism. These changes allow cancer cells to efficiently generate ATP, reduce power in the form of Nicotinamide adenine dinucleotide phosphate (NADPH) , and produce macromolecules necessary for cell division (Martinez-Outschoorn et al. 2017).

Key Metabolic Pathways Involved in Cancer

Glycolysis and the Warburg Effect

The Warburg effect, first described by Otto Warburg in the 1920s, is characterized by the preference of cancer cells for glycolysis over oxidative phosphorylation, even when oxygen is abundant. This metabolic shift allows cancer cells to convert glucose to lactate, producing ATP more rapidly than oxidative phosphorylation. Although glycolysis is less efficient regarding ATP yield per glucose molecule, it provides a quick supply of energy and intermediates for anabolic processes. The increased glycolytic flux in cancer cells supports the synthesis of nucleotides, amino acids, and lipids, which are essential for cell proliferation. Key enzymes in glycolysis, such as hexokinase 2 (HK2), pyruvate kinase M2 (PKM2), and lactate dehydrogenase A (LDHA), are often upregulated in cancer cells, facilitating the enhanced glycolytic activity. Additionally, oncogenes like c-Myc and hypoxia-inducible factor 1-alpha (HIF-1α) further promote glycolysis by upregulating the expression of glycolytic genes (Akter et al. 2024).

TCA Cycle Alterations

The TCA cycle, also known as the Krebs cycle, is a central metabolic pathway that generates high-energy molecules like NADH and $FADH_2$, which are used in oxidative phosphorylation to produce ATP. In cancer cells, the TCA cycle is often altered to support the biosynthetic and bioenergetic needs of the tumour. Mutations in enzymes such as isocitrate dehydrogenase (IDH1/2) can lead to the production of oncometabolites like 2-hydroxyglutarate, which can inhibit DNA and histone demethylation, contributing to the epigenetic reprogramming of cancer cells. Additionally, cancer cells may exhibit an increased reliance on glutamine as an alternative carbon source to replenish TCA cycle intermediates through a process known as anaplerosis. Glutamine is converted to glutamate and then to α-ketoglutarate, feeding into the TCA cycle and supporting the production of ATP and macromolecules (Pujana-Vaquerizo et al. 2024).

A Hike in Lipid Metabolism

Lipid metabolism is crucial for membrane biosynthesis, energy storage, and the production of signalling molecules. Cancer cells often show increased lipogenesis, driven by the upregulation of enzymes like ATP-citrate lyase (ACLY), acetyl-CoA carboxylase (ACC), and fatty acid synthase (FASN). These enzymes convert citrate to acetyl-CoA and subsequently to fatty acids, essential for forming cell membranes and lipid signalling molecules. Furthermore, cancer cells can enhance lipid uptake and storage by upregulating fatty acid transport proteins and lipoprotein lipase. This allows cancer cells to maintain a reservoir of lipids that can be mobilized for energy production and membrane synthesis during rapid proliferation.

Amino Acid Metabolism and Its Modulation

Amino acids serve as building blocks for proteins and are also involved in various metabolic processes. Cancer cells often exhibit altered amino acid metabolism to support their growth and survival. For instance, glutamine metabolism is upregulated in many cancers, providing a carbon and nitrogen source for biosynthetic processes. Glutamine is converted to glutamate and then to α ketoglutarate, feeding into the TCA cycle and supporting the production of nucleotides, amino acids, and lipids. Serine and glycine metabolism are often enhanced in cancer cells. These amino acids are involved in one-carbon metabolism, which is crucial for nucleotide synthesis and methylation reactions. The enzyme phosphoglycerate dehydrogenase (PHGDH), which diverts glycolytic intermediates into serine biosynthesis, is frequently overexpressed in cancer cells (Chen et al. 2024).

Impact on Tumour Growth and Progression

The metabolic reprogramming in cancer cells profoundly impacts tumour growth and progression. The enhanced glycolytic flux and altered TCA cycle support the rapid production of ATP and biosynthetic precursors, enabling cancer cells to sustain high proliferation rates. The increased lactate production from glycolysis also contributes to the acidic tumour microenvironment, promoting invasion, metastasis, and immune evasion. Moreover, reprogramming lipid and amino acid metabolism provides the necessary components for membrane biosynthesis, energy storage, and signalling, further supporting the aggressive behaviour of cancer cells. The importance of cancer metabolism stems from its crucial role in tumour development and progression. By comprehending the metabolic alterations in cancer cells, researchers can pinpoint potential therapeutic targets and devise strategies to interfere with these modified pathways, thereby hindering cancer growth and enhancing patient outcomes. (Vogel et al. 2024).

BIOPHYSICAL TECHNIQUES IN CANCER RESEARCH

Biophysical Techniques – in Brief

Biophysical techniques encompass various methods that apply physics principles to study biological systems. These techniques are essential for elucidating the structure, dynamics, interactions, and functions of biomolecules in living organisms. In cancer research, biophysical techniques allow for precisely characterizing cancer cells and their metabolic pathways at the molecular and atomic levels. The scope of biophysical techniques includes methods such as nuclear magnetic resonance (NMR) spectroscopy, mass spectrometry (MS), X-ray crystallography, cryo-electron microscopy (cryo-EM), fluorescence spectroscopy, and surface plasmon resonance (SPR). Each technique offers unique advantages for analyzing different aspects of cellular and molecular biology. For example, NMR and MS provide detailed information about the chemical composition and dynamics of metabolites, while X-ray crystallography and cryo-EM reveal the high-resolution structures of proteins and complexes. Fluorescence spectroscopy and SPR enable the study of molecular interactions and kinetics in real time. These techniques have been instrumental in advancing our understanding of biological processes by providing quantitative and qualitative data that are critical for deciphering complex biochemical pathways and interactions. Studying cancer metabolism is essential for identifying the altered metabolic pathways that drive tumour growth and progression. Biophysical techniques are essential in this field as they offer detailed insights into the molecular mechanisms driving metabolic reprogramming in cancer cells.

Nuclear Magnetic Resonance Spectroscopy

Principles and Methodology

Nuclear magnetic resonance (NMR) spectroscopy is a potent analytical technique used to determine the structure, dynamics, and interactions of molecules. The fundamental principle of NMR is based on the magnetic properties of atomic nuclei. When placed in a strong magnetic field, nuclei with a non-zero spin, such as hydrogen-1 (^{1}H) or carbon-13 (^{13}C), align with the field. Applying a radiofrequency pulse can excite these nuclei to higher energy states. As the nuclei relax back to their equilibrium state, they emit radiofrequency signals that can be detected and transformed into an NMR spectrum. The NMR spectrum provides information about the chemical environment of the nuclei, allowing for the identification of molecular structures. The position of the NMR signals (chemical shifts) indicates the electronic environment surrounding the nuclei, while the splitting patterns (J-coupling) reveal information about the connectivity between atoms. The intensity of the signals can be used to quantify the concentration of metabolites (Brindle 2024).

Methodologically, NMR spectroscopy involves several key steps:

Sample Preparation: The sample, which can be a pure compound, a mixture, or a biological sample (e.g., tissue extract or cell lysate), is prepared and placed in an NMR tube.

Magnetic Field Application: The sample is placed in a strong, homogeneous magnetic field, typically provided by a superconducting magnet.

Radiofrequency Pulses: A series of radiofrequency pulses is applied to the sample to excite the nuclei.

Signal Detection: The emitted signals as the nuclei relax back to their equilibrium state are detected by the NMR spectrometer.

Data Processing: The detected signals are processed using Fourier transformation to generate the NMR spectrum.

Advanced NMR techniques, such as two-dimensional (2D) NMR, can provide more detailed information about molecular structure and dynamics by correlating interactions between different nuclei (Zhao et al. 2024).

Applications in Cancer Metabolism Research

NMR spectroscopy has a wide range of applications in cancer metabolism research due to its ability to provide detailed insights into the metabolic alterations that occur in cancer cells. Here are some key applications:

Metabolite Profiling and Quantification: NMR spectroscopy is used to profile and quantify metabolites in cancer cells, tissues, and biofluids (e.g., blood, urine). By comparing the metabolite profiles of cancerous and non-cancerous samples, researchers can identify metabolic biomarkers associated with different types of cancer. For example, elevated levels of lactate and alanine are often observed in cancer cells due to increased glycolysis and altered amino acid metabolism.

Metabolic Flux Analysis: NMR can be used to trace the metabolic pathways in cancer cells by incorporating isotopically labelled substrates (e.g., ^{13}C-glucose, ^{13}C-glutamine). By analyzing the distribution of labelled atoms in various metabolites, researchers can map the flow of carbon through metabolic networks, providing insights into the activity of key pathways such as glycolysis, the TCA cycle, and lipid metabolism.

Dynamic Metabolic Changes: NMR spectroscopy allows for monitoring dynamic metabolic changes in response to external stimuli or therapeutic interventions. By performing time-resolved NMR experiments, researchers can observe how metabolic profiles change in real time, providing insights into the mechanisms of drug action and resistance in cancer cells.

Structural Elucidation of Metabolic Enzymes: High-resolution NMR can be used to determine the three-dimensional structures of metabolic enzymes involved in cancer metabolism. Understanding the structure of these enzymes and how they interact with substrates and inhibitors can inform the design of targeted therapies. For example, the structure of mutant IDH1, an enzyme involved in the TCA cycle, has been elucidated using NMR, providing a basis for developing specific inhibitors.

Tumour Microenvironment Analysis: NMR spectroscopy can be used to study the metabolic interactions between cancer cells and their microenvironment. By analyzing the metabolites present in the extracellular space and within cancer-associated fibroblasts, researchers can gain insights into how cancer cells modulate their environment to support tumour growth and invasion.

In Vivo Metabolic Imaging: Magnetic resonance spectroscopy (MRS), a technique related to NMR, allows for non-invasive in vivo metabolic imaging. By performing MRS on patients, researchers can obtain spatially resolved metabolic information about tumours, helping to diagnose cancer, assess tumour heterogeneity, and monitor treatment response. Thus, NMR spectroscopy is a versatile and strong tool in cancer metabolism research, offering detailed insights into the metabolic alterations that drive cancer progression. Its ability to profile and quantify metabolites, trace metabolic pathways, monitor dynamic changes, elucidate enzyme structures, and analyze the tumour microenvironment makes it invaluable for advancing our understanding of cancer biology and developing targeted therapies. As NMR technology continues to evolve, it will undoubtedly play an increasingly important role in the fight against cancer (Teng et al. 2024).

Mass Spectrometry

Principles and Methodology

Mass spectrometry (MS) is a powerful analytical technique used to measure the mass-to-charge ratio (m/z) of ions, providing detailed information about the molecular weight, structure, and composition of compounds. The core principle of MS involves ionizing chemical compounds to generate charged molecules or molecular fragments and then measuring their mass-to-charge ratios (Ma and Fernández 2024).

The methodology of MS generally follows these steps:

Sample Introduction: The sample, which can be a solid, liquid, or gas, is introduced into the mass spectrometer. This can be done directly or via chromatographic techniques such as gas chromatography (GC) or liquid chromatography (LC), which separate complex mixtures before MS analysis.

Ionization: The sample molecules are ionized to form charged particles. Common ionization techniques include:

Electron Ionization (EI): High-energy electrons are used to ionize gas-phase molecules, commonly used in GC-MS.

Electrospray Ionization (ESI): A high-voltage electric field is applied to a liquid sample to create charged droplets, commonly used in LC-MS for large, non-volatile molecules.

Matrix-Assisted Laser Desorption/Ionization (MALDI): A laser is used to ionize samples embedded in a matrix, often used for analyzing biomolecules like proteins and peptides.

Mass Analysis: Using a mass analyzer, the ions are separated based on their *m/z* ratio. Types of mass analyzers include:

Quadrupole: Uses oscillating electric fields to filter ions based on their m/z ratio.

Time-of-Flight (TOF): Measures the time it takes for ions to travel a fixed distance, with lighter ions reaching the detector faster.

Orbitrap: Traps ions in an electrostatic field and measures their oscillation frequencies.

Ion Trap: Traps ions in a three-dimensional electric field and sequentially ejects them based on their *m/z* ratio.

Detection: The separated ions are detected, and their intensities are recorded. Detectors convert ion impacts into electrical signals, generating a mass spectrum that displays the m/z ratios and relative abundances of the ions.

Data Analysis: The mass spectrum is analyzed to determine the molecular weight and structure of the sample components. Sophisticated software tools are used to interpret the data, identify compounds, and quantify their concentrations.

Applications in Metabolite Profiling

Metabolite profiling, or metabolomics, involves the comprehensive analysis of small molecules (metabolites) within a biological sample. MS is a key tool in metabolomics due to its high sensitivity, specificity, and ability to analyze a wide range of metabolites. Here are some key applications of MS in metabolite profiling:

Comprehensive Metabolite Identification: MS allows for the identification of thousands of metabolites in complex biological samples. By coupling MS with chromatographic techniques (LC-MS or GC-MS), researchers can separate and identify metabolites based on their *m/z* ratios and retention times. High-resolution MS and tandem MS (MS/MS) provide detailed structural information, enabling the identification of unknown compounds.

Quantitative Analysis: MS can quantify metabolites with high accuracy and sensitivity. By using internal standards and calibration curves, the concentrations of metabolites in biological samples can be precisely measured. This is essential for understanding the metabolic alterations associated with diseases such as cancer.

Pathway Analysis: MS-based metabolomics enables the mapping of metabolic pathways by identifying and quantifying biochemical reaction intermediates and end products. This helps researchers understand how metabolic networks are reprogrammed in cancer cells, revealing key pathways that support tumour growth and survival.

Biomarker Discovery: MS can identify potential biomarkers for cancer diagnosis and prognosis by comparing the metabolite profiles of healthy and diseased tissues. Metabolite biomarkers can provide insights into disease mechanisms, serve as diagnostic tools, and help monitor treatment response.

Dynamic Metabolic Flux Analysis: Stable isotope-labelled substrates (e.g., ^{13}C-glucose, ^{13}C-glutamine) can be combined with MS to trace the metabolic fate of these compounds in cells. This approach, known as metabolic flux analysis, allows researchers to quantify the flow of metabolites through different pathways, providing a dynamic view of cellular metabolism. It is particularly useful for studying how cancer cells alter their metabolic fluxes in response to oncogenic signals or therapeutic interventions.

Single-Cell Metabolomics: Recent advances in MS technology enable the analysis of metabolites at the single-cell level. This is important for understanding cellular heterogeneity within tumours and identifying metabolic subpopulations that may contribute to drug resistance or aggressive behaviour.

Mass spectrometry is a versatile and robust tool in metabolite profiling, offering detailed insights into the metabolic alterations associated with cancer. Its ability to identify, quantify, and analyze metabolites with high sensitivity and specificity makes it indispensable for understanding cancer metabolism. Through applications in comprehensive metabolite identification, quantitative analysis, pathway analysis, biomarker discovery, dynamic metabolic flux analysis, and single-cell metabolomics, MS plays a crucial role in advancing our knowledge of cancer biology and facilitating the development of targeted therapies (Grooms et al. 2024).

Stable Isotope Labelling by Amino Acids in Cell Culture

Stable isotope labelling by amino acids in cell culture (SILAC) is an intense technique used in quantitative proteomics and metabolic flux analysis. The fundamental principle of SILAC involves the incorporation of non-radioactive, isotopically labelled amino acids into the proteins of cultured cells. These labelled amino acids, typically containing stable isotopes such as ^{13}C or ^{15}N, replace their natural counterparts during protein synthesis. This labelling allows for the precise quantification and comparison of protein abundances and metabolic fluxes between different cell populations (Sun et al. 2024).

SILAC Methodology Involves Several Key Steps

Cell Culture: Cells are cultured in media containing isotopically labelled amino acids. Commonly used labelled amino acids include $^{13}C_6$-lysine and $^{13}C_6$-arginine, which are substituted for their naturally occurring forms (^{12}C-lysine and ^{12}C-arginine) in the growth medium. Parallel cultures are grown in either labelled or unlabelled media to generate differentially labelled cell populations.

Incorporation of Labelled Amino Acids: As cells grow and divide, they incorporate the labelled amino acids into their newly synthesized proteins. Complete labelling typically requires several cell divisions, ensuring that nearly all proteins contain the isotopically labelled amino acids.

Experimental Treatment: Once labelling is complete, cells can be subjected to various experimental treatments, such as drug exposure, genetic modifications, or environmental changes. This allows researchers to study the effects of these treatments on protein expression and metabolic activity.

Cell Lysis and Protein Extraction: After the experimental treatment, cells are lysed, and proteins are extracted from the labelled and unlabelled cell populations.

Protein Digestion and Mass Spectrometry: The extracted proteins are enzymatically digested into peptides, usually using trypsin. The resulting peptides are then analyzed using MS. The mass spectrometer can distinguish between peptides containing labelled and unlabelled amino acids based on their mass differences.

Data Analysis: The MS data are analyzed to identify and quantify the peptides from the labelled and unlabelled samples. The relative abundances of the peptides provide quantitative information about the differences in protein expression or metabolic activity between the experimental conditions.

Applications in Quantitative Proteomics and Metabolic Flux Analysis

SILAC has various applications in quantitative proteomics and metabolic flux analysis, offering valuable insights into cellular processes and disease mechanisms.

Quantitative Proteomics: SILAC is widely used in quantitative proteomics to compare protein expression levels between different cell states. For example, researchers can use SILAC to compare the proteomes of cancer cells with and without drug treatment, identifying proteins whose expression levels are altered by the treatment. This information can reveal potential drug targets and biomarkers for cancer therapy. SILAC-based quantitative proteomics has been instrumental in uncovering signalling pathways and regulatory networks involved in cancer progression and response to therapy.

Post-Translational Modifications: SILAC can also be used to study post-translational modifications (PTMs) such as phosphorylation, acetylation, and ubiquitination. By enriching for modified peptides and using SILAC, researchers can quantify changes in PTMs in response to various stimuli. This is crucial for understanding the regulation of protein function and the role of PTMs in disease.

Metabolic Flux Analysis: In metabolic flux analysis, SILAC allows for tracing metabolic pathways by incorporating labelled amino acids into cellular metabolites. Researchers can map the carbon and nitrogen flow through metabolic networks by analyzing the incorporation of labelled amino acids into metabolic intermediates and end products. This is particularly useful for studying how cancer cells reprogramme their metabolism to support rapid growth and survival. For instance, SILAC can be used to investigate how cancer cells utilize glutamine and glucose to fuel the TCA cycle and biosynthetic pathways.

Protein Turnover and Dynamics: SILAC can be employed to study protein turnover and dynamics by measuring the rates of protein synthesis and degradation. By switching cells from unlabelled to labelled media (or vice versa) and tracking the incorporation or depletion of labelled amino acids over time, researchers can determine the half-lives of proteins and understand how their stability is regulated. This information is critical for understanding the mechanisms of protein homeostasis and the impact of dysregulated protein turnover in diseases such as cancer and neurodegeneration.

Intercellular Communication: SILAC can be used to study intercellular communication by analyzing secreted proteins and exosomes. By labelling donor cells with isotopically labelled amino acids, researchers can track the transfer of labelled proteins to recipient cells, shedding light on mechanisms of cell–cell signalling and the role of the tumour microenvironment in cancer progression.

SILAC is a versatile technique in quantitative proteomics and metabolic flux analysis, providing precise and accurate measurements of protein abundances and metabolic pathways. Its applications in studying protein expression, post-translational modifications, metabolic fluxes, protein turnover, and intercellular communication make it an invaluable tool in cancer research and other fields of biology. By enabling the detailed analysis of cellular processes at the molecular level, SILAC contributes to our understanding of disease mechanisms and the development of targeted therapies (Ibáñez-Molero et al. 2024).

Positron Emission Tomography

Positron emission tomography (PET) is a dominant imaging technique used to visualize metabolic processes in the body. It provides quantitative data on tissues' physiological and biochemical functions, making it invaluable for diagnosing and monitoring diseases, including cancer, neurological disorders, and cardiovascular diseases (Montrazi et al. 2024).

Principles of PET

Radioactive Tracers: PET relies on radioactive tracers, also known as radiopharmaceuticals, which are molecules labelled with positron-emitting radionuclides. The most used tracer is fluorodeoxyglucose (FDG), a glucose analogue labelled with fluorine-18 (^{18}F). Other tracers can be used to target specific metabolic pathways or receptors.

Positron Emission: When the radionuclide decays, it emits a positron (the antimatter counterpart of an electron). This positron travels a short distance in the tissue and is annihilated upon encountering an electron. This annihilation produces two gamma photons, each with an energy of 511 keV, that are emitted in opposite directions (approximately 180 degrees apart).

Detection of Gamma Photons: PET scanners are equipped with a ring of detectors that surround the patient. These detectors capture the gamma photons emitted from the annihilation events. The simultaneous detection of two photons (coincidence detection) allows the scanner to localize the origin of the positron emission along a straight line.

Image Reconstruction: The data collected from the coincidence detections are used to reconstruct three-dimensional images of the tracer distribution within the body. This reconstruction is achieved using algorithms such as filtered back-projection or iterative reconstruction.

Quantification: The resulting PET images provide quantitative information about the tracer concentration in different tissues. This quantification can be expressed in standardized uptake values (SUVs), which normalize the tracer uptake to factors such as injected dose and patient body weight.

PET Methodology

Tracer Preparation: Radiopharmaceuticals are synthesized in specialized facilities, often using cyclotrons to produce the radionuclides. The tracers must be sterile and contaminant-free, with high radiochemical purity.

Patient Preparation: Patients are prepared for the PET scan based on the tracer and the metabolic process being studied. For example, when using FDG, patients may need to fast for several hours to reduce background glucose levels and enhance tracer uptake in target tissues.

Tracer Injection: The radioactive tracer is injected intravenously into the patient. The patient then rests for a period (e.g., 30–60 minutes) to allow the tracer to distribute and accumulate in the tissues of interest.

Image Acquisition: The patient is positioned in the PET scanner, and images are acquired over a specific period, typically ranging from a few minutes to an hour, depending on the study's goals and the tracer's pharmacokinetics.

Image Processing and Analysis: The acquired data are processed to reconstruct images, which are then analyzed to assess the distribution and concentration of the tracer. Quantitative measurements and visual assessments are performed to interpret the metabolic activity in the tissues.

Applications in Imaging Metabolic Activity

PET is widely used to image metabolic activity in various medical fields, providing crucial insights into physiological and pathological processes.

Cancer Imaging: PET is extensively used in oncology to detect and stage cancer, monitor treatment response, and detect recurrences. FDG-PET, which exploits the increased glucose metabolism of cancer cells, helps identify malignant lesions, assess tumour aggressiveness, and evaluate the efficacy of therapies such as chemotherapy and radiation. Other tracers can target specific tumour characteristics, such as hypoxia, proliferation, or receptor expression.

Neurological Disorders: PET is valuable in diagnosing and researching neurological conditions like Alzheimer's disease, Parkinson's disease, and epilepsy. For example, FDG-PET can assess brain glucose metabolism, revealing patterns of hypometabolism associated with neurodegenerative diseases. PET tracers targeting amyloid plaques and tau protein are used to study Alzheimer's pathology. In epilepsy, PET helps localize epileptic foci for surgical planning.

Cardiovascular Diseases: PET is used to evaluate myocardial perfusion and viability in patients with coronary artery disease. Tracers such as rubidium-82 (^{82}Rb) and nitrogen-13 ammonia (^{13}N-ammonia) assess blood flow and identify ischemic but viable myocardium, guiding revascularization decisions. FDG-PET can also evaluate myocardial inflammation and infection.

Metabolic Disorders: PET is used to study metabolic disorders by imaging specific pathways or organ functions. For instance, in diabetes research, FDG-PET can assess glucose uptake in insulin-sensitive tissues. PET tracers targeting fatty acid metabolism, amino acid transport, or endocrine receptors provide insights into metabolic diseases and endocrine disorders.

Research and Drug Development: PET is a compelling biomedical and pharmaceutical development tool. It allows researchers to study the pharmacokinetics and pharmacodynamics of new drugs, assess target engagement, and evaluate biological effects in vivo. PET imaging can accelerate drug development by providing early insights into drug efficacy and safety.

PET is a versatile and formidable imaging technique that provides detailed insights into metabolic activity in vivo. Its principles and methodology enable precise quantification of tracer distribution and metabolic processes, making it invaluable in oncology, neurology, cardiology, metabolic disorders, and biomedical research. By visualizing and quantifying metabolic alterations, PET contributes significantly to diagnosing, treating, and understanding various diseases, ultimately improving patient care and advancing scientific knowledge (Tricarico et al. 2024).

Fluorodeoxyglucose PET

Principles and Methodology

Fluorodeoxyglucose (FDG) PET is a specialized imaging technique that leverages the metabolic properties of glucose to detect and monitor various diseases, particularly cancer. FDG is a glucose analogue in which the hydroxyl group at the second carbon position is replaced by the positron-emitting radionuclide ^{18}F. The principles and methodology of FDG-PET involve several key steps:

Radioactive Tracer: FDG is synthesized by incorporating ^{18}F into a glucose molecule. Due to its structural similarity to glucose, FDG is taken up by cells via glucose transporters.

Cellular Uptake and Trapping: Once inside the cell, FDG is phosphorylated by hexokinase to FDG-6-phosphate. Unlike glucose-6-phosphate, FDG-6-phosphate is not further metabolized in the glycolytic pathway and is effectively trapped within the cell. This trapping mechanism allows for the accumulation of FDG in tissues with high glucose uptake.

Positron Emission and Annihilation: The ^{18}F in FDG undergoes positron emission decay, producing a positron that travels a short distance before annihilating with an electron. This annihilation event generates two gamma photons emitted in opposite directions.

Detection and Image Reconstruction: PET scanners detect these gamma photons using a ring of detectors. The simultaneous detection of the two photons (coincidence detection) allows the scanner to localize the origin of the positron emission. The data are then processed and reconstructed into three-dimensional images representing the distribution of FDG in the body.

Quantification: The intensity of the FDG signal in the images correlates with the amount of FDG uptake and, consequently, the glucose metabolic activity of the tissues. Standardized uptake values (SUVs) are often used to quantify FDG uptake, providing a normalized measure of tracer concentration (Tricarico et al. 2024).

Applications in Detecting Glucose Metabolism in Tumours

FDG-PET is widely used in oncology due to its ability to detect and quantify glucose metabolism, which is typically upregulated in cancer cells. The Warburg effect describes this phenomenon, where cancer cells preferentially metabolize glucose via glycolysis even under aerobic conditions, leading to increased glucose uptake and lactate production. FDG-PET exploits this metabolic alteration to detect and monitor tumours.

Cancer Detection and Diagnosis: FDG-PET is highly sensitive in detecting primary and metastatic tumours. Increased FDG uptake in malignant tissues contrasts with the lower uptake in normal tissues, allowing for the visualization of tumours. FDG-PET is particularly effective in identifying cancers with high metabolic activity, such as lung, colorectal, breast, and lymphoma.

Tumour Staging: Accurate staging of cancer is crucial for determining the appropriate treatment strategy. FDG-PET provides comprehensive whole-body imaging, enabling the detection of metastatic disease that might not be visible on conventional imaging modalities like CT or MRI. By identifying both the primary tumour and distant metastases, FDG-PET helps in accurate cancer staging.

Treatment Planning and Response Monitoring: FDG-PET is used to assess the metabolic activity of tumours before, during, and after treatment. Baseline FDG-PET scans help in planning treatment strategies by identifying the most metabolically active regions of the tumour. Subsequent scans during or after treatment can evaluate the response, differentiating between responsive and resistant disease. A significant reduction in FDG uptake indicates a positive response, whereas persistent or increased uptake suggests treatment resistance or disease progression.

Detection of Recurrence: FDG-PET is valuable in detecting tumour recurrence, especially when there are clinical or biochemical indications of relapse but conventional imaging is inconclusive. By identifying areas of increased metabolic activity, FDG-PET can detect recurrent disease earlier than structural changes become apparent.

Guiding Biopsies and Radiotherapy: FDG-PET can guide biopsies to the most metabolically active and likely malignant areas of a tumour, increasing diagnostic yield. It also aids in radiotherapy planning by delineating the most metabolically active tumour regions, ensuring precise targeting while sparing healthy tissues.

Prognostic Value: The level of FDG uptake in tumours has prognostic significance. Higher FDG uptake is often associated with more aggressive disease and a poorer prognosis. Thus, FDG-PET not only aids in diagnosis and treatment monitoring but also provides valuable prognostic information.

FDG-PET is a crucial imaging modality in oncology, capitalizing on the increased glucose metabolism characteristic of many cancers. FDG-PET enhances the detection, staging, and monitoring of tumours by providing detailed and quantitative images of metabolic activity. Its applications extend from initial diagnosis through treatment response assessment to recurrence detection, offering a comprehensive tool for managing cancer patients. Through its ability to visualize the metabolic landscape of tumours, FDG-PET continues to play a vital role in improving cancer care and patient outcomes (Tricarico et al. 2024).

Magnetic Resonance Imaging

Principles and Methodology

Magnetic resonance imaging (MRI) is a non-invasive imaging technique that uses effective magnetic fields and radio waves to generate detailed images of the body's internal structures. MRI is widely used in medical diagnostics because it provides high-resolution images of soft tissues without ionizing radiation (Dou et al. 2024).

Principles of MRI

Magnetic Field: MRI scanners generate a strong magnetic field, typically between 1.5 and 3.0 Tesla (T), although higher field strengths are used in research settings. This magnetic field aligns the nuclear spins of hydrogen atoms (protons) in the body's water and fat molecules.

Radiofrequency Pulses: Short bursts of radiofrequency (RF) pulses are transmitted into the body, temporarily disturbing the alignment of the hydrogen protons. When the RF pulse is turned off, the protons return to their original alignment, emitting RF signals.

Signal Detection: The emitted RF signals are detected by coils placed around the body part being imaged. These signals contain information about the location and environment of the protons.

Image Reconstruction: The detected signals are processed using Fourier transform techniques to construct cross-sectional images of the body. By manipulating the magnetic field gradients, images can be generated in various planes (axial, sagittal, coronal) and three-dimensional reconstructions.

MRI Methodology

Patient Preparation: The patient is positioned on a movable table that slides into the bore of the MRI scanner. Depending on the area being imaged, surface coils may be placed around the region of interest to improve signal reception.

Scanning Protocol: Specific imaging protocols are selected based on the clinical question. These protocols define the sequence of RF pulses, gradient pulses, and data acquisition parameters. Common sequences include T1-weighted, T2-weighted, and proton density-weighted images, each highlighting different tissue characteristics.

Image Acquisition: The patient must remain still during the scan to avoid motion artefacts. Breath-holding may be required for certain scans, particularly those involving the abdomen or chest.

Contrast Agents: In some cases, contrast agents such as gadolinium-based compounds are injected intravenously to enhance the visibility of blood vessels, tumours, and areas of inflammation.

Data Analysis: The acquired images are analyzed by radiologists to diagnose various conditions. Advanced software tools are used to quantify tissue properties, assess functional parameters, and create detailed anatomical maps.

Functional MRI and Its Applications in Cancer

Functional MRI (fMRI) extends the capabilities of conventional MRI by measuring and mapping brain activity and other physiological processes. In the context of cancer, fMRI provides valuable insights into tumour biology, treatment response, and the effects of therapy on normal tissue function (Daimiel Naranjo et al. 2024).

Blood Oxygen Level-Dependent (BOLD) fMRI: This technique measures changes in blood oxygenation associated with neural activity. Although primarily used in neuroscience, BOLD fMRI can be applied in cancer research to study brain tumours and the effects of treatments on brain function. By detecting changes in oxygenation, BOLD fMRI can map functional brain regions and assess the impact of tumours or surgical interventions on these regions.

Diffusion-Weighted Imaging (DWI): DWI is a fMRI technique that measures the diffusion of water molecules within tissues. In cancer, DWI detects and characterizes tumours based on their cellular density and structural integrity. Tumours often restrict water diffusion more than normal tissues, producing hyperintense signals on DWI. DWI is particularly useful in identifying malignant lesions, monitoring treatment response, and detecting early recurrence.

Dynamic Contrast-Enhanced MRI (DCE-MRI): DCE-MRI involves the rapid acquisition of images following the injection of a contrast agent. This technique measures the kinetics of contrast agent uptake and washout in tissues, providing information about tissue perfusion, capillary permeability, and extracellular volume. In cancer, DCE-MRI is used to evaluate tumour vascularity, assess the efficacy of anti-angiogenic therapies, and predict treatment response.

Magnetic Resonance Spectroscopy (MRS): MRS is a non-invasive technique that measures the concentrations of metabolites within tissues. By analyzing the spectral peaks of various metabolites, MRS can provide biochemical information about tumours, such as the levels of choline (a marker of cell membrane turnover) and lactate (a marker of anaerobic metabolism). MRS is used to differentiate benign from malignant lesions, guide biopsy sites, and monitor metabolic changes during treatment.

Arterial Spin Labelling (ASL): ASL is a non-invasive MRI technique that measures tissue perfusion by magnetically labelling arterial blood. In cancer, ASL can be used to assess tumour blood flow, providing insights into tumour angiogenesis and the effects of therapies targeting the vasculature.

Functional Connectivity MRI (fcMRI): fcMRI measures the temporal correlations of neural activity between different brain regions. In cancer research, fcMRI can be used to study the impact of brain tumours and treatments on neural networks, providing a better understanding of cognitive and neurological side effects.

MRI is a versatile and intense imaging modality that provides detailed anatomical and functional information about tissues. fMRI techniques, including BOLD fMRI, DWI, DCE-MRI, MRS, ASL, and fcMRI, extend the capabilities of conventional MRI, offering valuable insights into tumour biology and treatment effects. In cancer research and clinical practice, these techniques enable the non-invasive assessment of tumour characteristics, treatment response, and the impact of therapies on normal tissue function. By combining anatomical and functional information, MRI and fMRI contribute significantly to the diagnosis, management, and understanding of cancer (Reis et al. 2024).

Fluorescence Resonance Energy Transfer

Fluorescence resonance energy transfer (FRET) is a potent technique used to study molecular interactions at very close distances, typically within 1–10 nm. It involves the non-radiative energy transfer from a donor fluorophore to an acceptor fluorophore, providing valuable insights into protein interactions, conformational changes, and metabolic pathways (Peng et al. 2024).

Principles of FRET

Fluorophores: FRET requires two types of fluorophores: a donor and an acceptor. The donor fluorophore absorbs light at a specific wavelength and can transfer energy to the acceptor fluorophore if they are in close proximity.

Energy Transfer: When the donor fluorophore is excited by an external light source, it can transfer its excitation energy to the acceptor fluorophore through dipole–dipole interactions. This transfer of energy occurs without the emission of a photon and is highly dependent on the distance between the donor and acceptor.

Distance Dependency: The efficiency of energy transfer (E) is inversely proportional to the sixth power of the distance (r) between the donor and acceptor. This relationship is described by the Förster equation:

$$E = \frac{R_0^6}{R_0^6 + r^6}$$

where R_0 is the Förster distance, the distance at which energy transfer efficiency is 50%. R_0 depends on the spectral overlap between the donor emission and acceptor absorption, the donor's quantum yield, and the dipoles' relative orientation.

Detection: FRET can be detected by measuring changes in the fluorescence intensity or lifetime of the donor, the acceptor, or both. When FRET occurs, the donor's fluorescence decreases, and the acceptor's fluorescence increases. This can be measured using fluorescence microscopy, spectroscopy, or flow cytometry.

FRET Methodology

Selection of Fluorophores: Choose donor and acceptor fluorophores with appropriate spectral properties, ensuring significant overlap between the donor's emission spectrum and the acceptor's absorption spectrum.

Labelling: Proteins or other biomolecules of interest are labelled with the chosen fluorophores. This can be achieved using genetic fusion with fluorescent proteins (e.g., GFP and mCherry) or chemical conjugation with small-molecule fluorophores.

Sample Preparation: Prepare the biological sample with the labelled molecules. Depending on the experimental design, this can involve cell cultures, tissue samples, or in vitro systems.

Excitation and Detection: Excite the donor fluorophore with a suitable light source (e.g., laser or LED) and detect the resulting fluorescence emissions from both the donor and acceptor using appropriate filters and detectors.

Data Analysis: Analyze the fluorescence signals to determine FRET efficiency, which provides information about the proximity and interactions between the labelled molecules. Various methods, such as acceptor photobleaching, sensitized emission, or fluorescence lifetime imaging microscopy (FLIM), can be used to quantify FRET.

Applications in Studying Protein Interactions and Metabolic Pathways

FRET is extensively used in biological research to investigate protein interactions, conformational changes, and metabolic pathways. Its ability to detect molecular interactions at nanometre-scale distances makes it an invaluable tool for understanding complex biological processes (Lan et al. 2024).

Protein–Protein Interactions: FRET is widely used to study the interactions between proteins in living cells. By labelling two interacting proteins with donor and acceptor fluorophores, researchers can monitor the formation and dissociation of protein complexes in real time. This provides insights into signalling pathways, enzyme–substrate interactions, and the dynamics of protein networks. For example, FRET has been used to study the interactions between receptor tyrosine kinases and their downstream signalling partners, revealing signal transduction mechanisms.

Conformational Changes: FRET can detect conformational changes within a single protein or protein complex. By placing the donor and acceptor fluorophores at different positions within the protein, researchers can monitor changes in distance that occur during conformational transitions. This approach has been applied to study the conformational dynamics of ion channels, molecular motors, and enzymes. For instance, FRET has been used to investigate the conformational changes of G-protein-coupled receptors (GPCRs) upon ligand binding.

Metabolic Pathways: FRET is valuable for studying metabolic pathways by monitoring the interactions between metabolic enzymes and their substrates or cofactors. By labelling enzymes and metabolites with suitable fluorophores, researchers can visualize the spatial and temporal dynamics of metabolic processes. FRET-based sensors have been developed to measure intracellular concentrations of key metabolites, such as ATP, glucose, and calcium. These sensors provide real-time information about metabolic fluxes and cellular energy status.

Nucleic Acid Interactions: FRET is used to study the interactions between nucleic acids (DNA and RNA) and proteins. This includes the binding of transcription factors to DNA, ribonucleoprotein complex assembly, and chromatin structure dynamics. FRET has been employed to investigate the mechanisms of DNA repair, replication, and transcription regulation.

Drug Discovery and Screening: FRET-based assays are utilized in drug discovery to screen for compounds that modulate protein–protein interactions or enzyme activity. High-throughput FRET assays can identify potential drug candidates by measuring changes in FRET efficiency in response to chemical libraries. This approach has accelerated the identification of inhibitors for various therapeutic targets, including kinases, proteases, and protein–protein interactions.

FRET is a versatile technique for studying molecular interactions and conformational changes at nanometre-scale distances. Its principles and methodology enable the detection and quantification of protein–protein interactions, conformational dynamics, and metabolic processes in real time. FRET's applications in biological research have provided profound insights into cellular signalling, enzymatic activity, metabolic pathways, and nucleic acid interactions. By enabling the visualization of molecular interactions in their native cellular context, FRET has become an essential tool for advancing our understanding of complex biological systems and developing novel therapeutic strategies (Lan et al. 2024).

Infrared and Raman Spectroscopy

Principles and Methodology

Infrared (IR) and Raman spectroscopy are complementary techniques used to study the vibrational properties of molecules. These techniques provide detailed information about molecular structure, chemical bonds, and the composition of samples. Both methods are non-destructive and can be applied to a wide range of biological and chemical materials, making them valuable tools in various fields, including cancer research.

Principles of IR Spectroscopy

Vibrational Transitions: IR spectroscopy measures the absorption of IR light by molecules, which causes changes in their vibrational states. Each type of chemical bond and functional group absorbs IR radiation at specific wavelengths, leading to a unique absorption spectrum that acts as a molecular fingerprint (Zheng et al. 2024).

IR Regions: The IR spectrum is typically divided into three regions: near-IR (4000–14000 cm^{-1}), mid-IR (400–4000 cm^{-1}), and far-IR (10–400 cm^{-1}). The mid-IR region is most commonly used for molecular characterization because it includes the fundamental vibrations of most functional groups.

Absorption Bands: The absorption spectrum of a sample displays peaks corresponding to the vibrational frequencies of different bonds. These peaks can be analyzed to identify functional groups, determine molecular structure, and quantify specific compounds.

Principles of Raman Spectroscopy

Inelastic Scattering: Raman spectroscopy relies on the inelastic scattering of light, known as the Raman effect. When monochromatic light (usually from a laser) interacts with a molecule, most photons are elastically scattered (Rayleigh scattering). A small fraction of photons, however, undergoes inelastic scattering, resulting in a shift in energy corresponding to the vibrational modes of the molecule.

Raman Shifts: The energy difference between the incident photons and the scattered photons (Raman shift) provides information about the vibrational modes of the molecules. The Raman spectrum is a plot of intensity versus Raman shift (measured in wavenumbers, cm^{-1}).

Complementarity to IR: While both IR and Raman spectroscopy provide information about molecular vibrations, they are based on different selection rules. IR spectroscopy is sensitive to changes in dipole moment, while Raman spectroscopy is sensitive to changes in polarizability. This makes them complementary techniques, with Raman spectroscopy often revealing vibrational modes that are weak or inactive in IR spectra.

IR and Raman Spectroscopy Methodology

Sample Preparation: Samples can be in various forms, including solids, liquids, and gases. Minimal preparation is required, and both techniques can analyze samples directly, making them suitable for in situ and real-time analysis.

Instrumentation

IR Spectroscopy: An IR spectrometer consists of an IR light source, a sample holder, a monochromator to select specific wavelengths, and a detector to measure the transmitted or reflected IR radiation.

Raman Spectroscopy: A Raman spectrometer includes a laser source, a sample holder, a monochromator or spectrograph to disperse the scattered light, and a detector to measure the intensity of the Raman-scattered light.

Data Analysis: The obtained spectra are analyzed by comparing the observed peaks with reference spectra or by using computational methods to assign vibrational modes. This analysis provides information about molecular composition, structure, and interactions.

Applications in Characterizing Metabolic Changes

IR and Raman spectroscopy are extensively used in biomedical research to characterize metabolic changes, providing insights into disease mechanisms, diagnosis, and treatment monitoring.

Cancer Diagnosis and Classification: Both IR and Raman spectroscopy can differentiate between normal and cancerous tissues based on their distinct molecular signatures. Cancer cells often exhibit altered metabolism, leading to changes in the composition of lipids, proteins, nucleic acids, and carbohydrates. Spectroscopic techniques can detect these changes, aiding in the early diagnosis and classification of various cancers.

Metabolic Profiling: IR and Raman spectroscopy can be used to profile metabolic changes in cells and tissues. For instance, the techniques can detect alterations in lipid metabolism, protein expression, and nucleic acid content, which are indicative of metabolic reprogramming in cancer cells. By analyzing these metabolic profiles, researchers can gain insights into the biochemical pathways involved in tumour progression and identify potential biomarkers for cancer diagnosis and prognosis.

Monitoring Treatment Response: These spectroscopic techniques can monitor changes in the metabolic state of cancer cells in response to treatment. For example, successful chemotherapy or radiotherapy can lead to specific changes in the spectral profiles of tumour tissues, reflecting apoptosis or necrosis. By tracking these changes, clinicians can assess treatment efficacy and make informed decisions about therapy adjustments.

Studying Drug Mechanisms: IR and Raman spectroscopy can elucidate the mechanisms of action of anticancer drugs. By analyzing the interactions between drugs and cellular components, researchers can understand how drugs alter cellular metabolism and identify potential targets for therapeutic intervention. This information is crucial for developing more effective and targeted cancer treatments.

Single-Cell Analysis: Advances in spectroscopic techniques allow for the analysis of individual cells, providing high-resolution insights into cellular heterogeneity. Single-cell IR and Raman spectroscopy can identify variations in metabolic activity among different cell populations within a tumour, helping to understand tumour heterogeneity and the development of drug resistance.

Non-Invasive Diagnostics: Raman spectroscopy, in particular, offers the potential for non-invasive diagnostics through techniques such as surface-enhanced Raman scattering (SERS). SERS enhances the Raman signal by using metal nanoparticles, enabling the detection of low-abundance metabolites and biomarkers in biological fluids such as blood, urine, and saliva. This approach holds promise for developing non-invasive cancer diagnostic tools.

IR and Raman spectroscopy are intense and complementary techniques for characterizing metabolic changes in biological systems. Their principles and methodologies enable detailed analysis of molecular vibrations, providing insights into the composition, structure, and interactions of biomolecules. In cancer research, these techniques are invaluable for diagnosing cancer, profiling metabolic changes, monitoring treatment responses, studying drug mechanisms, and developing non-invasive diagnostic tools. By advancing our understanding of metabolic alterations in cancer, IR and Raman spectroscopy contribute significantly to improving cancer diagnosis, treatment, and patient outcomes (Li et al. 2024).

Extracellular Flux Analysis

Principles and Methodology

Extracellular flux analysis is a dynamic technique used to measure the metabolic activity of live cells, specifically focusing on cellular respiration and glycolysis. It provides real-time data on the consumption and production of key metabolic substrates and products, such as oxygen and protons, which are indicators of mitochondrial function and glycolytic activity, respectively.

Principles of Extracellular Flux Analysis

Oxygen Consumption Rate (OCR): OCR is an indicator of mitochondrial respiration. It measures the rate at which cells consume oxygen, which is primarily used in the electron transport chain to produce ATP. A decrease in OCR can indicate mitochondrial dysfunction or a shift towards glycolytic metabolism.

Extracellular Acidification Rate (ECAR): ECAR measures the rate of proton production, primarily due to the conversion of glucose to lactate during glycolysis. An increase in ECAR is indicative of heightened glycolytic activity, which is often observed in cancer cells due to the Warburg effect (Sorrells et al. 2024).

Extracellular Flux Analysis Methodology

Instrument Setup: The extracellular flux analyzer (e.g., Seahorse XF Analyzer) is equipped with sensors that simultaneously measure OCR and ECAR in multi-well plates. Each well contains a microchamber that traps a small volume of media above the cells, allowing for the detection of changes in oxygen and proton concentrations.

Cell Preparation: Cells are seeded into the wells of a specialized plate and allowed to adhere. The appropriate density of cells is critical to ensure accurate measurements without nutrient depletion or overcrowding effects.

Assay Medium: The growth medium is replaced with a specialized assay medium that is optimized for measuring metabolic fluxes. This medium lacks bicarbonate to prevent interference with pH measurements.

Calibration and Equilibration: The analyzer is calibrated using known standards, and the cells are equilibrated in the assay medium to stabilize their metabolic activity before measurements begin.

Sequential Injections: During the assay, various compounds can be sequentially injected into the wells to perturb cellular metabolism. Commonly used compounds include:

Oligomycin: Inhibits ATP synthase, leading to a decrease in OCR and an increase in ECAR, allowing the measurement of ATP-linked respiration.

Carbonyl Cyanide-*p*-trifluoromethoxyphenylhydrazone (FCCP): Uncouples oxidative phosphorylation by collapsing the proton gradient, maximizing OCR to measure maximal respiratory capacity.

Rotenone/Antimycin A: Inhibit complexes I and III of the electron transport chain, respectively, effectively shutting down mitochondrial respiration and allowing for the measurement of non-mitochondrial respiration.

2-Deoxy-D-glucose (2-DG): Inhibits glycolysis by blocking glucose uptake, used to assess glycolytic dependence and reserve capacity.

Data Acquisition and Analysis: The analyzer collects data on OCR and ECAR over time. The results are analyzed using dedicated software, which calculates various parameters such as basal respiration, maximal respiration, glycolytic capacity, and glycolytic reserve.

Applications in Measuring Cellular Respiration and Glycolysis

Extracellular flux analysis is extensively used in biomedical research to investigate cellular metabolism under various conditions, providing insights into metabolic reprogramming, disease mechanisms, and therapeutic interventions.

Cancer Metabolism: Cancer cells often exhibit altered metabolism characterized by increased glycolysis and reduced oxidative phosphorylation, known as the Warburg effect. Extracellular flux analysis allows researchers to quantify these metabolic changes in real time. By measuring OCR and ECAR, scientists can assess the metabolic phenotype of cancer cells, identify metabolic vulnerabilities, and evaluate the efficacy of metabolic inhibitors.

Drug Screening and Toxicology: This technique is valuable for screening potential therapeutic compounds that target cellular metabolism. By measuring changes in OCR and ECAR in response to drug treatment, researchers can determine the impact of compounds on mitochondrial function and glycolysis. This approach helps identify drugs that selectively target cancer cell metabolism without affecting normal cells, thereby reducing potential side effects.

Mitochondrial Dysfunction: Extracellular flux analysis is used to study mitochondrial diseases and conditions associated with mitochondrial dysfunction, such as neurodegenerative diseases, diabetes, and cardiovascular diseases. By measuring mitochondrial respiration, researchers can investigate these diseases' underlying mechanisms, identify mitochondrial dysfunction biomarkers, and evaluate the effects of therapeutic interventions aimed at restoring mitochondrial function.

Metabolic Pathway Analysis: The technique enables the detailed analysis of specific metabolic pathways. By using metabolic inhibitors and stimulators, researchers can dissect the contributions of various pathways to overall cellular metabolism. For example, the impact of glycolysis, glutaminolysis, and fatty acid oxidation on cellular energy production can be assessed by measuring OCR and ECAR under different conditions.

Immunometabolism: The metabolic activity of immune cells is crucial for their function and response to infections, cancer, and autoimmune diseases. Extracellular flux analysis allows the study of immune cell metabolism, including the metabolic reprogramming that occurs during activation, differentiation, and response to pathogens or tumours. This information is essential for developing strategies to modulate immune responses in various diseases.

Nutrient Sensing and Homeostasis: Cells respond to changes in nutrient availability by adjusting their metabolic activity. Extracellular flux analysis can be used to study how cells sense and adapt to different nutrient conditions. Researchers can gain insights into cellular nutrient-sensing mechanisms and their role in maintaining metabolic homeostasis by measuring OCR and ECAR in response to varying levels of glucose, amino acids, or fatty acids.

Extracellular flux analysis is a robust and versatile technique for studying cellular metabolism in real time. Its principles and methodology allow for the precise measurement of oxygen consumption and extracellular acidification, providing insights into mitochondrial function and glycolytic activity. Applications of this technique in cancer research, drug screening, mitochondrial dysfunction, metabolic pathway analysis, immunometabolism, and nutrient sensing have significantly advanced our understanding of cellular metabolism and its role in health and disease. By enabling the detailed analysis of metabolic changes, extracellular flux analysis is an essential tool in biomedical research and therapeutic development (Yamamoto et al. 2024).

APPLICATIONS OF BIOPHYSICAL TECHNIQUES IN CANCER DIAGNOSIS

Case Studies and Examples

Case Study: Breast Cancer Detection Using Raman Spectroscopy: Raman spectroscopy has been successfully used to differentiate between normal, benign, and malignant breast tissues. By analyzing the vibrational spectra, researchers identified specific biochemical changes in malignant tissues, such as increased protein and nucleic acid content and decreased lipid content. This technique demonstrated high sensitivity and specificity, suggesting its potential as a non-invasive diagnostic tool for breast cancer (Sezer et al. 2024).

Example: FDG-PET in Lung Cancer Diagnosis: FDG-PET has become a standard imaging technique for the diagnosis and staging of lung cancer. The high uptake of FDG in malignant lung tumours allows for accurate detection and differentiation from benign lesions. FDG-PET is also used to assess the response to chemotherapy and radiation therapy, helping to tailor treatment plans based on metabolic changes in the tumour (Wumener et al. 2024).

Case Study: Prostate Cancer Metabolomics Using NMR Spectroscopy: NMR spectroscopy has been applied to study the metabolic profiles of prostate cancer patients. Researchers identified altered levels of citrate, choline, and polyamines in prostate cancer tissues compared to healthy tissues. These metabolic changes serve as potential biomarkers for early detection and monitoring of prostate cancer progression (Zniber et al. 2024).

Case Study: Metabolic Profiling of Glioblastoma Using NMR Spectroscopy: Researchers used NMR spectroscopy to analyze the metabolic profiles of glioblastoma cells and identified elevated levels of 2-hydroxyglutarate (2-HG), a metabolite produced by mutant isocitrate dehydrogenase (IDH). This finding led to the development of IDH inhibitors, which are now being tested in clinical trials for the treatment of glioblastoma and other cancers with IDH mutations (Chai et al. 2024).

Example: FDG-PET in Evaluating Chemotherapy Response in Breast Cancer: FDG-PET has been widely used to monitor the response of breast cancer to chemotherapy. In one study, patients who showed a significant decrease in FDG uptake after the first cycle of chemotherapy had a higher likelihood of complete pathological response. This non-invasive imaging technique allowed for early assessment of treatment efficacy, guiding subsequent therapy decisions (Gebhart et al. 2024).

Case Study: SILAC in Identifying Therapeutic Targets in Lung Cancer: Using SILAC, researchers identified several proteins involved in the PI3K/AKT/mTOR pathway that were upregulated in non-small cell lung cancer (NSCLC). Inhibitors targeting this pathway showed promise in preclinical studies, leading to the development of targeted therapies that are currently being evaluated in clinical trials (Ortega Duran et al. 2024).

CHALLENGES AND LIMITATIONS

Technical Limitations

Biophysical Techniques, Although Powerful, Face Several Technical Challenges

Resolution and Sensitivity: Some techniques, like NMR spectroscopy and PET, struggle with resolution and sensitivity issues. For instance, NMR requires high concentrations of metabolites to generate detectable signals, which can be limiting for analyzing small biopsy samples or low-abundance metabolites.

Complexity and Cost: Advanced biophysical techniques often require sophisticated equipment and expertise. Instruments like mass spectrometers, PET scanners, and MRI machines are expensive to purchase and maintain, limiting their accessibility to well-funded research institutions.

Data Volume and Analysis: Techniques like MS and extracellular flux analysis generate vast amounts of data that require extensive computational resources and expertise to analyze. The complexity of these datasets can hinder the speed of data interpretation and the identification of meaningful patterns.

Sample Preparation: Proper sample preparation is crucial for accurate results. Contamination, degradation, and variability in sample handling can introduce significant errors in measurements, impacting the reliability of the data.

Interpretation of Biophysical Data

The Interpretation of Biophysical Data Presents Several Challenges

Biological Variability: Biological samples inherently exhibit high variability due to genetic, environmental, and temporal factors. Distinguishing meaningful biological signals from noise requires careful experimental design and robust statistical analysis.

Multifactorial Influences: Cancer metabolism is influenced by numerous factors, including genetic mutations, tumour microenvironment, and treatment history. Isolating the effects of specific variables can be challenging, making it difficult to draw definitive conclusions from biophysical data alone.

Correlation vs Causation: Biophysical techniques can identify correlations between metabolic changes and cancer progression, but establishing causation requires additional functional studies. This limitation necessitates complementary approaches to validate and extend biophysical findings.

Integrating Biophysical Techniques with Other Methods

Combining biophysical techniques with other methods enhances the comprehensiveness of cancer research but introduces integration challenges.

Data Integration: Integrating data from diverse biophysical techniques, such as NMR, MS, and PET, with genomic, transcriptomic, and proteomic data requires sophisticated bioinformatics tools and interdisciplinary expertise. Harmonizing data formats and ensuring compatibility between datasets are crucial for meaningful integration.

Interdisciplinary Collaboration: Effective integration of biophysical techniques with other methods necessitates collaboration across disciplines, including biology, chemistry, physics, and computational science. Coordinating these efforts can be logistically and intellectually challenging.

Standardization: Establishing standardized protocols for sample preparation, data acquisition, and analysis is essential for the reproducibility and comparability of results. However, achieving consensus on these standards across different techniques and research groups is a complex process.

Future Perspectives

Technological Advancements: Ongoing developments in instrumentation, such as higher-resolution NMR and MS, more sensitive PET tracers, and advanced imaging techniques, are expected to overcome some current technical limitations. These advancements will enhance the accuracy, resolution, and sensitivity of biophysical techniques.

Integration with Multi-Omics: Future research will likely focus on integrating biophysical data with multi-omics approaches (genomics, proteomics, metabolomics) to provide a more comprehensive understanding of cancer biology. This holistic view can uncover novel insights into the interplay between genetic mutations, protein expression, and metabolic reprogramming.

Personalized Medicine: Biophysical techniques will play a pivotal role in the development of personalized medicine. By providing detailed metabolic and molecular profiles of individual tumours, these techniques can guide the selection of targeted therapies, monitor treatment response, and predict resistance mechanisms.

Artificial Intelligence and Machine Learning: The application of artificial intelligence (AI) and machine learning (ML) in analyzing complex biophysical data is a promising future direction. These technologies can uncover hidden patterns, predict treatment outcomes, and suggest new therapeutic targets based on large datasets.

FUTURE DIRECTIONS AND INNOVATIONS

Emerging Biophysical Techniques

Single-Cell Metabolomics: Advances in single-cell analysis techniques are enabling the study of metabolic heterogeneity within tumours. Technologies like single-cell MS and single-cell RNA sequencing coupled with metabolic profiling provide insights into the metabolic states of individual cancer cells, revealing rare subpopulations and their roles in tumour progression and treatment resistance.

Quantum Sensors: Quantum sensing technologies are emerging as highly sensitive tools for detecting minute changes in metabolic activities. These sensors can measure magnetic fields, temperature, and chemical concentrations at the nanoscale, offering unprecedented resolution and sensitivity for studying cancer metabolism.

Advances in Existing Techniques

Enhanced Imaging Modalities: Advances in imaging technologies, such as super-resolution microscopy and hybrid imaging systems (e.g., PET-MRI), are improving the visualization of tumour biology at the cellular and molecular levels. These techniques provide detailed spatial and temporal information about tumour metabolism and the microenvironment.

Next-Generation Mass Spectrometry: Developments in MS, including higher resolution, faster scanning speeds, and improved ionization techniques, are enhancing the detection and quantification of metabolites and proteins. These advancements enable more comprehensive and accurate metabolic profiling of cancer cells and tissues.

Integration with Other Omics Technologies

Multi-Omics Integration: Combining biophysical techniques with genomics, transcriptomics, proteomics, and epigenomics provides a holistic view of cancer biology. Integrated multi-omics approaches can identify how genetic mutations drive metabolic reprogramming and how these changes impact protein expression and function.

Spatial Omics: Emerging spatial omics technologies, such as spatial transcriptomics and proteomics, allow the mapping of gene and protein expression within the spatial context of tissues. Integrating these techniques with imaging and metabolomics can reveal the spatial organization of metabolic processes within tumours.

Personalized Medicine Approaches

Precision Oncology: Biophysical techniques will be integral to precision oncology, where treatment strategies are tailored to the unique metabolic and molecular profiles of individual patients. By identifying specific metabolic vulnerabilities in tumours, these techniques can guide the selection of targeted therapies and monitor their effectiveness.

Real-Time Monitoring: Advances in non-invasive imaging and metabolite monitoring technologies will enable real-time assessment of treatment response. Techniques like liquid biopsy, coupled with biophysical analyses, can track metabolic changes in circulating tumour cells and cell-free DNA, providing dynamic insights into tumour evolution and therapy resistance.

SUMMARY OF KEY POINTS

In this comprehensive exploration of biophysical techniques in cancer research, we have covered several critical aspects. Firstly, understanding the intricacies of cancer metabolism and the necessity of studying metabolic reprogramming in cancer cells was highlighted. Metabolic pathways such as glycolysis, the TCA cycle, lipid metabolism, and amino acid metabolism play pivotal roles in tumour growth and progression. Biophysical techniques like nuclear magnetic resonance (NMR) spectroscopy, mass spectrometry (MS), stable Isotope labelling by amino acids in cell culture (SILAC), positron emission tomography (PET), and others have emerged as indispensable tools for probing these metabolic pathways. Each technique offers unique advantages and applications. NMR and MS provide detailed metabolic profiles, SILAC enables quantitative proteomics and metabolic flux analysis, and PET imaging with tracers like fluorodeoxyglucose (FDG) reveals glucose metabolism in tumours. Functional MRI and techniques like fluorescence resonance energy transfer (FRET) and infrared (IR) and Raman spectroscopy have broadened the scope of metabolic studies, facilitating the identification of metabolic biomarkers, therapeutic targets, and the monitoring of treatment responses. Despite these advancements, challenges persist. Technical limitations such as sensitivity, resolution, and cost, along with the complexity of data interpretation and integration with other methods, must be addressed. Nonetheless, the future of biophysical techniques in cancer research looks promising, with emerging technologies, advancements in existing techniques, and integration with other omics technologies paving the way for personalized medicine approaches.

THE SIGNIFICANCE OF BIOPHYSICAL TECHNIQUES IN CANCER METABOLISM

Biophysical techniques are at the forefront of cancer research, providing deep insights into the metabolic alterations that underpin cancer progression. These techniques have revolutionized our understanding of how cancer cells reprogramme their metabolism to support rapid growth, evade apoptosis, and adapt to hostile environments. By elucidating the specific metabolic pathways that are altered in cancer, biophysical techniques help identify novel therapeutic targets and biomarkers for early detection and diagnosis. Moreover, these techniques play a crucial role in drug development and testing. They enable the detailed analysis of drug–target interactions and the metabolic effects of therapeutic agents, facilitating the design of more effective and targeted treatments. The ability to monitor treatment response through non-invasive imaging and metabolic profiling ensures that therapies can be tailored to the individual patient, improving outcomes and minimizing side effects. The integration of biophysical techniques with other omics technologies enhances their utility further. Combining metabolic data with genomic, transcriptomic, and proteomic information provides a comprehensive view of cancer biology, revealing the complex interplay between genetic mutations, protein expression, and metabolic reprogramming. This holistic approach is essential for developing precision oncology strategies, where treatments are customized based on the unique metabolic and molecular profiles of individual tumours.

FUTURE OUTLOOK

The future of biophysical techniques in cancer research is bright, with several exciting developments on the horizon. Emerging technologies such as single-cell metabolomics and quantum sensors promise to provide unprecedented resolution and sensitivity in studying cancer metabolism. Advances in existing techniques, such as higher-resolution NMR and MS and enhanced imaging modalities, will further improve our ability to detect and analyze metabolic changes in cancer. Integration with multi-omics approaches and spatial omics technologies will offer deeper insights into the spatial and temporal dynamics of tumour metabolism, uncovering new therapeutic targets and biomarkers. The application of AI and ML in analyzing complex biophysical data will accelerate the discovery of novel patterns and therapeutic strategies.

Personalized medicine approaches, guided by biophysical techniques, will become increasingly prevalent. Real-time monitoring of treatment responses through non-invasive imaging and metabolite profiling will enable dynamic adjustments to therapy, ensuring the most effective and individualized treatment plans for patients. In conclusion, biophysical techniques are indispensable in the fight against cancer, providing critical insights into tumour metabolism, guiding the development of targeted therapies, and paving the way for personalized medicine. As these techniques continue to evolve and integrate with other scientific advancements, they will undoubtedly play a central role in improving cancer diagnosis, treatment, and ultimately, patient outcomes.

REFERENCES

Akter R, Awais M, Boopathi V, et al (2024) Inversion of the Warburg effect: Unraveling the metabolic nexus between obesity and cancer. *ACS Pharmacol Transl Sci* 7:560–569. https://doi.org/10.1021/acsptsci.3c00301

Brindle KM (2024) Imaging cancer metabolism using magnetic resonance. *npj Imaging* 2:1–5. https://doi.org/10.1038/s44303-023-00004-0

Chai X, Zhang Y, Zhang W, et al (2024) Tumor metabolism: A new field for the treatment of glioma. *Bioconjug Chem*. https://doi.org/10.1021/acs.bioconjchem.4c00287

Chen J, Cui L, Lu S, Xu S (2024) Amino acid metabolism in tumor biology and therapy. *Cell Death Dis* 15:42. https://doi.org/10.1038/s41419-024-06435-w

Daimiel Naranjo I, Bhowmik A, Basukala D, et al (2024) Assessment of hypoxia in breast cancer: Emerging functional MR imaging and spectroscopy techniques and clinical applications. *J Magn Reson Imaging*. https://doi.org/10.1002/jmri.29424

Dou Q, Grant AK, Coutinto de Souza P, et al (2024) Characterizing metabolic heterogeneity of hepatocellular carcinoma with hyperpolarized 13C pyruvate MRI and mass spectrometry. *Radiol Imaging Cancer* 6:e230056. https://doi.org/10.1148/rycan.230056

Ganapathy-Kanniappan S, Geschwind J-FH (2013) Tumor glycolysis as a target for cancer therapy: Progress and prospects. *Mol Cancer* 12:1–11. https://doi.org/10.1186/1476-4598-12-152

Gebhart G, Keyaerts M, Guiot T, et al (2024) Optimal [18F]FDG PET/CT cutoff for pathologic complete response in HER2-positive early breast cancer patients treated with neoadjuvant trastuzumab and pertuzumab in the PHERGain trial. *J Nucl Med* 65:708–713. https://doi.org/10.2967/jnumed.123.266384

Grooms AJ, Burris BJ, Badu-Tawiah AK (2024) Mass spectrometry for metabolomics analysis: Applications in neonatal and cancer screening. *Mass Spectrom Rev* 43:683–712. https://doi.org/10.1002/mas.21826

Ibáñez-Molero S, Pruijs JTM, Atmopawiro A, et al (2024) Phosphoprotein dynamics of interacting T cells and tumor cells by HySic. *Cell Rep* 43. https://doi.org/10.1016/j.celrep.2023.113598

Lan J, Liu L, Li Z, et al (2024) A multi-signal mitochondria-targeted fluorescent probe for simultaneously distinguishing biothiols and realtime visualizing its metabolism in cancer cells and tumor models. *Talanta* 267:125104. https://doi.org/10.1016/j.talanta.2023.125104

Li X, Yang G, Zhang S, Ren H (2024) Advancing clinical cancer care: Unveiling the power of surface-enhanced Raman spectroscopy. *J Raman Spectrosc* 55:429–444. https://doi.org/10.1002/jrs.6637

Ma X, Fernández FM (2024) Advances in mass spectrometry imaging for spatial cancer metabolomics. *Mass Spectrom Rev* 43:235–268. https://doi.org/10.1002/mas.21804

Martinez Outschoorn UE, Peiris-Pagés M, Pestell RG, et al (2017) Cancer metabolism: a therapeutic perspective. *Nat Rev Clin Oncol* 14:11–31. https://doi.org/10.1038/nrclinonc.2016.60

Montrazi ET, Sasson K, Agemy L, et al (2024) Molecular imaging of tumor metabolism: Insight from pyruvate- and glucose-based deuterium MRI studies. *Sci Adv* 10:eadm8600. https://doi.org/10.1126/sciadv.adm8600

Ortega Duran M, Shaheed S ul, Sutton CW, Shnyder SD (2024) A proteomic investigation to discover candidate proteins involved in novel mechanisms of 5-fluorouracil resistance in colorectal cancer. *Cells* 13:342. https://doi.org/10.3390/cells13040342

Peng F, Ai X, Sun J, et al (2024) Recent advances in FRET probes for mitochondrial imaging and sensing. *Chem Commun* 60:2994–3007. https://doi.org/10.1039/D4CC00018H

Pujana-Vaquerizo M, Bozal-Basterra L, Carracedo A (2024) Metabolic adaptations in prostate cancer. *Br J Cancer*. https://doi.org/10.1038/s41416-024-02762-z

Reis JC, Travado L, Heller AS, et al (2024) Greater perceived stress management skills and heightened brain metabolic activity in cortical and subcortical stress processing regions in metastatic breast cancer patients. *Brain Imaging Behav* 18:130–140. https://doi.org/10.1007/s11682-023-00821-2

Sezer G, Sahin F, Onses MS, Cumaoglu A (2024) Activation of epidermal growth factor receptors in triple negative breast cancer cells by morphine; analysis through Raman spectroscopy and machine learning. *Talanta* 272:125827. https://doi.org/10.1016/j.talanta.2024.125827

Sorrells JE, Park J, Aksamitiene E, et al (2024) Label-free nonlinear optical signatures of extracellular vesicles in liquid and tissue biopsies of human breast cancer. *Sci Rep* 14:5528. https://doi.org/10.1038/s41598-024-55781-4

Sun X, Wang S, Wong CCL (2024) Mass spectrometry–based proteomics technology in pancreatic cancer research. *J Pancreatol* 7:145. https://doi.org/10.1097/JP9.0000000000000152

Teng M, Li Z, Gu Y, et al (2024) Real-time monitoring of glucose metabolism and effects of metformin on HepG2 cells using 13C in-cell NMR spectroscopy. *Biochem Biophys Res Commun* 694:149383. https://doi.org/10.1016/j.bbrc.2023.149383

Tricarico P, Chardin D, Martin N, et al (2024) Total metabolic tumor volume on 18F-FDG PET/CT is a game-changer for patients with metastatic lung cancer treated with immunotherapy. *J Immunother Cancer* 12:e007628. https://doi.org/10.1136/jitc-2023-007628

Vogel FCE, Chaves-Filho AB, Schulze A (2024) Lipids as mediators of cancer progression and metastasis. *Nat Cancer* 5:16–29. https://doi.org/10.1038/s43018-023-00702-z

Wumener X, Zhang Y, Zang Z, et al (2024) The value of dynamic FDG PET/CT in the differential diagnosis of lung cancer and predicting EGFR mutations. *BMC Pulm Med* 24:227. https://doi.org/10.1186/s12890-024-02997-9

Yamamoto T, Nakayama J, Urabe F, et al (2024) Aberrant regulation of serine metabolism drives extracellular vesicle release and cancer progression. *Cell Rep* 43. https://doi.org/10.1016/j.celrep.2024.114517

Yu Y, Jiang Y, Glandorff C, Sun M (2024) Exploring the mystery of tumor metabolism: Warburg effect and mitochondrial metabolism fighting side by side. *Cell Signal* 120:111239. https://doi.org/10.1016/j.cellsig.2024.111239

Zhao Y, Ma C, Cai R, et al (2024) NMR and MS reveal characteristic metabolome atlas and optimize esophageal squamous cell carcinoma early detection. *Nat Commun* 15:2463. https://doi.org/10.1038/s41467-024-46837-0

Zheng C, Yu L, Zhao L, et al (2024) Label-free Raman spectroscopy reveals tumor microenvironmental changes induced by intermittent fasting for the prevention of breast cancer in animal model. *Spectrochim Acta A Mol Biomol Spectrosc* 317:124387. https://doi.org/10.1016/j.saa.2024.124387

Zniber M, Lamminen T, Taimen P, et al (2024) 1H-NMR-based urine metabolomics of prostate cancer and benign prostatic hyperplasia. *Heliyon* 10. https://doi.org/10.1016/j.heliyon.2024.e28949

2 Applications of Spectroscopy in Biology and Medicine

*Nidhi S, Jackson Rodrigues, Darshan Chikkanayakanahalli Mukunda, Shaik Basha, Shimul Biswas, and Krishna Kishore Mahato**

Department of Biophysics, Manipal School of Life Sciences, Manipal Academy of Higher Education, Manipal, Karnataka, India
*Corresponding Author

INTRODUCTION

Spectroscopy plays an important role in diagnosing biological and medical disorders by revealing information on the chemical composition, structure, and dynamics of biological molecules, thereby providing invaluable insights into various biological systems. There are different spectroscopy techniques that are in use to understand the properties and dynamics of biological macromolecules. These techniques are X-ray fluorescence spectroscopy, laser magnetic resonance imaging (MRI) spectroscopy, Raman spectroscopy, Fourier transform infrared (FT-IR) spectroscopy, reflectance spectroscopy, fluorescence spectroscopy, infrared (IR) spectroscopy, photoacoustic (PA) spectroscopy, and laser-induced breakdown spectroscopy (LIBS).

INFRARED SPECTROSCOPY

Infrared (IR) spectroscopy's electromagnetic spectrum analyzes molecules' absorption properties in the infrared region. An absorption frequency/wave number represents it on the x-axis, with the associated intensity/abundance on the y-axis. IR spectroscopy is a technique for identifying the different wavelengths of infrared light absorbed by molecules. This method is effective because of the specific frequency of light absorbed by different bonds within the molecules (Barth, A. 2007). However, the bond interaction and infrared radiation are based on the bonds' polarity. Polar bonds can interact with electromagnetic radiation in the infrared region. Polar molecules exhibit positive and negative charges permitting vibrational transitions upon electromagnetic excitation. When the vibrational energy changes, it alters the molecule's dipole moment. The polarity of the bond determines the strength of absorption. Bonds that are symmetrical and non-polar, such as those found in O=O and N≡N, are not absorbed by electromagnetic radiation (EMR) since they cannot interact with the electric vector of electromagnetic radiation (Richmond, G. L. 2002).

The basic principle of IR spectroscopy is the absorption of EMR by the molecules. These frequencies depend on factors like the shape and surface of the molecules, the vibronic coupling associated with them, and the atomic masses involved. When light hits the molecule, it can absorb energy, resulting in increased vibration of the functional region, fingerprint region, etc. The spectral region 1500–4000 cm^{-1} is considered a functional group region which indicates the different functional groups shown in Figure 2.1. The fingerprint area, which lies in the mid-IR region between 500 cm^{-1} and 1500 cm^{-1}, gives molecular characteristics of specific substances such as proteins (Kassem et al. 2023).

DOI: 10.1201/9781003545712-2

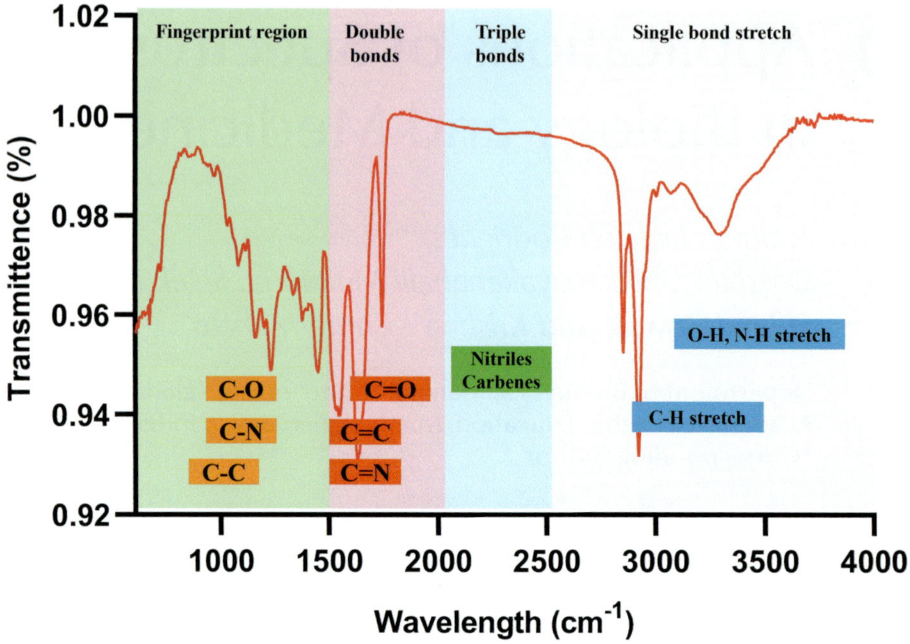

FIGURE 2.1 Chart showing the IR bands produced by the major functional groups and the fingerprint region specific to each compound. Adapted from Principles of the Fourier-Transform Infrared Spectroscopy (FT-IR). | Download Scientific Diagram. (n.d.). Retrieved October 16, 2023, from https://www.researchgate.net/figure/Principles-of-the-Fourier-Transform-Infrared-Spectroscopy-FT-IR_fig2_333498985.

FT-IR Spectroscopy

FT-IR is an analytical tool used for the structural characterization of unknown compounds. The underlying principle of all infrared spectroscopic methods is that the absorption of IR radiation occurs as it passes through a given sample, while the remaining, unabsorbed radiation is transmitted and is noted. The unique spectra produced by different molecules, based on their structural characteristics, enable the identification and differentiation of molecules. Like fingerprints or DNA, individuality is a unique characteristic (Figure 2.2).

Overall, FT-IR spectroscopy is an effective tool for analyzing the composition of a given sample, providing valuable insights into its molecular composition and structural properties. FT-IR is a better way to look at infrared light for a few reasons. First, you don't need to destroy the sample. Second, it's much faster than IR measurements. Third, it's more precise and can detect even tiny amounts of samples under study. FT-IR has many benefits due to the use of an interferometer as an infrared source. This allows for faster and more accurate measurements. An interferogram is a graph in contrast. The FT-IR spectrum determines their frequency based on time, making it possible to create a recognizable infrared spectroscopy spectrum. FT-IR spectroscopy has many applications, such as identifying the structure of unknown compounds, metal analysis, food analysis, organic synthesis, and pharmaceuticals. The technique can also be used to monitor chemical reactions and figure out how they work. FT-IR spectroscopy is very sensitive and reliable. It helps researchers learn more about the materials and compounds and their structure elucidation (Kassem et al. 2023).

Raman Spectroscopy

In 1930, the Nobel Prize in Physics was received by Sir C V Raman and his student K S Krishnan for the discovery of the Raman effect. This tool is based on light photon inelastic scattering at

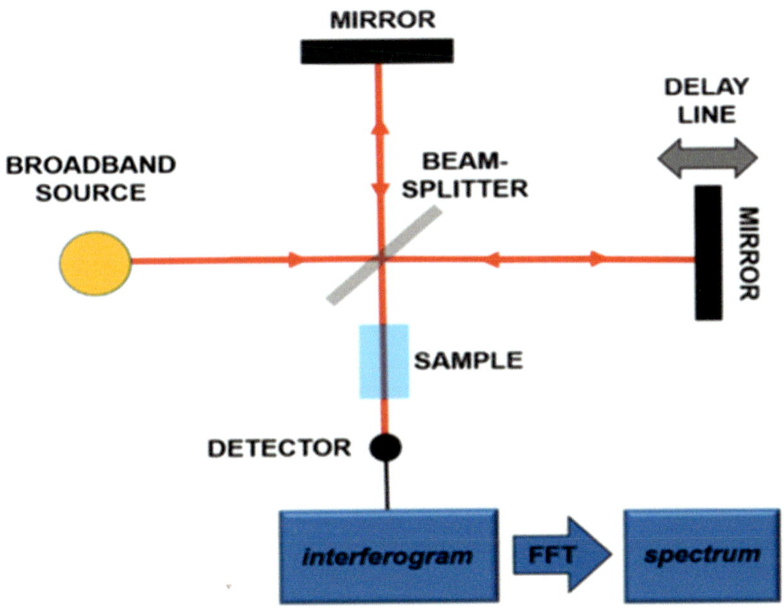

FIGURE 2.2 Schematic diagram of FT-IR spectrophotometer (principles of Fourier transform infrared spectroscopy (FT-IR)). Adapted figure from Barbara Patrizi et al. 2019 copyright © International Journal of Molecular Sciences, MDPI 2019.

suitable excitation wavelengths, which provides the vibrational properties of the biomolecules (Wan et al. 2017; Figure 2.3).

The Raman shift is the distinction between the dispersed light and incident light. When the light is scattered, the photon has lower energy than the energy of the incident light, known as Stokes scattering. When the kinetic energy of the scattered photon is greater than the energy of the incident photon, anti-Stokes scattering occurs. Due to differences in vibrational and rotational characteristics, the scattered photons display a different frequency to the incident photons. This variation leads to a change in wavelength, which is also studied through IR spectroscopy. However, Stokes signals, being stronger compared to anti-Stokes signals at low temperatures, are normally recorded. Raman spectroscopy uses polarized light sources (laser light) for sample excitation and charge-coupled device (CCD) detectors for signal detection (Kauffmann et al. 2019).

Fluorescence Spectroscopy

Fluorescence spectroscopy (FS) is the most common technique to investigate structural changes in macromolecules such as proteins, caused by stress, pH, heat, ionic strength, etc. It is an extremely accurate technique to study different biological fluorophores such as tryptophan, tyrosine, and phenylalanine and can monitor their spectral signatures constantly. When an excitation is triggered in an atom of the molecule, it will be moved from the ground state to the singlet state, and they eventually decline to the excited singlet state's lowest translational level of energy. To return the atom back to its ground state, energy will be less due to emitting photons. External conversion, internal conversion, and intersystem crossover are the three nonradiative relaxation steps that fluorescent molecules can go through without the electrical energy of excitation being converted into photons. The electrons proceed from a state of greater energy to a lower energy level during internal conversion, which occurs when the amount of energy difference between two states of electricity is comparatively modest. The energy will be transferred to the vibrational modes of the electric state. The

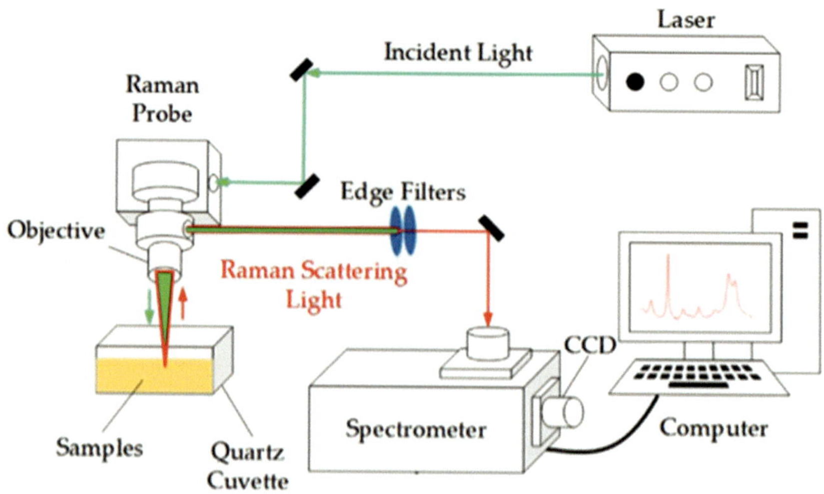

FIGURE 2.3 Schematic diagram of Raman spectrophotometer. Adapted figure from Wan et al., 2017 copyright © Energies, MDPI 2017.

intensity of fluorescence decreases as the temperature rises because vibrational activities are heat driven. Due to the collisional quenching caused by molecules of solute near the fluorophore undergoing external transformation, the energy is lost (dos Santos Rodrigues et al. 2023) (Figure 2.4).

The photons are emitted during phosphorescence when molecular transitions transition back to the ground state. Phosphorescence peaks manifest at longer wavelengths than fluorescence peaks because triplet states have less energy than singlet states. Phosphorescence lasts longer (10^{-4}–10^{-2} seconds) than fluorescence (10^{-9}–10^{-6} seconds) since these transitions are additionally prohibited. Prolonged lifetimes also result in intermolecular collisions, solvent migration during the quenching process, and heat degradation by oxygen. As a result, samples must be frozen to liquid nitrogen

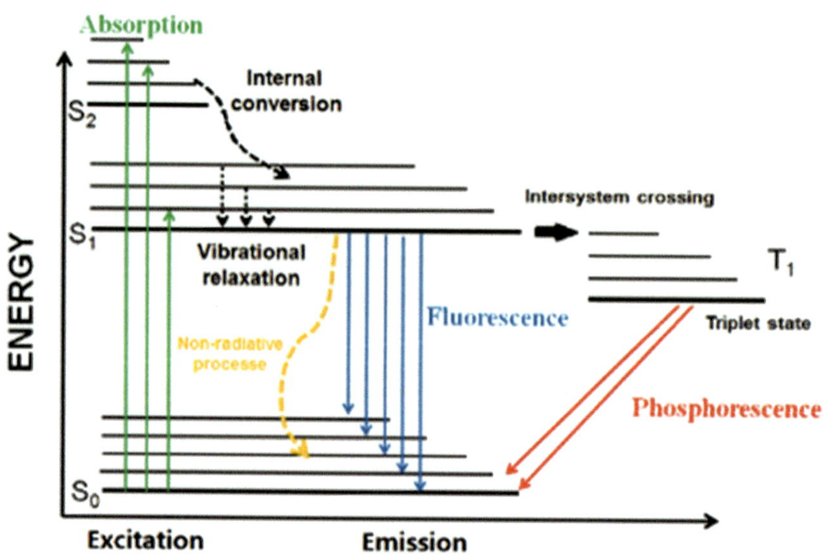

FIGURE 2.4 Jablonski diagram showing fluorescence. Adapted figure from Thorsten Schweizer et al., 2021 copyright © Automotive and Engine Technology, Spinger Nature 2021.

levels to see phosphorescence, which is frequently not visible at room temperature. The characteristics and intensity of the fluorescence generated by the substance's photoluminescence are the basis for fluorescent analysis. The analysis of the biological macromolecular structure and characteristics is one example of how it is now widely utilized for evaluating both the chemical and physical characteristics of an organism and its evolution (Application of Fluorescence Spectroscopy—Creative Proteomics n.d.).

Fluorescence spectroscopy can be used to examine solid powdered substances, crystals, films, liquids, and other objects. Depending on the sample, a quartz cell (liquid sample) or a solid sample holder (powder or sheet sample) can be used. Fluorescence analysis has the following advantages: (1) high sensitivity; (2) strong selectivity; (3) small sample volume; (4) additional physical parameters; and (5) ease of application. However, there are certain drawbacks, such as a limited application range and sensitivity to environmental conditions (various interference effects).

Fluorescence spectroscopy is primarily used to identify the presence of specific substances in specimens used for biotechnology and immunoassay analysis, such as the identification of substances known as deoxyribonucleic acid (DNA), immunoglobulin antigens, and other research elements (Camarca et al. 2021). For analysis and detection, various fluorescent probes are used. Biological nanofluorescent probes and biological non-nanofluorescent probes are the two main types. The advancement of nanotechnology has created a new discipline of fluorescence analysis. Nanomaterials offer outstanding spectral properties such as good fluorescence, wide excitation, and narrow emission, making them significant objects in fluorescent analysis that researchers are very interested in. Fluorescence spectroscopy is employed in different areas of biology such as protein fingerprinting, pharmacokinetics, drug analysis, pharmaceutics, etc. There are three different types of pharmaceutical fluorescence analysis: straight fluorescent evaluation, indirect fluorescence evaluation, and nanofluorescence assessment. Conventional fluorescence spectroscopy was originally used to examine the antimalarial drug, quinine. Fluorescence analysis has become widely employed in the investigation of antibacterial medications, analgesics, sedatives, and hemostatic pharmaceuticals since its invention (Orme et al. 2018; Devi et al. 2019).

Photoacoustic Spectroscopy

Photoacoustic spectroscopy relies on light-induced radiation to generate the acoustic waves measurable by ultrasonography detectors (Dogan et al. 2019). The chromophores/fluorophores within a specific sample absorb irradiation energy via the closely separated vibrational and rotational levels of energy of the excited states in various molecular orbitals (Buma et al. 2018). This is depicted in detail in pictorial form in the Jablonski graph, as shown in Figure 2.4. The absorption itself takes place in a time span of 10^{-15} seconds. Whenever there is absorption by a sample at a particular wavelength of light, a host of dramatic changes occurs within the sample. The electronic structure of each absorbing molecule rearranges itself, and the local molecular environment rushes to keep up the equilibrium. Fluorophores are often stimulated and shift to higher vibrational levels of the S1 or S2 singlet energy states (Priya et al. 2010; Rodrigues et al. 2021).

In a time span of picoseconds after absorption, the excited molecules relax and move down, reaching the first excited state's lowest vibrational state S1, and the vibronic energy is deposited into the surroundings as heat, usually causing a minimal expansion in the sample volume. At this step, there is a loss of energy with no light emission (non-radiative step) (Schmid et al. 2006; Sun et al. 2020). Since a significant number of vibrational cycles take place during the lifespan of the excited state, molecules almost invariably experience full vibrational relaxation. When the surplus vibrational energy collides with the excited state of the fluorescent molecules, it becomes heat, which is absorbed by nearby solvent molecules. Internal conversion, also known as vibrational relaxation, is a process where energy is lost without any form of radiation and is released as heat. This process occurs in the time range of picoseconds or less.

It is this internal conversion process that gives rise to a photoacoustic signal when the sample is irradiated by modulated light. The transitions from the S1 level to the S0 level may happen in the form of fluorescence or through intersystem crossing to T1 by phosphorescence as radiative relaxations. The nonradiative release of energy, which is in the form of heat, contributes to an acoustic pressure wave, containing an imprint of the molecular events that changed the excitation volume of the sample and can be detected by using suitable detectors like piezoelectric transducers. In the present study, the signal that we get is a pulsed laser-induced photoacoustic (PA) waveform (Gorey et al. 2019).

The various steps involved in the generation of a photoacoustic signal are represented sequentially, showing irradiation by a modulated or pulsed laser light that leads to optical absorption by the specimen molecules, resulting in the heating of the illuminated volume. The repeated heating and cooling of the illuminated volume give rise to thermal expansion and contraction, which produces the acoustic wave/PA signal within the sample. This PA waveform contains information on absorption by the specimen and changes in molecular conformations followed by pulsed/modulated excitations. Suitable analysis of the PA waveform can not only provide us with information regarding the time attributes of signal generation but also provide information about the optical properties of the samples, such as absorption coefficient, thermal diffusivity, acoustic velocity, etc. A mere PA waveform hosts multiple pieces of information and the first amplitude, its position, and time of appearance hide a plethora of information about the sample and its characteristics. Thus, PA spectroscopy, which offers a host of information, can be explored to evaluate various samples based on their intrinsic characteristics (Priya 2016).

Reflectance Spectroscopy

In reflection spectroscopy (RS), the spectral characteristics of surface-reflected radiation are studied in relation to the intensity of the incident primary energy. It's important to be aware of two exceptions: the first concerns the specular reflections from a smooth surface, and the second concerns the diffuse light reflected from a smooth surface. Between these two extremes are all conceivable variants. Two reflectance spectroscopy techniques that are fundamentally different are used for the two extreme cases: the first of them entails using the Fresnel equations to calculate graphic x (absorption index) and variables n (refractive index) as a function of wave A from the observed regular reflection. Fahrenfort has updated this old and problematic approach, which is typically incapable of producing particularly accurate findings, by switching the boundary of the air-sample stage with the boundary of phase over an insulator of refractive index (n) with the sample (n). When the specimen fails to absorb any radiation and its plane of incidence exceeds a predetermined set amount, total reflection occurs. Due to diffraction at the incident beam's edges, a small amount of energy passes into the less-dense phase when the two phases come into close optical contact. However, because the energy flux in both directions via the phase barrier is equal, the entire reflection occurs (Mirabella et al. 2020).

RS is a relatively new optically based endoscopic method that is quickly gaining traction. It is a safe and effective way of assessing tissue for ischemia. Many of the constraints involved with histopathology processing and diagnosis are resolved by this technology. RS is a straightforward technology for endoscopy. Because of a defined geometry of the light, the procedure is more trustworthy. This technique detects the intensity and the color of reflected light. Reflected light has the same wavelength as incident light, as opposed to autofluorescence spectroscopy, where different wavelengths are simultaneously absorbed and reflected to varying degrees (Wallace et al. 2009).

Whenever white light flashes over oxygenated hemoglobin, the bulk of the blue spectrum is absorbed and the red spectrum is reflected, giving the hemoglobin its unique hue. On the other hand, when exposed to white light, deoxygenated hemoglobin appears bluer because it absorbs more red light. RS is a technique that offers insights into hemoglobin concentration and oxygenation status in tissues. Because cancerous tissues encourage blood vessel formation, RS might potentially detect cancerous tissue by evaluating hemoglobin parameters. Light scattering is a type of spectroscopy known as reflectance spectroscopy that takes advantage of the elastic scattering capabilities of tissue structures. It measures the extent to which structures like nuclei, mitochondria, and collagen networks alter the angular trajectories of light. These concepts have since been expanded into various instruments that may have diagnostic uses (Shirkavand et al. 2023).

Laser-Induced Breakdown Spectroscopy

Laser-induced breakdown spectroscopy (LIBS) is a powerful method of assessment for materials identification and characterization (Ermilov et al. 2006). A laser pulse is focused on a liquid or solid surface, or into a specimen that contains a gas, liquid, or cloud of aerosolized particles (Esenaliev et al. 1997). This produces a microplasma containing both excited molecular and ionic constituents from the bulk sample and ablated particles (Figure 2.5).

These excited-state species produce light with distinct wavelengths that may be measured with a spectrometer and processed by a computer (Oraevsky et al. 1994; Oraevsky et al. 2018). LIBS can identify all elements without the requirement for considerable sample preparation because each element has a distinct emission spectrum. LIBS can also be performed remotely and in real-time, making it an excellent alternative for practical applications (LIBS, n.d.).

Magnetic Resonance Imaging Spectroscopy

Magnetic resonance spectrometry (MRS) is the arrangement of an atom surrounded by electrons with nuclei in different molecules encountered by an electromagnetic field. It leads to slightly distinct resonance frequencies and impulses. It is basically the identical technology to nuclear magnetic resonance, or NMR, in chemical analysis, but it is also known as MRS in the field of medicine to avoid any misunderstandings or issues related to the word "nuclear." MR spectroscopy may collect data from any nucleus with a non-zero spin, which is known as an "NMR-active" nucleus.

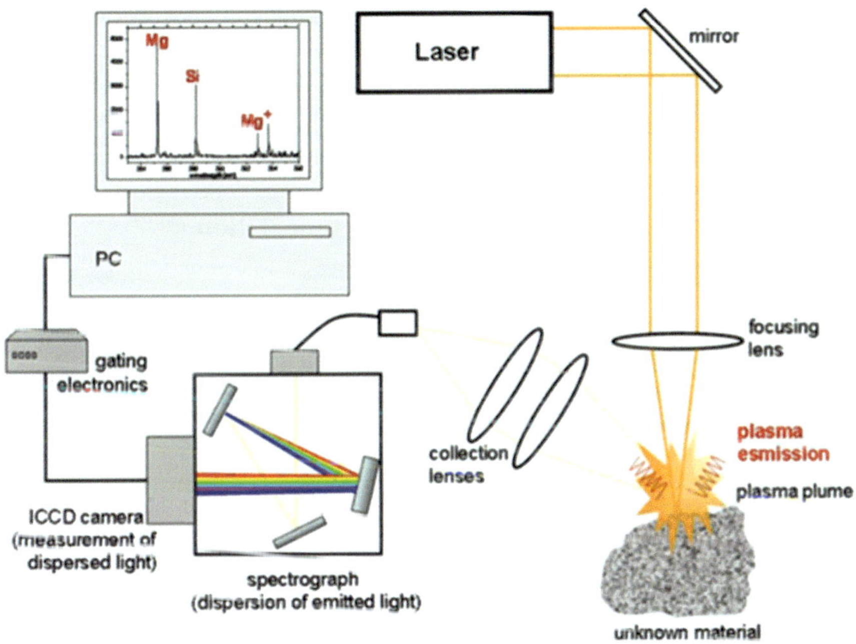

FIGURE 2.5 Schematic diagram of laser-induced breakdown spectroscopy (LIBS). Adapted figure from Thorsten Gaviola et al., 2021 copyright © 2021 Elsevier Ltd.

Protons, carbon-13, and phosphorus-31 are the most observed NMR-active nuclei. Proton spectra are commonly employed in clinical practice because they allow for precise measurement of profiles of metabolites in vivo. Phosphorus-31 (P-31) is commonly used to calculate the ratio of adenosine triphosphate (ATP) to phosphocreatine, as well as additional metabolites, and can be used to calculate the cell's energy change. Water suppression is an important step in MRS processes to avoid water from influencing the spectrum and rendering other spectra undetectable. Inversion recovery or chemical shift selective (CHESS) procedures can be used to accomplish this. If water suppression fails, a general gradient to the baseline can occur, affecting the height distribution of peaks. Magnetic resonance spectroscopy (MRS) employs a variety of pulse patterns, the most basic of which is a 90-degree radiofrequency (RF) pulse (Tognarelli et al. 2015) (Figure 2.6).

This pattern, which does not include any curves, entails the RF coil receiving the signal right after the single RF pulse is emitted. Imaging sequences, such as the spin echo sequence, can also be utilized for spectroscopy. There is a major difference, however, between imaging and spectroscopic sequences. A read-out gradient is not employed in spectroscopy while the RF coil receives the electrical signal from the individual or object under inspection. Instead of providing spatial or positional data, frequency information is employed to discriminate between distinct chemical substances. This is possible because, depending on the molecule and its position, the electron cloud around diverse compounds in chemicals shields the resonance-sensitive atoms of spectroscopic interest differently. The shielding causes the atoms' measured frequency of resonance to change slightly, allowing confirmation with MRS (Dager et al. 2008).

X-RAY FLUORESCENCE SPECTROSCOPY

The excitation state of the specimen's atoms serves as the foundation for the X-ray fluorescence (XRF) technique. A nucleus's inner shell electrons are directed by X-rays produced by a light source, which discharges electrons from an atom. The energy difference between the two electron outer coverings is equal to the energy of the fluorescent lighting. As a result, the intensity serves as atom-specific and determines atoms in the specimen. The material being studied will emit a range of X-rays with various energies because it contains several atoms. An energy dispersive XRF sensor's semiconductor detector gathers fluorescent emissions. The power of the energy that enters determines the electrical signals that the X-rays produce in the detectors. The signals are recorded using an integrated analyzer. This technique quickly analyzes each X-ray, one at a time. The detector of a contemporary XRF device will process 1 million counts per second. The spectrum can

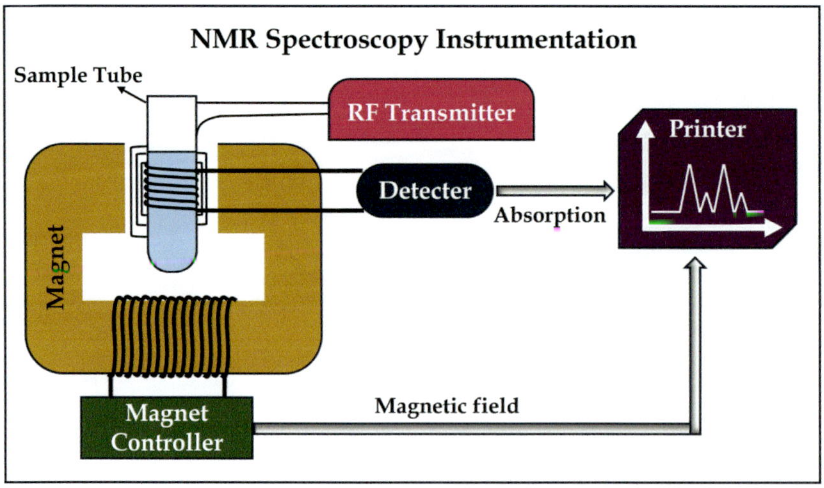

FIGURE 2.6 Schematic diagram of magnetic resonance imaging spectroscopy.

yield enough information, even with a small observation time frame, to compute intensity levels, which can then be utilized to determine the specimen's morphology. Better information is produced by using a longer measurement duration, which raises precision and peak-to-background proportions and, as a result, increases detection limits. More counts must be acquired in order to analyze a component that is present in the sample due to its sensitivity. If the specimen contains a higher concentration of a substance and the detector can manage a higher count percentage, it is reasonably simple; however, if the element concentrations are lower and the detection system can only tolerate a lower count rate, it becomes more challenging.

An X-ray fluorescence (XRF) spectrophotometer is a device that employs X-rays to analyze the chemical composition of minerals, rocks, and particles. It is based on the wavelength-dispersive spectroscopic idea, similar to an electron microprobe (EPMA), but with restrictions on smaller spot size examination. As a result, it is mostly used in bulk investigations of minerals. The preparation of samples is simple and inexpensive, and X-ray spectrometers are reliable and straightforward to use, making XRF a popular approach for studying both major and trace elements in mineral deposits and silt.

The motion of atoms when they contact high-energy, short-wavelength photons such as X-rays is the basis for XRF examination of rocks and soils. When a particle is excited, it can become ionized, and if the electromagnetic field is strong enough to dislodge a firmly bound inner electron, an atom becomes unstable. Since the outside electron's orbit has less binding energy than the interior electron's orbital, the outer electron replaces the absence of the interior electron and releases energy as a result. This energy is emitted in the form of fluorescence radiation, which has lower power than the main incident X-rays. The resulting fluorescence X-rays are obtained by identifying the proportions of the elements found in the material utilizing the distinctive energy of the produced photon, especially with regard to the transition that takes place between electron orbital configurations.

Techniques for vibratory spectroscopy are flexible tools with diverse applications. Biological investigation, IR spectroscopy, and FT-IR are examples. Radioactive MRI resonance, X-ray spectroscopy, magnetic resonance imaging, laser-induced degradation spectroscopy, and the use of imaging and other techniques are employed.

UV-visible spectroscopy can be used to quantify nucleic acids, proteins, and metabolites while also providing information on their quantity, structure, and interactions. Fluorescence spectroscopy allows for the research of molecular dynamics, changes in conformation, and binding interactions while studying the fluorescence properties of biomolecules. IR spectroscopy can aid in the diagnosis of disorders in biofluids such as blood or urine by recognizing functional groups and analyzing the chemical composition of biological samples. Raman spectroscopy provides information on vibrations within molecules that aid in the identification of substances, particularly in the detection and research of cancer medicines. While mass spectrometry (MS) precisely detects molecular masses to aid in identifying illnesses and the development of drugs, the use of nuclear magnetic resonance (NMR) spectroscopy is critical in understanding physiological molecules' 3D shapes and dynamics. Optical coherence tomography (OCT), which is beneficial in fields such as ophthalmology and cardiology, uses near-infrared light to create high-resolution pictures of biological tissues. X-ray spectroscopy, including crystallography, aids in comprehending the microscopic and molecular makeup of biomolecules. Electron spin characteristics in molecules are examined using electron paramagnetic resonance (EPR) analysis in structural biology and biochemistry. Magnetic resonance imaging (MRI), a critical screening tool, employs NMR techniques to generate exact images of interior structures. The LIBS (laser-induced breakdown spectrometry) is a tool to characterize the substances contained in biological material in forensic research, archaeological research, and environmental monitoring. Functional imaging tools like MRI and positron emission tomography (PET) scan used to analyze brain function and metabolism, assisting in the diagnosis and understanding of neurological illnesses. These numerous spectroscopic approaches, when combined, expand the comprehension of biology and improve medical operations. They are critical to the comprehension of biology and have numerous applications, including observing the environment, drug research, and medical diagnostics. To better grasp the complicated interactions between light and proteins, it is critical to understand the Jablonski diagram when evaluating spectroscopic data collected from biological samples. It describes the electronic energy levels and transitions that occur throughout the absorption and emission processes. This understanding is critical for the growth of biological applications in a variety of areas. A Jablonski diagram is the simplest approach to illustrate energy transitions among the molecule's electronic and vibrational properties that include radiative and non-radiative energy. This shows how light interacts with molecules to affect their electrically charged levels and transitions. It facilitates the visualization of absorption, internal transformation, resonant unwinding, luminescence, and phosphorescence processes. To depict the various potential energy levels, the lines in a Jablonski diagram are stacked in two columns, with energy on the vertical axis. The various levels are arranged horizontally based on the variety of electronic states. It consists of the ground state (S0), which has the lowest energy, and excited states (S1, S2, etc.) where particles may shift when photons are absorbed. The ascending transition to higher-energy levels is represented by absorption (A), which is accompanied by internal conversion (IC) and relaxation of vibration (VR), where surplus energy is dissipated. Some molecules fluoresce (F), generating lower-energy photons when they return to their neutral state, whereas others phosphoresce (P), releasing light over a longer duration from long-lived excited states, as illustrated in Figure 2.7.

Vibrational spectroscopy techniques are versatile tools with numerous applications, such as biochemical investigation, fluorescence spectroscopy (FS), Raman spectroscopy (RS), infrared (IR) spectroscopy, X-ray spectroscopy (XRD), nuclear magnetic resonance (NMR), laser-induced breakdown spectroscopy (LIBS), functional imaging (FI), etc. Proteins, nucleic acids, and metabolites can all be quantified using UV-visible spectroscopy, which also offers information about their concentration, structure, and interactions. When examining the fluorescence characteristics of biomolecules, fluorescence spectroscopy allows the investigation of molecular dynamics, conformational changes, and binding interactions. By identifying functional groups and analyzing the chemical makeup of biological samples, IR spectroscopy can help diagnose diseases in biofluids like blood or urine. RS offers data on molecular vibrations that help identify compounds, especially in the diagnosis and development of drugs for cancer. While MS precisely measures molecular masses to aid in disease diagnosis and drug discovery, NMR is crucial for

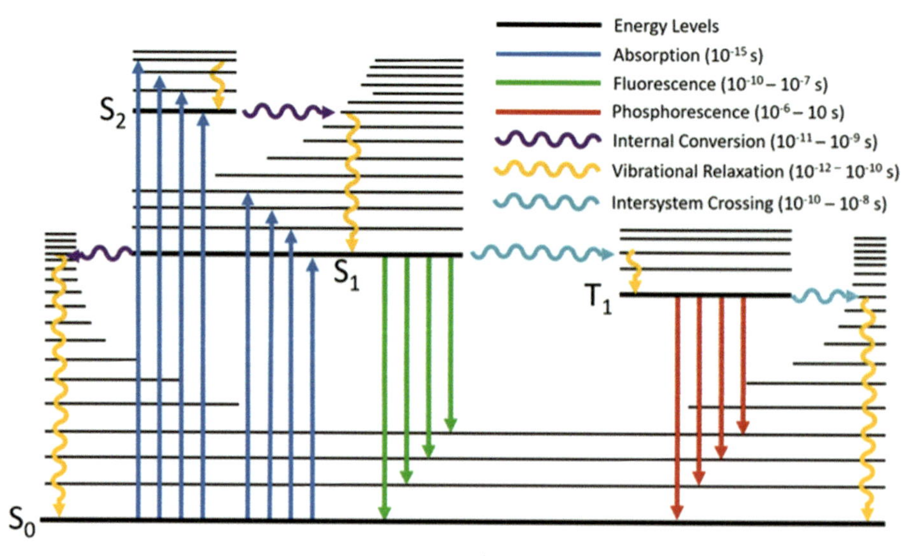

FIGURE 2.7 A typical Jablonski diagram showing the possible radiative and non-radiative transitions (Reprinted from Insero et al., 2023, copyright @ 2023 SAGE Publications Ltd).

understanding the 3D structures and dynamics of biological molecules. OCT, which is useful in areas like ophthalmology and cardiology, produces high-resolution images of biological tissues using near-infrared light. Understanding the atomic and molecular structures of macromolecules is facilitated by X-ray spectroscopy, including crystallography. For structural biology and biochemistry, electron spin properties in molecules are investigated by electron paramagnetic resonance (EPR) spectroscopy. NMR principles are used by MRI, a crucial diagnostic tool, to create precise images of internal structures. LIBS is a technique used in forensic science, archaeology, and environmental monitoring to determine the elements present in biological samples. Functional imaging methods like PET and fMRI study brain function and metabolism, aiding in the diagnosis and comprehension of neurological disorders. Together, these various spectroscopic techniques advance our understanding of biology and boost medical procedures. It is essential to expand our understanding of biology, environmental monitoring, drug development, and medical diagnosis (Haschke 2014).

ASSESSMENT OF BREAST CANCER USING SPECTROSCOPIC TECHNIQUES

The most frequently discovered malignancy and the main reason for cancer-related deaths globally is female breast cancer. It accounts for 11.7% of all cancer cases. It is the most frequently recognized carcinoma in women and the main cause of cancer mortality, accounting for one out of every four cancer cases and one out of every six cancer deaths, ranking first in most nations (Sung et al. 2021). Previous studies have demonstrated that early detection and treatment of breast cancer may significantly reduce the long-term death rate from the disease (Migowski 2015).

When a woman is diagnosed with breast cancer, she has two surgical options: mastectomy and breast-conserving surgery (BCS) (Kamath et al. 2011). Mastectomy is the surgical removal of the whole breast. BCS involves the removal of cancerous or abnormal tissue from the breast, as well as some surrounding healthy tissue, retaining as much normal breast as possible. The main benefit of BCS is that the patient retains most of their breast (Keynes et al. 1928). Breast-conserving surgery preserves normal breast tissue while keeping the breast's cosmetic appearance and is a less extensive surgery (Kosik et al. 2019). Therefore, it has been demonstrated that breast-conserving surgeries improves patients' quality of life (Curran et al. 1998; Nissen et al. 2001; Arndt et al. 2008). Literature states that women who undergo breast-conserving surgery have the same long-term survival percentages as those who have mastectomy surgery (van Dongen 2000; Christiansen et al. 2018; Ratosa et al. 2021).

BCS was originally described by Sir Geoffrey Keynes in 1924 and clinically accepted by the 1950s and 1960s. Today, breast cancers in their early stages are treated with BCS followed by radiotherapy. When it comes to the surgical resection of the tumor, proper localization is crucial to guarantee that all malignant tissue is eliminated for improving local control and lowering local recurrence rates. The distance between the cancer tissue and the boundary of the removed tissue is a strong factor in tumor recurrence. The ideal outcome is a negative tumor margin or clear margin, as shown in Figure 2.8(a) where no malignant tissue is identified at the edge of the resected lesion, implying that the diseased tissue is completely confined inside the excised tissue. It is recommended for women with ductal carcinoma in situ (DCIS) to have at least 2 mm of normal tissue between the tumor and the margin of the excised tissue. Figure 2.8(b) depicts a positive margin, in which malignant tissue is identified at the edge of the resected tissue, suggesting that the patient has residual cancer and that a second operation may be necessary. Since 20–40% of BCS margins come out to be positive (Jacobs 2008; Atkins et al. 2012), additional surgeries are required to remove more tissue. The high rate of re-excision results in higher expenditures, more patient anxiety, and a higher risk of post-surgical problems. As a result, better imaging technologies are required to ensure clear margins during breast-conserving surgery.

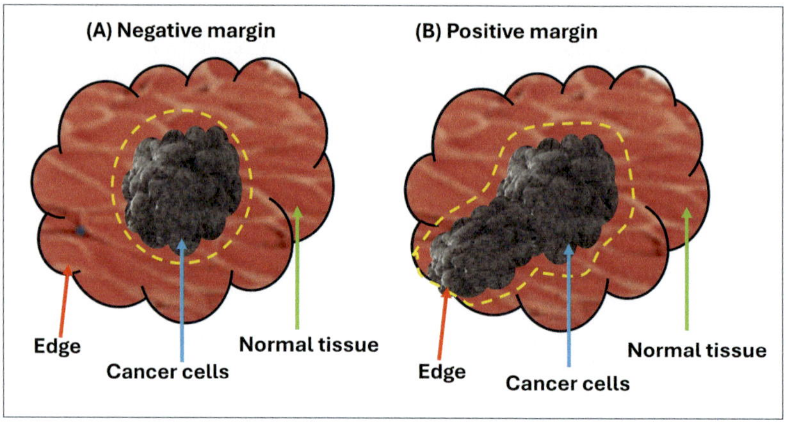

FIGURE 2.8 Negative and positive margin of normal and tumor tissue.

CURRENT MARGIN ASSESSMENT TECHNIQUES

In clinical practice, histology, frozen section analysis, and imprint cytology are used as intraoperative margin detection techniques in BCS. The histology technique is more labor-intensive and slower than other available approaches. However, it is currently considered the gold standard for comparison with other available detection approaches. As a part of margin classification, tissue is fixed, processed, sectioned, stained with hematoxylin and eosin, and interpreted microscopically. Even though this approach is accurate, it takes several days to complete after the surgery. Another round of surgery may be needed if the outcome is a positive or close margin. This leads to stress, additional cost, and surgical infection in some cases due to re-excision.

The frozen section is considered the treatment of choice in some hospitals. As part of the analysis, a small piece of tissue is frozen, sectioned, stained, and interpreted under the microscope. Around 20–30 minutes is required for this entire procedure, and the patient is under anesthesia during this period. In the breast, which contains dense tissues, freezing may generate artifacts. Also, the tissue used for frozen sectioning is not feasible for histopathologic margin assessment. Frozen section interpretations are very costly and require additional staff to cut the specimens. Due to this, a very small part of the specimen margin is frozen and analyzed. Frozen sectioning is not very effective for larger tumors and DCIS. This approach has a specificity of 95% and a sensitivity of 83%. It provides data similar to standard histology but much faster. Due to high cost, technical issues, and low sampling rate, it is not a recommended approach to the margin issue in BCS (Esbona et al. 2012).

Imprint cytology is a simple pathology-based margin detection technique that provides similar information as the frozen section but in a significantly shorter time. Also, it causes less damage to the tissue; therefore, there is no impact on postoperative analysis. As a part of the procedure, a glass slide is placed against the edge of each surface of a specimen and pressed. The tumor cells, which are malignant, would stick to the clean glass, but adipose cells wouldn't. Afterward, the imprint cytology specimen is analyzed to find the accurate surgical margin and the stage of cancer. Unlike spot checks in frozen sectioning, imprint cytology evaluates the whole surface of the resected tissue. Meta-analysis studies show that imprint cytology has a sensitivity of 72–91% and 95–97% specificity. For accurate and rapid analysis, cytopathology specialists, along with the regular surgical staff, are needed. Imprint cytology is limited to surface measurement. It may not be a recommended approach if the surgical margin is greater than 0 mm. Given the limitations of current modalities, there is a need for a simple, cost-effective, and sensitive tool for tumor margin detection. Optical absorption techniques like fluorescence and photoacoustic spectroscopy have shown better outcomes in the detection of breast tumor margin positivity.

AUTOFLUORESCENCE

The advantage of autofluorescence spectroscopic techniques is that they use non-ionizing radiation without extrinsic fluorophores, making the technique feasible for a clinical trial. Because pathologic techniques of margin evaluation are not standardized, there is limited agreement on what constitutes an appropriate negative margin. As a result, patient care varies depending on the surgeon's acceptance of adequate margins and the requirement for re-excision. Developing an objective margin evaluation approach for a precise broad local excision would be extremely beneficial to patients in preventing recurrence and increasing breast conservation surgery rather than mastectomy. The applicability of fluorescence spectroscopy in classifying oral pathology has been demonstrated in an earlier study (Raghushaker et al. 2021). The current study is an attempt to classify and distinguish tumor, margin, and normal tissue by procuring fluorescence spectral data analysis. The aim is to employ the autofluorescence spectroscopy technique in differentiating the breast tumor margin ex vivo.

PHOTOACOUSTIC SPECTROSCOPY

Photoacoustic (PA), also known as optoacoustic (OA), is a non-invasive hybrid technology that combines optical excitation with ultrasonic detection. In the last two decades, photoacoustic has gained popularity in the biomedical field, such as disease identification and characterization techniques, and it is now making its way into clinical practices (Kruger et al. 1994, 1995). The technique is a new way to identify and characterize diseases that holds great promise in the biomedical field. PA has been extensively used in vivo for oncological imaging as it can locate the tumor lodged within the deep tissue and its vasculature since it is highly sensitive and specifically detects (Mallidi et al. 2011; Gargiulo et al. 2019). Photoacoustic works on the principle of the photoacoustic effect (Bell 1881). When samples are subjected to pulsed/modulated laser lights, photons to propagate diffusely inside. Absorption of photons causes slight localized heating of the sample, leading to the thermoelastic expansion that generates pressure (ultrasound) (Lengenfelder et al. 2019; Li et al. 2020). Pressure is transmitted as photoacoustic waves, detected by an ultrasonic detector, and is used to acquire information about the sample (Wang & Hu 2012; Upputuri & Pramanik 2015). The photoacoustic effect is associated with changes in acoustic waves induced by variances in light absorption rates in various tissue samples, allowing the test object's diverse structures to be distinguished (Wilson et al. 2013). The first mechanism of the photoacoustic is based on light induced by specimen imaging, and it is done using a longer wavelength where the tissue absorption is minimal. Then the PA imaging system is connected to ultrasound (US) as they share the same detector and electronics. Hence the combined PA and US system can provide anatomical and functional information. Besides imaging, the PA is also used to study the oxygen content of the blood to study hypoxia in tumors (Mehrmohammadi et al. 2013).

Photoacoustic spectroscopy, which uses non-ionizing radiation and is sensitive to metabolic changes, can shorten diagnostic time and enhance diagnostic accuracy. The photoacoustic technique can collect functional and molecular information from various tissues. One of the most significant advantages of this technique over other methods is that it may offer tissue information without needing an exogenous molecular imaging agent, instead of relying solely on endogenous contrast (Zackrisson et al. 2014). The photoacoustic technique has been reported to have been used in breast tumor margin assessment by a multispectral photoacoustic tomography system (Li et al. 2015). Najafzadeh et al. used this technique to delineate the brain tumor margin. It categorizes photoacoustic signals based on tumor positivity by focusing on protein electromagnetic absorption.

According to prior research, distinct contrasting agents will produce diverse intensities of peaks using different light intensities. Angiogenesis is a prominent biomarker in cancer tissues and is required for the development and nutrition of rapidly developing cancer cells. Furthermore, breast tumor cells only develop up to 2 mm in length due to poor nutrition produced by disordered vasculature and systemic capillary leak, resulting in hypoxic circumstances. As a result, photoacoustic spectroscopy can be used to detect breast cancer in vitro more accurately.

The purpose of this investigation is to see if photoacoustic spectroscopy can be used to identify breast margins. We used this concept to record the photoacoustic spectra of tumor and counter normal breast tissue samples ex vivo. Furthermore, support vector machine (SVM) techniques based on principal component analysis (PCA) were developed and used to increase classification reliability (Priya et al. 2010; Rodrigues et al. 2021).

ASSESSMENT OF BREAST TUMOR SURGICAL MARGIN USING PHOTOACOUSTIC TECHNIQUES

Li et al. determined the margin condition of resected human breast tissue (R. Li et al. 2015). The power of the laser is 30 mJ/sec with a wavelength range of 670–2300 nm. In this study, a 128-element transducer with a central frequency of 21 MHz was employed to view fat and hemoglobin (Heijblom et al, 2013). The results demonstrated that photoacoustic tomography (PAT) might be utilized to distinguish between tumor and normal tissue with a sensitivity of 100% and a specificity of 75%. Scientists created a tailored PAT system that scanned lumpectomy tissues from 66 individuals in 10 minutes. The device produced a two-dimensional picture with a depth of 6 mm and an axial resolution of 200 μm. The results indicated that 90% specificity and 85.5% sensitivity were reached (R. Li et al. 2018).

Sangha et al. (2022) devised a novel multimodality imaging technique to assess the feasibility of PAT for characterizing breast tumor margins. Six breast cancer samples from different individuals were resected and scanned using a PAT system to get three-dimensional computational data. PAT was used in conjunction with a 128-element 40 MHz transducer and an Nd:YAG laser with a wavelength range from 670–2300 nm and a pulse rate of 5 ns. PAT images were equivalent to histology in terms of morphological and compositional information. Using PAT pictures, researchers were able to see hemoglobin-rich areas at 1100 nm and lipid-rich adipose tissue at 1210 nm. A previously published hypothesis reported that higher Hb contrast implies highly vascularized malignant locations. However, this study found that the normal region of the breast has higher Hb contrast than tumor parts. This was because blood vasculature collapsed after breast biopsies due to a lack of continuous blood supply. Furthermore, the Hb distribution is most likely altered because of tissue manipulation. Kosik et al. evaluated intraoperative photoacoustic screening (iPAS) using the near-infrared laser (680–950 nm, 20 Hz pulse repetition) directed on three human lumpectomy specimens. An arc-shaped 25-channel transducer array detected the PA signals. While imaging at a resolution of 2.5 mm, the results revealed a difference in lipid content between normal and cancerous breast tissue.

COMMERCIALLY ACCESSIBLE PHOTOACOUSTIC SPECTROSCOPY-BASED DEVICES

CANON PHOTOACOUSTIC MAMMOGRAPHY

Three prototypes of photoacoustic mammography (PAM) devices for breast cancer diagnostics were developed by Fukutani et al. (2011). The initial prototype invented by Fukuzaki et al., known as PAM-01, employed a flat scanning detection system and a light delivery system that could be directed both upward and caudally. Based on the fundamental design of the first PAM prototype, the second PAM prototype (PAM-02) was created. The main distinction between the two prototypes is the addition of US imaging. PAM-02 could take photo sonic and ultrasonic images at the same time. A continuous array of transducers was used to capture the ultrasound echo imaging (Asao et al. 2016). The third PAM prototype (PAM-03), with hemispherical sensors, was created with the assistance of Opt Sonics, Inc. (USA) to reduce the impact of a limited field of vision and improve visualization of the 3D structure of vessels.

The PAM-03 system was used to measure the Hb saturation rate of blood capillaries using a Q-switched alexandrite laser with tunable frequencies from 755 to 795 nm. A hemispherical cup holding water that matched the patient's body temperature held the patient's breast in the imaging setup. A hemispherical array of detectors with 512 elements was embedded in the inner surface of the holding cup (3 mm diameter each, 2 MHz central frequency). The detector array was scanned continuously in a spiral pattern inside a plane to improve picture quality at the center of the breast cup. The maximum light energy was adjusted to less than 10 mJ/cm^2, which is only half of the American National Standards Institute (ANSI) maximum permitted exposure level. The PA signal was detected using a piezoelectric zirconate titanate transducer. PAM-03 produced 3D photoacoustic images of the breast vascular networks in the breast (0.57 mm spatial resolution). However, the imaging region was limited to around 30 mm in depth, and the measurement time is 4 minutes for a scan with two wavelengths. Subsequently, the patient's body movements degrade the quality of the image and the accuracy of Hb saturation even with body motion compensation (Toi et al. 2017; Shiina et al. 2018).

The research comprised 22 instances of breast cancer. The most significant observation was that using a hemispherical array to record 3D pictures allows for greater resolution of blood vessel information than would otherwise be attainable with an MRI. They also compared DCIS and IBS patients in terms of clinical findings. The majority of carcinoma blood vessels were directed to the lesion in IBS patients, and 93% of vessels were ruptured or quickly constricted at the margin. There was no substantial difference between DCIS and IBC patients, although DCIS cases had intratumorally spotty signals, whereas IBC cases had intratumor vascular-like linear signals (Toi et al., 2017).

Twente Photoacoustic Mammography

The first-generation photoacoustic mammoscope was revealed by Manohar et al. (PAM 1) in 2005 (Manohar et al. 2005). Furthermore, the study's preliminary findings were published in 2007. Using PA in planar mode and one wavelength, the PAM 1 system was able to visualize a breast lesion in vivo with high imaging contrast (Manohar et al. 2004). PAM 1 had a small field of view (90 × 80 mm^2 detector surface area); therefore, it could not cover the entire breast in one measurement (Heijblom et al. 2015, 2016). PAM 1 had a small field of view (90 × 80 mm^2 detector surface area); hence complete breast coverage could not be achieved in a single measurement (Manohar et al. 2016). To overcome this limitation, Schoustra et al. developed an upgraded second-generation 3D Twente photoacoustic mammoscope 2 (PAM 2). A specialized transducer array, i.e., arc, was used in the upgraded second-generation 3D Twente photoacoustic mammoscope 2 (PAM 2) to avoid limited view effects (Neuschler et al. 2018).

The patient's breasts are suspended via an opening inside the imaging apparatus. Nine small laser beams are pointed towards the portion of the mammary gland closest to the walls of the chest, while the bottom beam of the laser is centered on the breast nipple. The single element of the ultrasound detector system has a 3.5 × 3.5 mm^2 surface area that contains 384 elements (32 elements × a total of 12 arcs) (Nyayapathi et al. 2021). Each arc array was composed of 32 elements. There was a total of 12 arrays, resulting in a total of 384 elements. These constituents have a center regularity of 1 MHz with fractional bandwidth of 100%. The PAM-2 device has a dual illumination system comprising a Q-switched Nd.YAG (1064 m) and alexandrite laser (755 nm) with pulse durations of 5 ns and 60 ns, respectively, a pulse repetition rate of 10 Hz in "single wavelength mode" and 20 Hz in "double wavelength mode." The elevational resolution is estimated to be 0.96 mm, while the lateral resolution is 1.06 mm, and the imaging depth was assessed to be 22 mm (Schoustra et al. 2019).

Manohar et al., using PAM 1, showed high-intensity region PA images in four out of five cases of patients with malignancies, which they ascribe to vascular cancer distribution (Manohar et al.,

2007). A study conducted by Hibler et al. demonstrated the ability of PAM 1 to distinguish between benign cysts and malignant tumors using 1064 nm excitation. In the study from 2016, 33 breast cancer patients were scanned with a 1064 nm excitation wavelength, and photoacoustic signals were detected using a 588-element ultrasonic array in 10 minutes utilizing an updated prototype of PAM 1. The findings revealed that one or more cancers could be seen with high certainty in 31 cases (Heijblom et al. 2016). With PAM-02, 3D images of great anatomical structures with elevated detail in the vessel network were observed. A structure comparable to lactiferous ducts ending at the nipple was discovered in the breast of a healthy volunteer (Schoustra et al. 2019). Using the supporting cups, the same research group conducted a study on healthy human participants. They established through various experiments that using a cup reduces breast movement, which leads to better breast placement in the field of view, thus improving image quality (Schoustra et al. 2021).

OptoSonics PAM

Photoacoustic mammography (PAM) is a 3D breast imaging technique developed by Kruger et al. that employs a hemispherical detector to examine the breast. The device included an alexandrite laser that emitted light at 756 nm for light energy of 10 mJ/cm^2 and a pulse duration of 75 ns. A 512-element hemispherical ultrasonic transducer was utilized to constantly scan the hemisphere array in a spiral fashion (radius – 127 mm, 2 MHz central frequency, bandwidth – 70% bandwidth), ensuring that all PAc data was collected at equal distances and helping to increase the field of view (FOV) (24 cm).

Four healthy volunteers' breasts were scanned using a 96 mm spiral approach to a depth of 53 mm for the research. Because of the 2 MHz transducer, the spatial resolution was calculated to be 0.42 mm. In four human volunteers with breast sizes ranging from C to DD cups, the vasculature could be clearly seen through the tissue in the breasts, including on the chest wall (Kruger et al. 2013).

Laser Optoacoustic Ultrasonic Imaging System

Laser optoacoustic ultrasonic imaging system (LOUISA-3D) was introduced in 1996 for the detection of breast cancer progression (Menezes et al. 2019). Oraevsky et al. (2001) a 32-element PVDF transducer with a 120-degree arc in a 2001 investigation for 2D imaging of the breast. The lesion's depth resolution was 0.4 mm, and the side-to-side resolution was around 1 mm, based on where the lesion was in relation to the transducer array. In 2009, Ermilov et al. (2009) designed a laser optoacoustic imaging system for the detection of breast cancer (LOIS64). The device was created to use an arc-shaped array of 64 ultrawideband acoustic transducers to image an individual mammary segment in craniocaudal or mediolateral translation. LOUIS-3-DM, which integrates multiwavelength optoacoustic tomography and laser-induced ultrasonic tomography, was introduced in 2015 by the same group (Ermilov et al. 2016).

LOUIS-3-DM, a clinical prototype 3-D laser OA ultrasonic imaging system assembly (LOUISA-3D), was developed. This novel device is capable of imaging the whole breast using an alexandrite laser at two different wavelengths (757 and 797 nm). Each wavelength has a 50–100 ms delay, which allows the functional imaging of vasculature. The module was made up of an arc-shaped fiber, a hemispherical mold for the breast, and an ultrasonic transducer that rotated around the breast. The 96-element transducer has a frequency range of 50 kHz to 6 MHz. Additionally, the system is comprised of an ultrasound subsystem that consists of 192 element transducers with a central frequency of 7 MHz. The ultrasound subsystem produced 2D slices of breast structure that can be easily layered with corresponding OA slices from 3D functional pictures (Zhang et al. 2010, 2011).

A patient with a breast tumor and one healthy person were scanned for the pre-clinical trial. A tiny tumor that was not evident on ultrasound was detected, presumably indicating a high system sensitivity of the optoacoustic subsystem to small but actively developing malignant lesions with high-density angiogenesis microvasculature. Deoxygenated veins and oxygenated arteries were seen in a healthy volunteer. The major breast vasculature (0.5–1 mm) was evident at depths of up to 40 mm with a resolution of 0.3 mm (Figure 2.8). The LOUISA-3D pilot clinical validation results revealed that the technology is ready for a statistically meaningful clinical feasibility study (Oraevsky et al. 2018).

The Single Breath-Hold PA Computed Therapy

The limitations of MRI and mammography may be partially overcome by the promising method known as photoacoustic computed tomography (PACT). By combining diffuse optical tomography's functional optical contrast and ultrasonography's high spatial resolution, PACT is able to reach tissues in depth. Over the last two decades, many breast PACTs with greater temporal and spatial resolution and adequately deep non-ionizing penetration have been developed. However, these devices have several drawbacks, such as a narrow field of view and a lengthy scanning time (Toi et al. 2017).

Lin and his team improved PACT technology by developing single breath-hold PA computed therapy (SBH-PACT) to solve the traditional system's limitations. Patients may lie supine on a platform with their breasts placed via a hole in the table into a water tank. The SBH-PACT system employed a 1064 nm Quanta Ray laser beam (10 z pulse repetition rate, 8–12 ns pulse width) to attain maximum optical penetration in breast tissue. The laser fluence was 20 mJ m^{-2} which is less than the ANSI safety limit for laser exposure. The PA signal was detected by 512 elements with a diameter of 220 mm, a full ring US transducer with a 2.25 MHz central frequency, and a bandwidth of 95%, which enables full-view fidelity and high image quality. SBH-PACT attained in-vivo imaging depths of up to 4 cm and in-plane resolutions of 255 m, which were about four times higher than contrast-enhanced MRI. The system is coupled with data acquisition circuits (DAQ), which help to obtain the entire breast image within 15 s. In this preliminary study (Lin et al. 2018), SBH-PACT was utilized to investigate one normal participant and seven patients with breast tumors. The image created by scanning a healthy 27-year-old woman displayed a thorough depiction of the complete breast's blood vascular anatomy. The approach identified eight of the nine tumors by highlighting blood vessel structures, as well as dense breasts that were not identified by mammography.

Seno Medical Imagio System

Seno Medical Instruments developed the OA/US dual-modality system, which could offer physicians more confidence in identifying and treating normal and cancerous masses. The CE mark was granted to Seno's Imagio breast imaging system in April 2014, and FDA clearance was granted in 2021, clearing the route for future clinical translation.

The PA Seno medical imaging system includes a portable ultrasonic and PA probe with a built-in fiber optic cable. The system consists of two lasers with a 5 ms delay delivered with a repetition rate of 5 Hz: a 1064 nm Nd:YAG laser with a pulse duration of 15 ns/pulse and a 757 nm Alexandrite laser with a pulse period of 50 ns. With a central frequency of 7.5 MHz, the linear transducer array comprises 128 elements. The axial resolution of the alexandrite and Nd: YAG lasers was reported to be 0.47 mm and 0.42 mm, respectively. The lateral resolution of the alexandrite and Nd: YAG lasers was determined to be 0.81 mm and 0.73 mm, respectively. Structures like blood vessels, for example, were scanned to a depth of 40 mm (Oraevsky et al. 2018).

The morphology and hemoglobin concentration, as well as oxygenation within and around breast masses, were shown using this OA/US modality. Malignant tumors promote neo-angiogenesis and drain oxygen from hemoglobin to tissue. In PA/US mode, to construct a functional image of the PA signal that was recorded, RGBA color mappings were used. The resulting image showed the functional images of total hemoglobin differentiating malignant from benign tumors.

Dual Scan Mammoscope

In 2019, Nyayapathi and colleagues revolutionized breast imaging by introducing the dual scan mammoscope, a cutting-edge PACT device that enables precise imaging of angiographic features through mammogram-like images. The system was equipped with an Nd:YAG laser at 1064 nm (pulse width: 10 ns, 10 Hz pulse repetition rate) used for clinical imaging procedures. The two linear transducers had curved 128 elements, and the central frequency was 2.25 MHz. The amount of energy irradiated on the live organism was calculated to be 21 mJ/cm^2. The depth of the imaging system was reported to be 7 cm, and the lateral and elevation resolutions were 0.97 and 1.05 mm, respectively (Sang et al. 2006). The dual scan mammoscope (DSM) system, like the Twente PAM, records data in CC view, making it easier to correlate to X-ray mammograms, with which radiologists are most familiar (Nyayapathi et al. 2020).

Zheng et al. developed second-generation DSM (DSM-2), which offered a larger FOV through laterally shifted round trip scanning, allowing the coverage of the entire breast tissue (171.2 mm) within 38 s without losing scanning speed or image quality (Wang et al. 2009; Wong et al. 2017). By enhancing the firing voltage and increasing the number of firing angles, the quality of the images was improved. The technology could execute quasi-static compression, allowing tissue stiffness to be used as an additional parameter in tumor classification (Zheng et al. 2021).

In vivo, clinical trials were used to show the DSM2's performance by imaging two participants with breast cups. The data were represented as a depth-encoded "Maximum Intensity Projection (MIP)" with a color scale ranging from blue (shallow region) to red (deep region) after image reconstruction. Findings of cup sizes C and D were obtained by the transducers (top and bottom). In a top-down view, the final merged pictures depict the entire distribution of the vascular structure.

The aim of this study was to use photoacoustic spectroscopy on ex vivo tissues to detect the margin of breast tumors. To simulate clinical conditions, tumor and normal tissue were placed next to each other in a quartz cuvette, creating a margin at the junction of the two tissue types. Photoacoustic spectral acquisition was used to analyze the cuvette. The microcontroller-controlled translation stage was used to move the quartz cuvette containing the tumor and normal tissue to the photoacoustic spectroscopy system for analysis. Using the translation stage, the sample was excited longitudinally every 1 mm. After acquiring photoacoustic spectra from the tumor and the junction of normal and tumor tissue, data analysis was done by preprocessing, transformation, feature extraction, and principal component analysis. PCA analysis showed scattered data points suggesting further analysis of the photoacoustic spectra. Hence, the input feature matrix for photoacoustic spectroscopy was created with labels and fed into a machine-learning algorithm using the MATLAB classification learner app. The algorithm achieved an accuracy of 88.7%, a specificity of 93.75%, and a sensitivity of 100%. The study indicates that photoacoustic spectroscopy can detect the margin of breast tumors with suitable modifications to the instrumentation.

ACKNOWLEDGMENT

The authors would like to thank Manipal School of Life Sciences, Manipal Academy of Higher Education, Manipal, for infrastructure facilities. Authors would also like to thank DBT-BUILDER (*BT/INF/22/SP43065/2021*) for providing infrastructure facilities at Manipal School of Life Sciences (MSLS), MAHE, Manipal. Authors would also like to thank Director of MSLS for his encouragements and support.

REFERENCES

Application of Fluorescence Spectroscopy—Creative Proteomics. (n.d.). Retrieved October 16, 2023, from https://www.creative-proteomics.com/pronalyse/application-of-fluorescence-spectroscopy.html

Arndt, V., Stegmaier, C., Ziegler, H., & Brenner, H. (2008). Quality of life over 5 years in women with breast cancer after breast-conserving therapy versus mastectomy: A population-based study. *Journal of Cancer Research and Clinical Oncology, 134*(12), 1311–1318. https://doi.org/10.1007/s00432-008-0418-y

Asao, Y., Hashizume, Y., Suita, T., Nagae, K., Fukutani, K., Sudo, Y., Matsushita, T., Kobayashi, S., Tokiwa, M., Yamaga, I., Fakhrejahani, E., Torii, M., Kawashima, M., Takada, M., Kanao, S., Kataoka, M., Shiina, T., & Toi, M. (2016). Photoacoustic mammography capable of simultaneously acquiring photoacoustic and ultrasound images. *Journal of Biomedical Optics, 21*(11), 116009. https://doi.org/10.1117/1.JBO.21.11.116009

Atkins, J., Mushawah, F. al, Appleton, C. M., Cyr, A. E., Gillanders, W. E., Aft, R. L., Eberlein, T. J., Gao, F., & Margenthaler, J. A. (2012). Positive margin rates following breast-conserving surgery for stage I-III breast cancer: Palpable versus nonpalpable tumors. *Journal of Surgical Research, 177*(1), 109–115. https://doi.org/10.1016/j.jss.2012.03.045

Barth, A. (2007). Infrared spectroscopy of proteins. *Biochimica et Biophysica Acta (BBA)-Bioenergetics, 1767*(9), 1073–1101.

Bell, A. G. (1881). The production of sound by radiant energy. *Science, 2*(49), 242–253. https://doi.org/10.1126/science.os-2.49.242

Buma, T., Conley, N. C., & Choi, S. W. (2018). Multispectral photoacoustic microscopy of lipids using a pulsed supercontinuum laser. *Biomedical Optics Express, 9*(1), 276. https://doi.org/10.1364/BOE.9.000276

Camarca, A., Varriale, A., Capo, A., Pennacchio, A., Calabrese, A., Giannattasio, C., ... & Staiano, M. (2021). Emergent biosensing technologies based on fluorescence spectroscopy and surface plasmon resonance. *Sensors, 21*(3), 906.

Christiansen, P., Carstensen, S. L., Ejlertsen, B., Kroman, N., Offersen, B., Bodilsen, A., & Jensen, M.-B. (2018). Breast conserving surgery versus mastectomy: overall and relative survival—A population based study by the Danish Breast Cancer Cooperative Group (DBCG). *Acta Oncologica, 57*(1), 19–25. https://doi.org/10.1080/0284186X.2017.1403042

Curran, D., van Dongen, J. P., Aaronson, N. K., Kiebert, G., Fentiman, I. S., Mignolet, F., & Bartelink, H. (1998). Quality of life of early-stage breast cancer patients treated with radical mastectomy or breast-conserving procedures: Results of EORTC trial 10801. *European Journal of Cancer, 34*(3), 307–314. https://doi.org/10.1016/S0959-8049(97)00312-2

Dager, S. R., Corrigan, N. M., Richards, T. L., & Posse, S. (2008). Research applications of magnetic resonance spectroscopy to investigate psychiatric disorders. *Topics in Magnetic Resonance Imaging, 19*(2), 81–96.

Devi, D. R., Mekala, V., & Rao, B. G. (2019). Pharmacognostic evaluation, fluorescence and TLC analysis of Peltophorum pterocarpum (Dc.) backer ex heyne leaves. *Int. J. Herb. Med, 7*(2), 13–18.

Dogan, B. E., Menezes, G. L. G., Butler, R. S., Neuschler, E. I., Aitchison, R., Lavin, P. T., Tucker, F. L., Grobmyer, S. R., Otto, P. M., & Stavros, A. T. (2019). Optoacoustic imaging and Gray-scale US features of breast cancers: Correlation with molecular subtypes. *Radiology, 292*(3), 564–572. https://doi.org/10.1148/radiol.2019182071

dos Santos Rodrigues, F. H., Delgado, G. G., da Costa, T. S., & Tasic, L. (2023). Applications of fluorescence spectroscopy in protein conformational changes and intermolecular contacts. *BBA Advances, 3*, 100091.

Ermilov, S. A., Conjusteau, A., Mehta, K., Lacewell, R., Henrichs, P. M., & Oraevsky, A. A. (2006). 128-channel laser optoacoustic imaging system (LOIS-128) for breast cancer diagnostics (A. A. Oraevsky & L. v. Wang, Eds.; p. 608609). https://doi.org/10.1117/12.658158

Ermilov, S. A., Khamapirad, T., Conjusteau, A., Leonard, M. H., Lacewell, R., Mehta, K., Miller, T., & Oraevsky, A. A. (2009). Laser optoacoustic imaging system for detection of breast cancer. *Journal of Biomedical Optics, 14*(2), 024007. https://doi.org/10.1117/1.3086616

Ermilov, S. A., Su, R., Conjusteau, A., Anis, F., Nadvoretskiy, V., Anastasio, M. A., & Oraevsky, A. A. (2016). Three-dimensional optoacoustic and laser induced ultrasound tomography system for preclinical research in Mice. *Ultrasonic Imaging, 38*(1), 77–95. https://doi.org/10.1177/0161734615591163

Esbona, K., Li, Z., & Wilke, L. G. (2012). Intraoperative imprint cytology and frozen section pathology for margin assessment in breast conservation surgery: A systematic review. *Annals of Surgical Oncology, 19*(10), 3236–3245. https://doi.org/10.1245/s10434-012-2492-2

Esenaliev, R. O., Karabutov, A. A., Tittel, F. K., Fornage, B. D., Thomsen, S. L., Stelling, C., & Oraevsky, A. A. (1997). Laser optoacoustic imaging for breast cancer diagnostics: Limit of detection and comparison with x-ray and ultrasound imaging (B. Chance & R. R. Alfano, Eds.; pp. 71–82). https://doi.org/10.1117/12.280213

Fukutani, K., Someda, Y., Taku, M., Asao, Y., Kobayashi, S., Yagi, T., Yamakawa, M., Shiina, T., Sugie, T., & Toi, M. (2011). Characterisation of photoacoustic tomography system with dual illumination (A. A. Oraevsky & L. v. Wang, Eds.; p. 78992J). https://doi.org/10.1117/12.873690

Gargiulo, S., Albanese, S., & Mancini, M. (2019). State-of-the-art preclinical photoacoustic imaging in oncology: Recent advances in cancer theranostics. *Contrast Media and Molecular Imaging, 2019*. https://doi.org/10.1155/2019/5080267

Gorey, A., Jacob, P. M., Abraham, D. T., John, R., Manipadam, M. T., Ansari, M. S., Chen, G. C. K., & Vasudevan, S. (2019). Differentiation of malignant from benign thyroid nodules using photoacoustic spectral response: A preclinical study. *Biomedical Physics & Engineering Express, 5*(3), 035017. https://doi.org/10.1088/2057-1976/ab101c

Haschke, M. (2014). Laboratory micro-X-ray fluorescence spectroscopy. *Cham: Springer International Publishing, 10*, 978–983.

Heijblom, M., Piras, D., Brinkhuis, M., van Hespen, J. C. G., van den Engh, F. M., van der Schaaf, M., Klaase, J. M., van Leeuwen, T. G., Steenbergen, W., & Manohar, S. (2015). Photoacoustic image patterns of breast carcinoma and comparisons with Magnetic Resonance Imaging and vascular stained histopathology. *Scientific Reports, 5*(1), 11778. https://doi.org/10.1038/srep11778

Heijblom, M., Piras, D., Maartens, E., Huisman, E. J. J., van den Engh, F. M., Klaase, J. M., Steenbergen, W., & Manohar, S. (2013). Appearance of breast cysts in planar geometry photoacoustic mammography using 1064-nm excitation. *Journal of Biomedical Optics, 18*(12), 126009. https://doi.org/10.1117/1.JBO.18.12.126009

Heijblom, M., Piras, D., van den Engh, F. M., van der Schaaf, M., Klaase, J. M., Steenbergen, W., & Manohar, S. (2016). The state of the art in breast imaging using the Twente Photoacoustic Mammoscope: Results from 31 measurements on malignancies. *European Radiology, 26*(11), 3874–3887. https://doi.org/10.1007/s00330-016-4240-7

Insero, G., Fusi, F., & Romano, G. (2023). The safe use of lasers in biomedicine: Principles of laser-matter interaction. *Journal of Public Health Research, 12*(3), 22799036231187077.

IR Spectroscopy and FTIR Spectroscopy: How an FTIR Spectrometer Works and FTIR Analysis. (n.d.). Analysis & Separations from Technology Networks. Retrieved October 16, 2023, from http://www.technologynetworks.com/analysis/articles/ir-spectroscopy-and-ftir-spectroscopy-how-an-ftir-spectrometer-works-and-ftir-analysis-363938

Jacobs, L. (2008). Positive margins: The challenge continues for breast surgeons. *Annals of Surgical Oncology, 15*(5), 1271–1272. https://doi.org/10.1245/s10434-007-9766-0

Kamath, S. D., Ray, S., & Mahato, K. K. (2011). Photoacoustic spectroscopy of ovarian normal, benign, and malignant tissues: A pilot study. *Journal of Biomedical Optics, 16*(6), 067001. https://doi.org/10.1117/1.3583573

Kassem, A., Abbas, L., Coutinho, O., Opara, S., Najaf, H., Kasperek, D., … & Tiquia-Arashiro, S. (2023). Applications of Fourier Transform-Infrared spectroscopy in microbial cell biology and environmental microbiology: advances, challenges, and future perspectives. *Frontiers in Microbiology, 14*, 1304081.

Kauffmann, T. H., Kokanyan, N., & Fontana, M. D. (2019). Use of Stokes and anti-Stokes Raman scattering for new applications. *Journal of Raman Spectroscopy, 50*(3), 418–424.

Keynes, G. (1928). Radium treatment of primary carcinoma of the breast. *The Lancet, 212*(5473), 108–111. https://doi.org/10.1016/S0140-6736(00)83605-3

Kosik, I., Brackstone, M., Kornecki, A., Chamson-Reig, A., Wong, P., Araghi, M. H., & Carson, J. J. L. (2019). Intraoperative photoacoustic screening of breast cancer: A new perspective on malignancy visualisation and surgical guidance. *Journal of Biomedical Optics, 24*(05), 1. https://doi.org/10.1117/1.JBO.24.5.056002

Kruger, R. A., Kuzmiak, C. M., Lam, R. B., Reinecke, D. R., Rio, S. P. del, & Steed, D. (2013). Dedicated 3D photoacoustic breast imaging. https://doi.org/10.1118/1.4824317

Kruger, R. A., & Liu, P. (1994). Photoacoustic ultrasound: Pulse production and detection in 0.5% Liposyn. *Medical Physics, 21*(7), 1179–1184. https://doi.org/10.1118/1.597399

Kruger, R. A., Liu, P., Fang, Y. "Richard," & Appledorn, C. R. (1995). Photoacoustic ultrasound (PAUS)—Reconstruction tomography. *Medical Physics, 22*(10), 1605–1609. https://doi.org/10.1118/1.597429

Lengenfelder, B., Mehari, F., Hohmann, M., Heinlein, M., Chelales, E., Waldner, M. J., Klämpfl, F., Zalevsky, Z., & Schmidt, M. (2019). Remote photoacoustic sensing using speckle-analysis. *Scientific Reports, 9*(1), 1057. https://doi.org/10.1038/s41598-018-38446-x

Li, R., Lan, L., Xia, Y., Wang, P., Han, L. K., Dunnington, G. L., Obeng-Gyasi, S., Sandusky, G. E., Medley, J. A., Crook, S. T., & Cheng, J.-X. (2018). High-speed intra-operative assessment of breast tumour margins by multimodal ultrasound and photoacoustic tomography. *Medical Devices & Sensors, 1*(3), e10018. https://doi.org/10.1002/mds3.10018

Li, R., Wang, P., Lan, L., Lloyd, F. P., Goergen, C. J., Chen, S., & Cheng, J.-X. (2015). Assessing breast tumor margin by multispectral photoacoustic tomography. *Biomedical Optics Express, 6*(4), 1273. https://doi.org/10.1364/BOE.6.001273

Li, X., Kang, L., Zhang, Y., & Wong, T. T. W. (2020). High-speed label-free ultraviolet photoacoustic microscopy for histology-like imaging of unprocessed biological tissues. *Optics Letters, 45*(19), 5401. https://doi.org/10.1364/OL.401643

LIBS. (n.d.). Retrieved October 16, 2023, from https://www.uni-muenster.de/Planetology/en/ifp/research/geologischeplanetologie/LIBS.html

Lin, L., Hu, P., Shi, J., Appleton, C. M., Maslov, K., Li, L., Zhang, R., & Wang, L. v. (2018). Single-breath-hold photoacoustic computed tomography of the breast. *Nature Communications, 9*(1), 2352. https://doi.org/10.1038/s41467-018-04576-z

Mallidi, S., Luke, G. P., & Emelianov, S. (2011). Photoacoustic imaging in cancer detection, diagnosis, and treatment guidance. *Trends in Biotechnology, 29*(5), 213. https://doi.org/10.1016/J.TIBTECH.2011.01.006

Manohar, S., Kharine, A., van Hespen, J. C. G., Steenbergen, W., & van Leeuwen, T. G. (2004). Photoacoustic mammography laboratory prototype: Imaging of breast tissue phantoms. *Journal of Biomedical Optics, 9*(6), 1172. https://doi.org/10.1117/1.1803548

Manohar, S., Kharine, A., van Hespen, J. C. G., Steenbergen, W., & van Leeuwen, T. G. (2005). The Twente Photoacoustic Mammoscope: System overview and performance. *Physics in Medicine and Biology, 50*(11), 2543–2557. https://doi.org/10.1088/0031-9155/50/11/007

Manohar, S., & Razansky, D. (2016). Photoacoustics: A historical review. *Advances in Optics and Photonics, 8*(4), 586. https://doi.org/10.1364/AOP.8.000586

Manohar, S., Vaartjes, S. E., van Hespen, J. C. G., Klaase, J. M., van den Engh, F. M., Steenbergen, W., & van Leeuwen, T. G. (2007). Initial results of in vivo non-invasive cancer imaging in the human breast using near-infrared photoacoustics. *Optics Express, 15*(19), 12277. https://doi.org/10.1364/OE.15.012277

Mehrmohammadi, M., Yoon, S. J., Yeager, D., & Emelianov, S. Y. (2013). Photoacoustic imaging for cancer detection and staging. *Current Molecular Imaging, 2*(1), 89. https://doi.org/10.2174/2211555211302010010

Menezes, G. L. G., Mann, R. M., Meeuwis, C., Bisschops, B., Veltman, J., Lavin, P. T., van de Vijver, M. J., & Pijnappel, R. M. (2019). Optoacoustic imaging of the breast: Correlation with histopathology and histopathologic biomarkers. *European Radiology, 29*(12), 6728–6740. https://doi.org/10.1007/s00330-019-06262-0

Migowski, A. (2015). A detecção precoce do câncer de mama e a interpretação dos resultados de estudos de sobrevida. *Ciência & Saúde Coletiva, 20*(4), 1309–1309. https://doi.org/10.1590/1413-81232015204.17772014

Mirabella, F. M. (2020). Principles, theory, and practice of internal reflection spectroscopy. In *Internal Reflection Spectroscopy* (pp. 17–52. CRC Press.

Neuschler, E. I., Lavin, P. T., Tucker, F. L., Barke, L. D., Bertrand, M. L., Böhm-Vélez, M., Destounis, S., Dogan, B. E., Grobmyer, S. R., Katzen, J., Kist, K. A., Makariou, E. v., Parris, T. M., Young, C. A., & Butler, R. (2018). Downgrading and upgrading Gray-scale ultrasound BI-RADS categories of benign and malignant masses with optoacoustics: A pilot study. *American Journal of Roentgenology, 211*(3), 689–700. https://doi.org/10.2214/AJR.17.18436

Nissen, M. J., Swenson, K. K., Ritz, L. J., Farrell, J. B., Sladek, M. L., & Lally, R. M. (2001). Quality of life after breast carcinoma surgery: A comparison of three surgical procedures. *Cancer, 91*(7), 1238–1246.

Nyayapathi, N., Lim, R., Zhang, H., Zheng, W., Wang, Y., Tiao, M., Oh, K. W., Fan, X. C., Bonaccio, E., Takabe, K., & Xia, J. (2020). Dual Scan Mammoscope (DSM)—A new portable photoacoustic breast imaging system with scanning in craniocaudal plane. *IEEE Transactions on Biomedical Engineering, 67*(5), 1321–1327. https://doi.org/10.1109/TBME.2019.2936088

Nyayapathi, N., Zhang, H., Zheng, E., Nagarajan, S., Bonaccio, E., Takabe, K., Fan, X. C., & Xia, J. (2021). Photoacoustic dual-scan mammoscope: Results from 38 patients. *Biomedical Optics Express, 12*(4), 2054. https://doi.org/10.1364/BOE.420679

Oraevsky, A. A., Clingman, B., Zalev, J., Stavros, A. T., Yang, W. T., & Parikh, J. R. (2018). Clinical optoacoustic imaging combined with ultrasound for coregistered functional and anatomical mapping of breast tumors. *Photoacoustics, 12*, 30–45. https://doi.org/10.1016/j.pacs.2018.08.003

Oraevsky, A. A., Esenaliev, R. O., Jacques, S. L., & Tittel, F. K. (1994). Laser-based optoacoustic imaging in biological tissues. *Laser-Tissue Interaction V; and Ultraviolet Radiation Hazards, 16*. https://doi.org/10.1117/12.182927

Oraevsky, A. A., Karabutov, A. A., Solomatin, S. v., Savateeva, E. v., Andreev, V. A., Gatalica, Z., Singh, H., & Fleming, R. D. (2001). Laser optoacoustic imaging of breast cancer in vivo (A. A. Oraevsky, Ed.; p. 6). https://doi.org/10.1117/12.429300

Oraevsky, A. A., Su, R., Nguyen, H., Moore, J., Lou, Y., Bhadra, S., Anastasio, M., Forte, L., & Yang, W. (2018). Full-view 3D imaging system for functional and anatomical screening of the breast. In A. A. Oraevsky & L. v. Wang (Eds.), *Photons Plus Ultrasound: Imaging and Sensing 2018* (p. 248). SPIE. https://doi.org/10.1117/12.2318802

Orme, M. E., Andalucia, C., Sjölander, S., & Bossuyt, X. (2018). A comparison of a fluorescence enzyme immunoassay versus indirect immunofluorescence for initial screening of connective tissue diseases: systematic literature review and meta-analysis of diagnostic test accuracy studies. *Best Practice & Research Clinical Rheumatology, 32*(4), 521–534.

Principles of the Fourier-Transform Infrared Spectroscopy (FT-IR). | Download Scientific Diagram. (n.d.). Retrieved October 16, 2023, from https://www.researchgate.net/figure/Principles-of-the-Fourier-Transform-Infrared-Spectroscopy-FT-IR_fig2_333498985

Priya, M. (2016). Pulsed laser induced photoacoustic spectroscopy PLPAS of the changes in breast tumor microenvironment and tumor growth an ex vivo experimental study. University. https://shodhganga.inflibnet.ac.in:8443/jspui/handle/10603/357024

Priya, M., Satish Rao, B. S., Chandra, S., Datta, A., Nayak, S. G., & Mahato, K. K. (2015). Monitoring breast tumor progression by photoacoustic measurements: A xenograft mice model study. *Journal of Biomedical Optics, 20*(10), 105002. https://doi.org/10.1117/1.JBO.20.10.105002

Raghushaker, C. R., Rodrigues, J., Nayak, S. G., Ray, S., Urala, A. S., Satyamoorthy, K., & Mahato, K. K. (2021). Fluorescence and photoacoustic spectroscopy-based assessment of mitochondrial dysfunction in oral cancer together with Machine Learning: A Pilot Study. *Analytical Chemistry* Dec 14, *93*(49), 16520–16527. doi: 10.1021/acs.analchem.1c03650. Epub 2021 Nov 30. PMID: 34846862.

Ratosa, I., Plavc, G., Pislar, N., Zagar, T., Perhavec, A., & Franco, P. (2021). Improved survival after breast-conserving therapy compared with mastectomy in stage I-IIA breast cancer. *Cancers, 13*(16), 4044. https://doi.org/10.3390/cancers13164044

Richmond, G. L. (2022). Molecular bonding and interactions at aqueous surfaces as probed by vibrational sum frequency spectroscopy. *Chemical Reviews, 102*(8), 2693–2724.

Rodrigues, J., Amin, A., Raghushaker, C. R., Chandra, S., Joshi, M. B., Prasad, K., Rai, S., Nayak, S. G., Ray, S., & Mahato, K. K. (2021). Exploring photoacoustic spectroscopy-based machine learning together with metabolomics to assess breast tumor progression in a xenograft model ex vivo. *Laboratory Investigation, 101*(7), 952–965. https://doi.org/10.1038/s41374-021-00597-3

Sangha, G. S., Hu, B., Li, G., Fox, S. E., Sholl, A. B., Brown, J. Q., & Goergen, C. J. (2022). Assessment of photoacoustic tomography contrast for breast tissue imaging using 3D correlative virtual histology. *Scientific Reports, 12*(1), 2532. https://doi.org/10.1038/s41598-022-06501-3

Schmid, T. (2006). Photoacoustic spectroscopy for process analysis. *Analytical and Bioanalytical Chemistry, 384*(5), 1071–1086. https://doi.org/10.1007/s00216-005-3281-6

Schoustra, S. M., op 't Root, T. J. P. M., Pompe van Meerdervoort, R. P., Alink, L., Steenbergen, W., & Manohar, S. (2021). Pendant breast immobilisation and positioning in photoacoustic tomographic imaging. *Photoacoustics, 21*, 100238. https://doi.org/10.1016/j.pacs.2020.100238

Schoustra, S. M., Piras, D., Huijink, R., op 't Root, T. J. P. M., Alink, L., Kobold, W. M., Steenbergen, W., & Manohar, S. (2019). Twente Photoacoustic Mammoscope 2: System overview and three-dimensional vascular network images in healthy breasts. *Journal of Biomedical Optics, 24*(12), 1. https://doi.org/10.1117/1.JBO.24.12.121909

Schweizer, T., Kubach, H., & Koch, T. (2021). Investigations to characterize the interactions of light radiation, engine operating media and fluorescence tracers for the use of qualitative light-induced fluorescence in engine systems. *Automotive and Engine Technology, 6*(3), 275–287.

Shiina, T., Toi, M., & Yagi, T. (2018). Development and clinical translation of photoacoustic mammography. *Biomedical Engineering Letters, 8*(2), 157–165. https://doi.org/10.1007/s13534-018-0070-7

Shirkavand, A., Babadi, M., Fashtami, L. A., & Mohajerani, E. (2023). Application of Optical Spectroscopy in Diagnosing and Monitoring Breast Cancers: A Technical Review. *Clinical Spectroscopy*, 100027.

Sun, A., Guo, H., Gan, Q., Yang, L., Liu, Q., & Xi, L. (2020). Evaluation of visible NIR-I and NIR-II light penetration for photoacoustic imaging in rat organs. *Optics Express, 28*(6), 9002. https://doi.org/10.1364/OE.389714

Sung, H., Ferlay, J., Siegel, R. L., Laversanne, M., Soerjomataram, I., Jemal, A., & Bray, F. (2021). Global cancer statistics 2020: GLOBOCAN estimates of incidence and mortality worldwide for 36 cancers in 185 countries. *CA: A Cancer Journal for Clinicians, 71*(3), 209–249. https://doi.org/10.3322/caac.21660

Tognarelli, J. M., Dawood, M., Shariff, M. I., Grover, V. P., Crossey, M. M., Cox, I. J., ... & McPhail, M. J. (2015). Magnetic resonance spectroscopy: principles and techniques: lessons for clinicians. *Journal of Clinical and Experimental Hepatology, 5*(4), 320–328.

Toi, M., Asao, Y., Matsumoto, Y., Sekiguchi, H., Yoshikawa, A., Takada, M., Kataoka, M., Endo, T., Kawaguchi-Sakita, N., Kawashima, M., Fakhrejahani, E., Kanao, S., Yamaga, I., Nakayama, Y., Tokiwa, M., Torii, M., Yagi, T., Sakurai, T., Togashi, K., & Shiina, T. (2017). Visualisation of tumor-related blood vessels in human breast by photoacoustic imaging system with a hemispherical detector array. *Scientific Reports, 7*(1), 41970. https://doi.org/10.1038/srep41970

Tsang, V. T. C., Li, X., & Wong, T. T. W. (2020). A review of endogenous and exogenous contrast agents used in photoacoustic tomography with different sensing configurations. *Sensors, 20*(19), 5595. https://doi.org/10.3390/s20195595

Upputuri, P. K., & Pramanik, M. (2015). Performance characterisation of low-cost, high-speed, portable pulsed laser diode photoacoustic tomography (PLD-PAT) system. *Biomedical Optics Express, 6*(10), 4118. https://doi.org/10.1364/BOE.6.004118

van Dongen, J. A. (2000). Long-term results of a randomised trial comparing breast-conserving therapy with mastectomy: European Organization for Research and Treatment of Cancer 10801 trial. *Journal of the National Cancer Institute, 92*(14), 1143–1150. https://doi.org/10.1093/jnci/92.14.1143

Wallace, M. B., Wax, A., Roberts, D. N., & Graf, R. N. (2009). Reflectance spectroscopy. *Gastrointestinal Endoscopy Clinics of North America, 19*(2), 233–242.

Wan, F., Du, L., Chen, W., Wang, P., Wang, J., & Shi, H. (2017). A novel method to directly analyze dissolved acetic acid in transformer oil without extraction using Raman spectroscopy. *Energies, 10*(7), Article 7. https://doi.org/10.3390/en10070967

Wang, L. v. (2009). Multiscale photoacoustic microscopy and computed tomography. *Nature Photonics, 3*(9), 503–509. https://doi.org/10.1038/nphoton.2009.157

Wang, L. v., & Hu, S. (2012). Photoacoustic tomography: In vivo imaging from organelles to organs. *Science, 335*(6075), 1458–1462. https://doi.org/10.1126/science.1216210

Wilson, K. E., Wang, T. Y., & Willmann, J. K. (2013). Acoustic and photoacoustic molecular imaging of cancer. *Journal of Nuclear Medicine, 54*(11), 1851–1854. https://doi.org/10.2967/jnumed.112.115568

Wong, T. T. W., Zhang, R., Hai, P., Zhang, C., Pleitez, M. A., Aft, R. L., Novack, D. v., & Wang, L. v. (2017). Fast label-free multilayered histology-like imaging of human breast cancer by photoacoustic microscopy. *Science Advances, 3*(5). https://doi.org/10.1126/sciadv.1602168

Zackrisson, S., van de Ven, S. M. W. Y., & Gambhir, S. S. (2014). Light in and sound out: emerging translational strategies for photoacoustic imaging. *Cancer Research* Feb 15, *74*(4), 979–1004. doi: 10.1158/0008-5472.CAN-13-2387. Epub 2014 Feb 10. PMID: 24514041; PMCID: PMC3944207.

Zhang, X., Zhang, H. F., Puliafito, C. A., & Jiao, S. (2011). Simultaneous in vivo imaging of melanin and lipofuscin in the retina with photoacoustic ophthalmoscopy and autofluorescence imaging. *Journal of Biomedical Optics, 16*(8), 080504. https://doi.org/10.1117/1.3606569

Zhang, Y., Cai, X., Choi, S.-W., Kim, C., Wang, L. v., & Xia, Y. (2010). Chronic label-free volumetric photoacoustic microscopy of melanoma cells in three-dimensional porous scaffolds. *Biomaterials, 31*(33), 8651–8658. https://doi.org/10.1016/j.biomaterials.2010.07.089

Zheng, E., Zhang, H., Goswami, S., Kabir, I. E., Doyley, M. M., & Xia, J. (2021). Second-generation dual scan mammoscope with photoacoustic, ultrasound, and elastographic imaging capabilities. *Frontiers in Oncology, 11*. https://doi.org/10.3389/fonc.2021.779071

3 Advances and Challenges with Oral Cancer Diagnosis

*Deva Nanda A R, Jackson Rodrigues, Darshan Chikkanayakanahalli Mukunda, Shaik Basha, Shimul Biswas, and Krishna Kishore Mahato**

Department of Biophysics, Manipal School of Life Sciences, Manipal Academy of Higher Education, Manipal, Karnataka, India
*Corresponding Author

INTRODUCTION

Oral cancer (OC) is a malignancy of the oral cavity (Zhang & Wang, 2022) and is considered a condition of growing concern globally as the sixth most common malignancy (Dhanuthai et al., 2018), with 274,300 cases reported every year (Moafa et al., 2022). Epidemiological classifications include oral, lip, and oropharyngeal cancer (Zhang & Wang, 2022), along with a propensity for lymph node metastasis (Rivera, 2015). Cancers of the mouth and oropharyngeal origin are the 16th most prevalent malignancies globally (Dhanuthai et al., 2018). Oral cancers are 2–3 times more incident in males than in the female population (Rivera, 2015). The most frequently occurring neoplasm is oral squamous cell carcinoma (OSCC), in 90–95% of cases resulting in death with a 1:2 ratio of women to men (Lajolo et al., 2022). The South-East Asian region and Europe have been identified as having the greatest incidence and mortality of OC (Rivera, 2015).

Exposure to carcinogens, alterations in the cytogenetic processes, and other biological factors promote carcinogenesis and malignancy (Rivera, 2015). The transformed cells survive, proliferate, and pass on these properties to their cellular clones (Rivera, 2015).

Generally classified as a preventable disease, the principal causes of oral cancer include the synergistic effects of the consumption of alcohol and tobacco (Dhanuthai et al., 2018). Smokers have a three times higher risk of developing OC than non-smokers, leading to immunosuppression of the oral cavity, which promotes gingivitis, periodontitis, and progresses to OC (Rivera, 2015).

The pre-carcinogens present in tobacco, like nitrosamines, benzopyrenes, and aromatic amines (Rivera, 2015), induce formation of DNA adducts which cause deleterious alterations in the oncogenes and tumour-suppressor genes leading to carcinogenesis (Dhanuthai et al., 2018). A malfunction of the enzymes alcohol dehydrogenase (ADH) and aldehyde dehydrogenase (ALDH) is involved in alcohol oxidation. An elevated release of free radicals and lowered immunity that influence carcinogenesis is seen in both cases (Dhanuthai et al., 2018). The highest burden of oral cancer is on the Asian continent due to the traditional practices of chewing betel quid and areca nut (Moafa et al., 2022). The slaked lime used combined with these induces the penetration of arecoline into the lamina propia to induce fibrosis (Wetzel & Wollenberg, 2020). Other risk factors include persistent inflammations, subpar hygiene, infectious viruses, occupation-associated vulnerability, malnourishment, low intake of fruits and vegetables, and genetics (Dhanuthai et al., 2018). Mostly, the viral infection by human papillomavirus (HPV) produces viral proteins E6 and E7 that induce the degradation of tumour-suppressor gene products of the p53 and pRb genes (Neville et al., 2009) mostly affecting the tonsils and the base of tongue (Rivera, 2015). Oral mucosal trauma arising from the

 DOI: 10.1201/9781003545712-3

teeth and prosthetic devices, mucosal inflammation, oral microbiome, and genetic predisposition in people with faulty DNA repair mechanisms show an increased risk in the development of OC (Dhanuthai et al., 2018). Lip cancers are caused mostly by actinic cheilitis (inflammatory lesion on the lower lip) due to excessive sun exposure (Wetzel & Wollenberg, 2020).

The occurrence of pre-malignant lesions is indicative of potential cancer progression. These clinical presentations include leukoplakia, erythroplakia, oral lichen planus, and oral submucous fibrosis (Montero & Patel, 2015). The lesions are classified as oral potentially malignant disorders (OPMDs) (Cherry et al., 2019). Oral leukoplakias typically occur on the alveolar mucosa, mucosal layer of the buccal cavity, tongue, roof and floor of the mouth, and have a prevalence of 1–3%, affecting men more, particularly in the age window of 50–70 years (Dhanuthai et al., 2018) can be either homogenous or heterogenous based on the morphological characteristics, with homogenous lesions being less risky compared to non-homogenous ones (Warnakulasuriya et al., 2007). Proliferative verrucous leukoplakia (PVL) is a rarely occurring type of multifocal non-homogenous leukoplakia, with no history of alcohol or tobacco usage, usually seen in individuals in their sixth decade of life (Wetzel & Wollenberg, 2020). Erythroplakia are red, smooth, velvety lesions which have a higher risk of transformation than other conditions, with a transformation rate of 14–50% (Wetzel & Wollenberg, 2020). These commonly occur on the soft palate, underneath the tongue, floor of the mouth, and on tonsil pillars (Wetzel & Wollenberg, 2020). Erythroleukoplakia has a mixed red and white appearance, is poorly demarcated, and can be accompanied by pain or soreness (Wetzel & Wollenberg, 2020). These patches appearing on the oral cavity "cannot be characterized as being caused by any other conditions" as defined by WHO, and excision through surgery is most often the only definitive way for management and treatment most of the time (Montero & Patel, 2015). Oral submucous fibrosis is a chronic mucosal disorder due to the loss of fibroelasticity of the affected tissue, mostly due to increased betel quid usage, with a male predilection (Wetzel & Wollenberg, 2020). Oral lichen planus is a chronic inflammation mostly seen in women, attributed to diverse factors including genetic predisposition (Nosratzehi, 2018). The identification of such morphologically altered precursor lesions can suggest the potential malignancy of oral mucosa and help in surveillance over time to prevent progression (Cherry et al., 2019).

The five-year survival rates go from 20–30% for cancers of floor of mouth to 40–60% for cancers of the tongue and alveolar processes (Pałasz et al., 2017). Almost 50% of OCs are subjected to diagnostic delay as either the patients are asymptomatic or they do not seek medical expertise unless the symptoms worsen (Dhanuthai et al., 2018). Diagnosis at early stages is delayed in marginalized populations, with low health awareness and inability to follow up (Thankappan et al., 2021). The delay can also be due to incorrect medical diagnosis and treatment leading to a worsening prognosis and increasing risk for progression of the malignancy (Dhanuthai et al., 2018). Therefore, the earliest diagnosis of oral cancer is required for a better prognosis. With poor rates of survival and high chances of recurrence up to 50% (Towle et al., 2021), the chances of spread and worsening of the disease can be lowered with earlier intervention. A definitive diagnosis is impossible in many cases as patients don't present with characteristic oral mucosal anomalies (Awan, 2015). The conventional techniques must be replaced with better methodologies to differentiate the lesions from normal tissue. They are clinically subtle, heterogenous in their presentation; some are not readily detectable to the naked eye (Figures 3.1–3.3).

CONVENTIONAL TECHNIQUES

CONVENTIONAL ORAL EXAMINATION

The conventional method of screening is performed by illuminating the oral cavity with a bright incandescent light source to inspect for abnormalities that could lead to cancer (Warnakulasuriya & Kerr, 2021). An extensive physical examination of the head, neck, oral, and pharyngeal regions is done by visual inspection and palpation. Digital neck node area palpation is performed to evaluate regional metastasis. This method is a routine screening procedure because it is of a less invasive and painless nature (Raghushaker et al., 2020). A study by Walsh et al. shows more than 70% sensitivity and specificity of more than 90% for conventional oral examination (COE) when compared to other techniques. COE can be applied alongside other index tests for increased sensitivity (Walsh et al., 2013). The technique is inefficient in distinguishing pre-malignant lesions from other benign lesions (Raghushaker et al., 2020). The inaccessibility to visually restrained areas of the oral cavity and lesions present underneath a normal-looking mucosa can be missed, which could further progress to malignancies later (Raghushaker et al., 2020). Even though direct

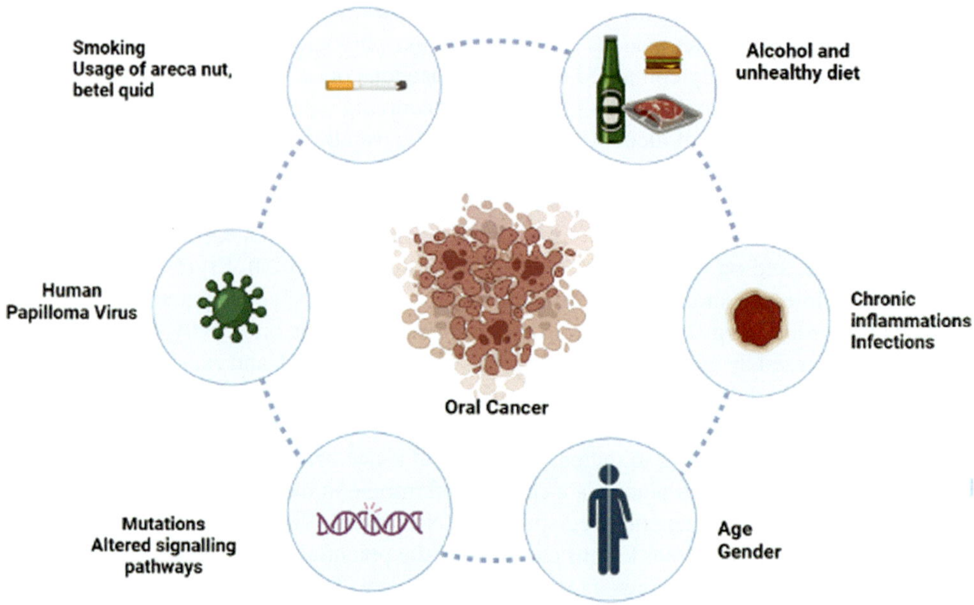

FIGURE 3.1 Major factors causing oral cancer.

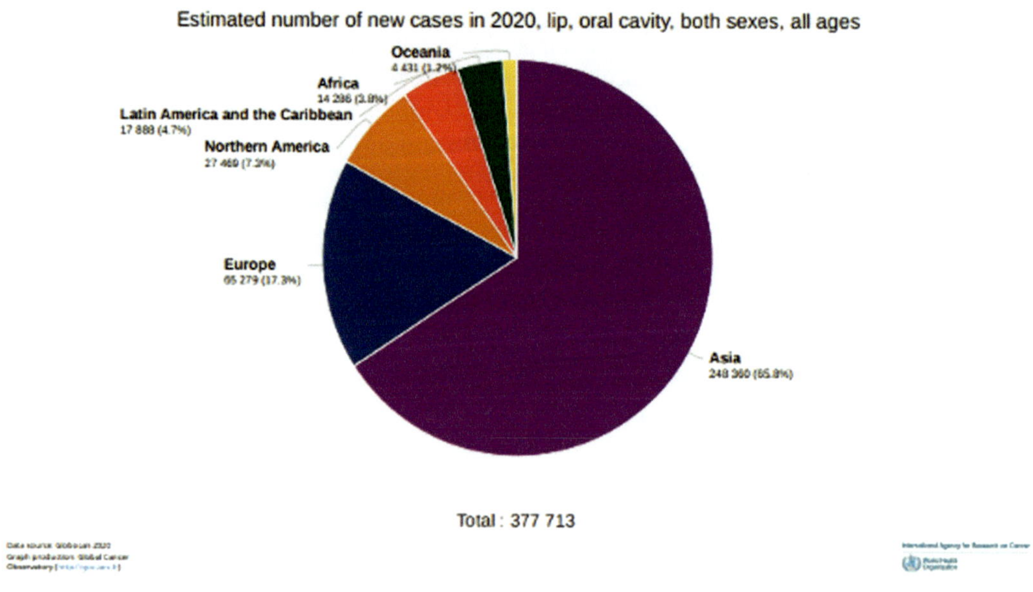

FIGURE 3.2 Incidence rate of cancers of lip and oral cavity as of 2020 (GLOBOCAN).

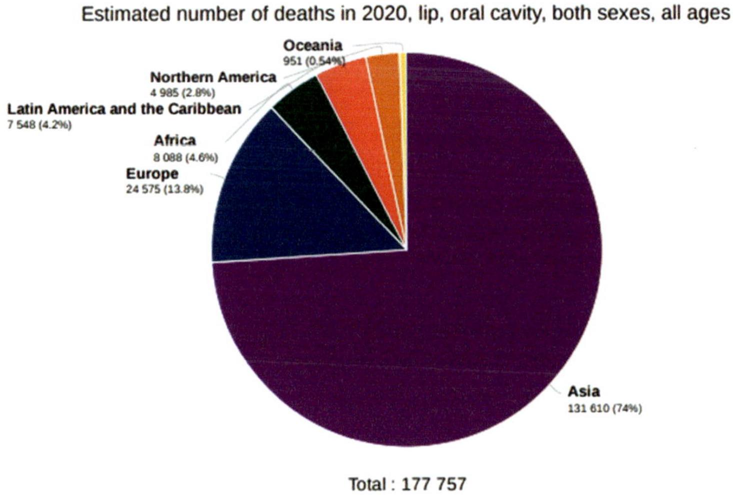

FIGURE 3.3 Mortality rates of cancers of lip and oral cavity as of 2020 (GLOBOCAN).

visualization is enabled, the evaluation of disease extension is not possible with this method (Arya et al., 2012). The oral cavity can be easily examined, but diagnosing metastases in local lymph nodes and organs is difficult, which can be used to make conclusive decisions on resection, further treatment, implantations, adverse consequences of therapy, and disease recurrence (Pałasz et al., 2017). This calls for the importance of histopathological analysis of the tissue to understand its true nature.

Biopsy of the Oral Tissue

Performing oral tissue biopsy to interrogate the abnormalities helps clinicians distinguish between normal and potentially malignant lesions and conduct the correct diagnosis of the disease (Raghushaker et al., 2020). The two methods usually followed are surgical biopsy or brush cytology.

The process of tissue removal, staining, and microscopic examination is involved in surgical biopsy and is considered the gold standard for oral cancer diagnosis (Singh et al., 2016). Even though this process relies on histological risk stratification, there are complications associated with the invasive procedure of tissue removal, chances of sampling errors, incorrect interpretations of the findings, and the likelihood of tissue undergoing biochemical changes after removal (Singh et al., 2016). As the usage of scalpel or punch biopsy is not welcomed by patients most of the time, an easier alternative uses brush cytology to scrape off cells from the affected area using an oral brush (Raghushaker et al., 2020). Oral brush biopsy has been effective when modified along with molecular markers like tenascin and keratins (Reddy et al., 2012) or DNA image cytometry (Kaur et al., 2016) and has an overall sensitivity of 40–90% and specificity of 68–100% (Raghushaker et al., 2020). Despite being considered a gold standard method, a conflict exists in almost 12% of cases until final histopathology, occurring due to the heterogeneity of the tumour leading to sampling error (Vonk et al., 2021). The procedure is discomforting and invasive for the patient and requires expertise to be performed (Kaur et al., 2021).

The samples are subjected to genomic and transcriptomic studies to profile the mutations, altered expressions, signalling, epigenetics of genes, deregulations of miRNA, and patterns of methylation, along with suitable validation studies which are useful in finding biomarkers (Rai et al., 2018).

Staining Techniques

In vivo staining using Toluidine Blue, Lugol's Iodine, or Rose Bengal, is a fast, inexpensive, and effective method to visualize potential biopsy sites that are otherwise non-apparent. The in vivo method helps detect and define the lesions' margins by binding to biological entities in cells to produce colours (Elimairi et al., 2017).

One of the most effective screening techniques is the use of Toluidine Blue stain. The topical application of this widely used stain differentiates the normal mucosa by staining only the carcinoma, benign lesions, melanoma, fibrosarcoma, and lymphosarcoma (Myers) (Helsper, 1972). This can be used as a preliminary step to mark the margins before excision, to visualize a satellite lesion near a larger one, and to differentiate traumatic ulcers or inflammatory ulcers from cancer (Helsper, 1972). It works by binding to the nucleic acids (DNA, RNA), which are more abundant in malignancy (Helsper, 1972).

Lugol's Iodine staining, being glycophilic, stains the glycogen in the cytoplasm, changing the colour to brown reversibly (Elimairi et al., 2017). The regions of abnormalities on mucosa are low in glycogen due to elevated glycolysis levels and no cellular differentiation; therefore, they remain partially or unstained (Elimairi et al., 2017). In a study conducted by Kurita and Kurashina (Kurita et al., 2012), this method could be used to mark the margins precisely to ascertain the extension of the lesion for surgical excision. This method is seen to have 94.7% sensitivity and 83.8% specificity (Chaudhari et al., 2013) and was observed to have 90% accuracy when used along with Toluidine Blue (Nagaraju et al., 2010). Even though vital staining causes little or no discomfort, the mucosal barrier can hinder the dyes, making it difficult to inspect the deep-seated dysplasia (Vijayakumar et al., 2019; Table 3.1).

CHEMILUMINESCENCE-BASED DEVICES

Chemiluminescence is the light emission process upon excitation of a molecule from its ground state due to the energy from a chemical reaction (Dodeigne et al., 2000; Kim et al., 2022). The structural and metabolic changes are subjected to malignancy alter the mucosal refractive index, thereby reflecting the white light, whereas normal mucosa will absorb it (Raghushaker et al., 2020). The ViziLite® (Zila Pharmaceuticals, Phoenix, AZ) is an Food and Drug Administration (FDA)-approved technology that works on the principle of chemiluminescence. A modified form that combines this with Toluidine Blue staining is ViziLite Plus. Along with the acidophilic dye that stains DNA or RNA, which is more prevalent in the abnormal tissue, it uses a 1% acetic acid wash to break the glycoprotein barrier so that the light can be absorbed (Dodeigne et al., 2000). Normal tissues' dehydrated cytoplasm will absorb the tool's blue light at a wavelength of 430–580 nm, whereas the abnormal tissue will reflect the light, appearing "acetowhite" (Raghushaker et al., 2020). Sensitivity ranges from 0–100%, whereas specificity is generally low, from 14.2% to 81.5% (Dodeigne et al., 2000). Due to the application of acetic acid, the saliva secretion levels incline, which increases mucosal reflectance (Awan et al., 2011).

Microlux (Zila Pharmaceuticals, Inc.) emits a 440 nm blue light from a battery-powered LED fibreoptic source. This method was seen to have a sensitivity of 94.3% and a specificity of 99.6%, and Microlux/DL+TB has better potential to mark the borders with a sensitivity of 88.7% and a specificity of 99.3% (Ibrahim et al., 2014).

This method should be administered only in the primary stages and is not for conclusive testing (Chakraborty et al., 2019). Commonly encountered drawbacks include accurately pinpointing the site of the lesion, more background noise, difficulty in quantifying the outputs, and choosing the appropriate light source (Chakraborty et al., 2019).

TABLE 3.1
Commonly Used Staining Methodologies for Detection of Oral Cancer

Staining method	Principle	Interpretation	Applications	Advantages	Sensitivity	Specificity	Limitations	References
Toluidine Blue Staining	Acidophilic metachromatic dye of the thiazine group; high affinity for acidic compounds; stains cells with greater nucleic acid content	Dark blue + pale blue – uncertain No colour uptake – negative	Screening of oral cancer and precancer; post treatment surveillance of OPMDs	High detection rate of OPMDs	38–100%	9–100%	High false positive results; variability in staining techniques; subjectivity in colour interpretation	Epstein et al. (1992); Gandolfo et al. (2006); Guneri et al. (2013); Awan et al. (2012); Lingen et al. (2008)
Rose Bengal Staining	Tetrachloro and tetraiodo derivative of fluorescein	Stains desquamated epithelial cells, degenerated or dead cells but not epithelial cells	To diagnose ocular surface disorders; detection of OPMDs and oral cancer	Conjugated gold nanorod with rose Bengal dye advocated for optical detection of malignant cells	90–100%	73.7–89.09%	Requires further research to be used in a clinical setting	Murabe (2014); Du et al. (2007); Mittal et al. (2012); Wang et al. (2014)
Methylene Blue Staining	Stains nucleic acids	Deep and focal staining on the lesion	High sensitivity	Used for screening oral malignancy in high risk cases	90–91.4%	66.6–69%	Low specificity	Riaz et al. (2013); Chen et al. (2007a); Chen et al. (2007b); Riaz et al. (2013)
Lugol's Iodine	Reaction of iodine with glycogen present in cytoplasm visualized by colour change; Malignant cells fail to exhibit this iodine-starch reaction owing to enhanced glycolysis and loss of cellular differentiation	Normal mucosa stains brown or mahogany; dysplastic and cancerous tissues appear pale as they do not take up the stain	Screening tool for OPMDs and oral cancer	Effective delineation of the lesions, finding correct biopsy sites, cost effective	87.5–94.7%	83.8–84.2%	Minor irritation and severe pain reported in some cases	Petruzzi et al. (2010); Chaudhari et al. (2013); Chen et al. (2007b); Elimairi et al. (2017)

IMAGING TECHNIQUES

Radiologic scans of the head/neck/chest regions to stage the tumour are performed using computed tomography (CT), magnetic resonance imaging (MRI), ultrasound (US), and positron emission tomography (PET) (Wong & Wiesenfeld, 2018). CT is a widely used tool to stage head and neck tumours based on metastasis and vessel infiltration (Pérez et al., 2015). These are greatly helpful in understanding the nature of the lesion before the treatment procedures are decided.

Radiography

Orthopantomogram is an X-ray of the oral cancer performed in finding the stage and spread of the disease by visualizing the maxilla and mandible tissues and bones (Borse et al., 2020). Intra-oral tube radiography is used to capture radiograms of the oral cavity, using a rod-like X-ray source and a receptor setup with an intensifying screen and a film placed in the oral cavity in a cassette (Keur & Macdonald, 1986).

The panoramic view of the anterior part of the oral cavity is examined for abnormalities but requires additional radiograms (Keur & Macdonald, 1986). Plain radiographs with a sensitivity of 75% and specificity of 100% are routinely recommended by clinicians to help diagnose carcinoma based on tissue morphological changes, even though they are not efficient in assessing soft tissues and bones (Pałasz et al., 2017).

Magnetic Resonance Imaging

MRI is a great tool to visualize the extent of tumour node metastasis and lymphadenopathy with precision (Singh et al., 2017). It can be a reliable preoperative measure to stage the malignancy and proceed with the suitable therapy (Singh et al., 2017). This method does not use harmful radiation on the patients and can excellently perform soft-tissue discrimination (Singh et al., 2017).

The MRI contrast agents are also said to be less immunogenic (Singh et al., 2017). Therefore, they can be used along with other techniques like biopsy for precision screening and decision-making (Borse et al., 2020). Nae et al. studied 46 patients with a mean age 59.7 years who were subjected to preoperative imaging modalities like MRI and CT to assess medullary involvement. MRI was shown to have a sensitivity, specificity, and accuracy of 88%, 83%, and 85% respectively (Nae et al., 2019).

Ultrasonography

Another commonly used, easy, and affordable technique is ultrasonography. In the process of intra-oral ultrasonography, ultrasonic images of the oral cavity can contour the margins of the tumour by placing a covered transducer to manipulate the surface of the lesion by the application of sterile gel, which can vertically measure the distance from the surface to the maximum depth, thereby measuring the thickness of the tumour (Nae et al., 2019; Shintani et al., 2001). This is a non-invasive method, capable of detecting very thin lesions as well as accurately measuring the spread of the tumour in a rapid and reproducible manner (Shintani et al., 2001). Even though intraoral ultrasonography gives a better yield than the extraoral method, it cannot differentiate the inflammations around the tumour from the tumour itself (Shintani et al., 2001). It can be used to predict lymph node metastasis as it can differentiate between cystic and solid lesions and benign and malignant masses (Waech et al., 2021).

Computed Tomography

CT is a cross-sectional imaging modality in OC in preoperative and post-surgical scenarios. The chances and severity of relapse can be studied after surgical invasion (Baba et al., 2018). Lymph node metastases are a major contributor to decreasing survival rates by 50% (Ho et al., 2017). This method is influenced by the contrast agent and looks for the involvement of soft tissues and bones in metastases (Figueiredo et al., 2010). The chances of disease recurring can also be assessed with CT after the surgery (Baba et al., 2018).

Multispectral Optoacoustic Tomography

The accurate detection of small lymph nodes is mostly compromised in current techniques, which highlights the importance of a non-invasive imaging method called multispectral optoacoustic tomography (MSOT) (Vonk et al., 2022). Tissues absorb short laser pulses of different wavelengths and, in response, produce ultrasound waves that can detect the intrinsic chromophores and contrast agents present in the depths of the tissue using a tomographic detector (Vonk et al., 2022). Near-infrared fluorescent tracers and tumour-specific contrast agents help identify molecules with distinct spectral properties (Stoffels et al., 2015). Cross sections of less than 1 ms are obtained in real time using MSOT (Stoffels et al., 2015). When the two systems (optoacoustic and ultrasound) are brought together, they provide lymph node features in a single step, unlike other techniques (Stoffels et al., 2015). As this procedure can be repeatedly performed, it is possible to obtain the cross-sections at certain intervals to better study the spread (Stoffels et al., 2015).

Optical Coherence Tomography

Another optical method that has sought attention is optical coherence tomography (OCT). The interactions of the incident light with the tissue chromophores are measured using an interference signal obtained as a difference in the intensity of light from the sample path beam and the reference path beam generated by splitting the white incident light (Reddy & Praveen, 2017). Therefore, the high-resolution images generated can be compared to the biopsy samples and are called optical biopsy (Reddy & Praveen, 2017).

BIOMARKERS AND BIOSENSORS

The analysis of molecules of biological importance present in the malignant transformation of tissues, called biomarkers, has gained attention as a minimally invasive method that addresses the cellular and molecular level changes in tissues (Kaur et al., 2021). Sources of biomarkers can vary from urine, blood, or saliva samples, from which circulating tumour cells (CTCs), circulating tumour DNA (ct-DNA), RNA (like mRNA of IL8, DUSP1, HA3, SAT, etc.), microRNA, proteins (altered expression of cytokines, telomerase, catalase, transferrin, etc., indicative of malignant transformation), metabolic biomarkers like valine, lactic acid, and phenylalanine (Kaur et al., 2021; Goldoni et al., 2021). The presence of these is characteristic of the site of occurrence, the progression, chances of recurrence, and the lifespan of the patient (Goldoni et al., 2021). When the biomarker properties are addressed along with other clinical measures, the rate of nodal metastases will go down from 18% to 11% in the preliminary stages of OSCC (Chai et al., 2016; Figure 3.4).

Serum Biomarkers

Earlier, liquid biopsy was restricted to blood samples, where the circulating miRNAs secreted from exosomes indicated varying cellular functions (Mazumder et al., 2019). In a study by Chai et al., biomarkers present in abundance in serum were identified using proteomics analysis and compared between node positive and node-negative metastasis cases of OSCC (Chai et al., 2016). The use of blood samples is limited as some degree of invasiveness is involved, and there is a risk of coagulation, careful sample handling, processing, and storage requirements (Mazumder et al., 2019). Schiegnitz et al. discovered that the pro-inflammatory cytokines IL-6, IL-8, and sIL-2R levels increased in OSCC patients compared to oral premalignant lesion (OPL) patients, and that elevated levels of IL-6 correlated with higher T-grade and lymph node metastasis (Schiegnitz et al., 2018).

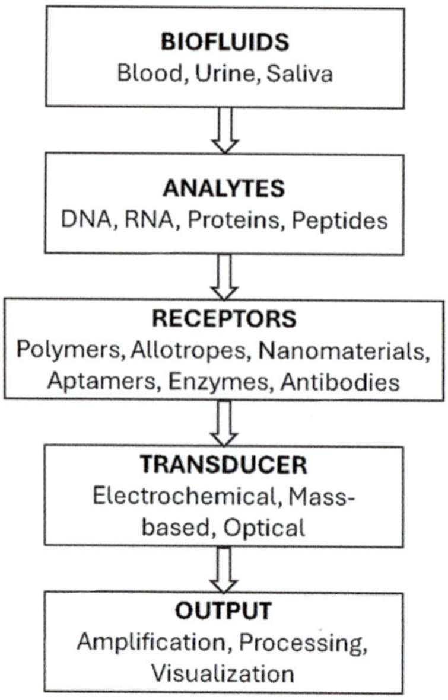

FIGURE 3.4 Schematic of the principle of biosensing.

Salivary Biomarkers

The ease of collection and investigation makes saliva a commonly used specimen, the constituents showing the characteristics of the present state of the body (Cheng et al., 2014). "Salivaomics" is an umbrella term for the genomic, transcriptomic, metabolomic, proteomic, and microbiome profiles of salivary composition (Goldoni et al., 2021). There are at least 100 salivary biomarkers reported so far, mostly released into the blood by tumours directly or by alterations in the tumour microenvironment (Goldoni et al., 2021). Apart from its pivotal role in digestion, antimicrobial properties, buffering, and lubrication, the alterations in salivary constituents can be used to detect oral abnormalities at an early stage, as the collection is comparatively easier and less invasive (Goldoni et al., 2021).

The unusual methylation of DNA also serves as biomarkers and is indicative of malignant transformation even before the condition is evident (Adeoye et al., 2022). Differential methylation patterns of cancer-specific genes altogether provide better diagnostic test accuracy (DTA) than single ones (Adeoye et al., 2022). A statistical evaluation done to differentiate OSCC from normal and OPMD states shows the sensitivity and specificity of these markers in saliva to be 72.0% and 88.4%, and that in oral swabs to be 71.2% and 97.1% respectively (Adeoye et al., 2022). Even though the specimen is advantageous because it is available in large quantities and can be repeated in a minimally invasively way (Goldoni et al., 2021), the testing process of salivary biomarkers is still not in a standard format. Therefore, it requires immediate action and post-test validations as the constitution of saliva can vary and be influenced by the volume and various factors from person to person (Cheng et al., 2014). It must be noted that several metabolites present in the saliva also undergo changes in their levels (Table 3.2).

A biosensor is designed to assess the levels of a biomarker in the sample, as the transducer will convert the signal of biomarker interaction with the corresponding probe molecules into a measurable outcome (Goldoni et al., 2021). It consists of a bioreceptor probe that recognizes the biomolecule, a transducing unit, and another unit to process the signals that are connected to a setup that reads these (Singhal et al., 2021). The probes used here can be cells, genetic material, or proteinaceous compounds (Singhal et al., 2021). The probe recognizes the marker in the vicinity from the fluid under analysis using a biomolecule or a substance that mimics one, which determines the efficiency of the device (Singhal et al., 2021). The interactivity of the analyte and the probe gives signals that are captured by the transducer, which checks for any differential presentations and transforms them into a form that can be quantified (Singhal et al., 2021). Ever-advancing technologies are making it possible to develop platforms that can function based on mass, optical, electrochemical, or photoelectrochemical properties (Goldoni et al., 2021). This is paving the way for the development of new methodologies for point of care (POC) devices for the timely detection and intervention of OC (Goldoni et al., 2021). Biosensors scale down many techniques into a single operational device with outputs manageable even by even inexperienced technicians (Singhal et al., 2021).

DIAGNOSTICS BASED ON AUTOFLUORESCENCE PROPERTIES

Optical Fluorescence Imaging-Based Techniques

The conventional imaging techniques are helpful in effective diagnosis and treatment yet cannot provide optimum resolution and detect early changes before carcinogenesis, requiring advanced modalities like optical fluorescence imaging (OFI) (Vonk et al., 2021). Unlike the other practices, OFI is a good predictor of OSCC and oral epithelial dysplasia (OED) as it compiles clinical information obtained from lesion detection, assessment, and management (Tiwari et al., 2020).

Fluorescence imaging makes use of low-energy but longer wavelength emissions from fluorophores in cells upon excitation at certain wavelengths (Vonk et al., 2021). Some naturally occurring fluorophores include tryptophan, collagen, elastin, keratin, and haemoglobin, NADH (Balasubramaniam et al., 2015) that are excited by 375–440 nm light (Awan, 2015) to have an alteration in their concentrations and natural light-scattering and absorption characteristics, helping to differentiate normal tissues from malignancy (Balasubramaniam et al., 2015). Alterations in the physiology and metabolism of the tissue affect the internal fluorescence properties as they encounter light (López-Jornet et al., 2011). When tissues are irradiated with blue light of 400-460 nm, they respond by producing autofluorescence, which is immediately detected (López-Jornet et al., 2011). The fluorescence is nothing but the photons with less energy released by the excited fluorophores (Yan et al., 2017). Due to depletion of collagen from the connective tissue in neoplasia and an increase in mitochondrial fluorophore concentrations (NADH and FAD) in dysplasia, there is a significant difference in fluorescent intensity when compared to normal mucosa (Balasubramaniam et al., 2015) Those cells that have undergone inflammation are seen to have lost their fluorescence properties (Balasubramaniam et al., 2015) The evaluations based on autofluorescence are useful for pre-operative decision-making and can be effectively incorporated into mass screening processes (Balasubramaniam et al., 2015).

One of the pioneering experiments by Leonard et al. induced fluorescence by intravenously supplying hematoporphyrin to observe red fluorescence across all 29 lesions studied (Leonard & Beck, 1971). Fluorescence can be induced if chemical compounds called photosensitizers are applied on the surface of the tissue and irradiated (Ebihara et al., 2003). The accumulation of these agents upon exposure to sunlight becomes toxic and has been substituted by inducing the production of these agents inside the tissues (Ebihara et al., 2003), as in the case of protoporphyrin IX, a precursor of heme, it is made to accumulate by providing external sources of 5-aminolevulinic acid, which alters the feedback mechanism of heme, unlike how it occurs in endogenous synthesis (Ebihara et al., 2003). The tumour-affected tissues are seen to have an abundance of protoporphyrin IX (PpIX), forming the basis for photodynamic therapy (Smith, 2002). The fluorescence intensity due to protoporphyrin accumulation upon sensitization using 5-ALA is seen to be 8 15 fold greater in malignancy than in normal mucosa (Smith, 2002). Another study conducted by Leunig et al. incubated tissues in ALA and illuminated with a laser to observe that the normal tissue gave green fluorescence, whereas tumour tissue showed red fluorescence. The characteristic fluorescence was attributed to the accumulation of PpIX. The main drawbacks of such agents are the limited access to the depths of the tissue and the effects of coexisting conditions like morphological differences or

TABLE 3.2
Major Salivary Biomarkers and Their Properties

S. No	Biomarker	Properties	Indication	Regulation in Oral Cancer	Reference
1	Micro RNA	Coordinate cellular and tissue programming by activation or repression of translation in local cells or by affecting transcription of cell of origin	The tissue of origin, stage, progression, prognosis, response to treatment	Increases in OSCC	Schwarzenbach et al. (2014); Calin and Croce (2006)
2	Cell-free DNA	Released to blood from healthy or tumour cells due to various cellular processes. cfDNA from apoptosis is fragmented	cfDNA from necrotic tumour cells are long (180–200 base pairs; half-life of 15 minutes to 13 hours); apoptotic cells have shorter fragments	Not completely reliable as an early diagnostic marker as the cell has to reach necrosis to generate cfDNA	Goldoni et al. (2021)
3	Messenger RNA	Enclosed in apoptotic bodies, released in exosomes or macrovesicles	Regulations of the transcriptome	IL-8, IL-1-β, S100P – high upregulation H3F3A – moderate upregulation OAZ1, DUSP1, SAT – low upregulation	Goldoni et al. (2021); Arantes et al. (2018)
4	Cytokines	Present in oral fluids	Predicts poor survival, inflammation, immune responses and differentiation	Higher expression of proinflammatory and proangiogenic cytokines in patients	Katakura, A. et al. (2007)
5	Telomerase	Ribonucleoprotein to elongate repeat sequences of telomeric ends of chromosomes	Escape from cellular senescence and other proliferation restricting conditions	75% OSCC patients had increased levels	Zhong et al. (2005)
6	Cyfra21-1	Monoclonal antibody that exclusively reacts with Cytokeratin-19	Cytokeratin-19 expressed in simple epithelial carcinoma	Increased in OSCC and PML	Rajkumar et al. (2015)
7	Glutathione	Epidemiological marker for chemoprevention	Risk of development of OSCC, oxidative stress	Increased	Almadori et al. (2007)
9	Metabolic markers	Metabolic profiles of biofluids	Global perturbation demonstrating the genetic/physiological stimulations	Alanine and n-eicosanoic acid increased; valine, gamma-aminobutyric acid, phenylalanine decreased	Wei et al. (2011)

multiple microbial infestations (Smith, 2002). But several imaging techniques have advanced based on this phenomenon, improving the detection and diagnostics of oral cancer.

Loss of autofluorescence (LAF) can happen in tissues due to various factors like morphological or biochemical changes, elevated metabolic rates, angiogenesis, and alterations in the size and texture of the nucleus (Kordbacheh et al., 2016; Figure 3.5).

VELscope

One commonly used autofluorescence-based diagnostic imaging technique is the Visually Enhanced Lesion Scope (VELscope) (LED Medical Diagnostics Inc, Barnaby, Canada). VELscope, a chair-side machine with a blue light of special wavelength 400–460 nm, can penetrate the epithelial layer and basal layer to reach the stroma (Balasubramaniam et al., 2015). Real-time visualization is possible with the help of a narrow-band filter present in the device (Awan, 2015). With less than two minutes for detection, the VELscope indicates the difference in fluorescence between normal and abnormal mucosa due to the changes in fluorophore content and identification of tumour margins (Balasubramaniam et al., 2015). Green autofluorescence is retained in normal mucosa, whereas abnormalities are presented as dark in colour due to loss of fluorescence (Balasubramaniam et al., 2015; Awan, 2015). It can be used as an appurtenance to other methods to tell apart the premalignant lesions from the benign ones and also to demarcate the margins for operative invasions (Farah et al., 2012).

It was found that VELscope could determine five otherwise undetected lesions, and the diagnosis of the biopsy site was also modified (Farah et al., 2012). It could also give augmented visualization of the lesions by 34.74% (Farah et al., 2012). A range of sensitivity around 92–100% is mostly obtained, with specificities as low as 15% (Koch et al., 2011) to 92.3% (Koch et al., 2011; Paderni et al., 2011). Interpretation of the results requires expertise as the fluorescence pattern can be similar among varying perturbations of the oral cavity (Farah et al., 2012). Combining a VELscope with usual practices was useful as they gave a positive predictive value (PPV) of 29% and could identify unidentifiable lesions (Farah et al., 2012). VELscope should be put to use only after thoroughly assessing the potential presence of oral complications and their fluorescence signature, following COE (Farah et al., 2019). In spite of having high sensitivity, the specificity rates are lower for the device as it recognizes non-malignant inflammations and infections (Farah et al., 2019).

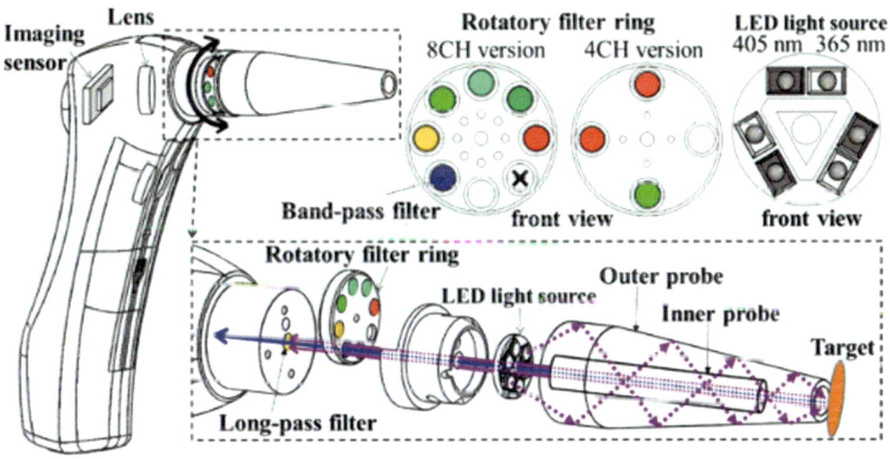

FIGURE 3.5 The components of a light-induced autofluorescence multispectral imager (adapted from Yan et al., 2021a).

Identafi

The visualization of the oral cavity can be performed using Identafi (DentalEZ, PA, USA), a multi-spectral instrument containing three lights (Lalla et al., 2015). The usage of white LED ensures better visibility compared to the commonly used extraoral light (Lalla et al., 2015). When normal mucosal layers are illuminated with violet light (405 nm), blue fluorescent emission is observed, whereas the irregular regions will appear darker due to LAF (Lalla et al., 2015). The basal vascular network of abnormalities can be determined using the 545-nn violet light by causing tissue reflectance (Lalla et al., 2015). Light naturally present in the surroundings can impact to alter the functionality of the violet light source, but it doesn't affect the green light (Lalla et al., 2015). Therefore, a modification of 25% intensity in the light source could contribute to improving the efficiency of the light sources (Lalla et al., 2015). The morphological assessment of the blood vessels facilitates the understanding of the degree of severity of the disease (Lalla et al., 2016). Well-trained expertise is required to differentiate between lesions and dysplasia; therefore, this method is suggested to be used by a dental practitioner in conjunction with traditional methods, to effectively evaluate the oral mucosa sequentially and then combine the outcomes (Lalla et al., 2016). This device is quite expensive and has to be used in dark environments for more precision, and the probability of falsely considering hyperpigmented regions as diseased also to be taken into account (Sharma et al., 2021; Figure 3.6).

ORALOOK

ORALOOK (HITS PLAN, Tokyo, Japan) is another lightweight device that captures images in real time (Morikawa et al., 2019). The light source, a blue light of 410 nm, is set perpendicularly with respect to the lesion at about a distance of 10 cm (Morikawa et al., 2019). The captured image can be processed easily, as the device has a filter (520–560 nm) that detects apple-green coloured fluorescence from the tissue (Morikawa et al., 2019). Real-time image retrieval and single-handed use make this modality attractive (Morikawa et al., 2019).

IllumiScan

Another handheld device that uses optical imaging is IllumiScan (Shofu, Kyoto, Japan), which saves the captured digital information to the camera (Kikuta et al., 2018). Compactly designed, this device offers convenience in operation and can even capture images of the keratinized mucosa (Kikuta et al., 2018). The device is usually placed at a distance of 15 cm (Kikuta et al., 2018). The emission is at 425 nm, and the absorption range is 490–550 nm (Taguchi et al., 2022). The rest of the input is deleted before the image is presented on the monitor (Taguchi et al., 2022).

GOCCLES

Glasses for Oral Cancer Curing Light Exposed Screening (GOCCLES®) is a device developed to render cheap and effortless fluorescence detection (Moro et al., 2015). The FDA-approved device

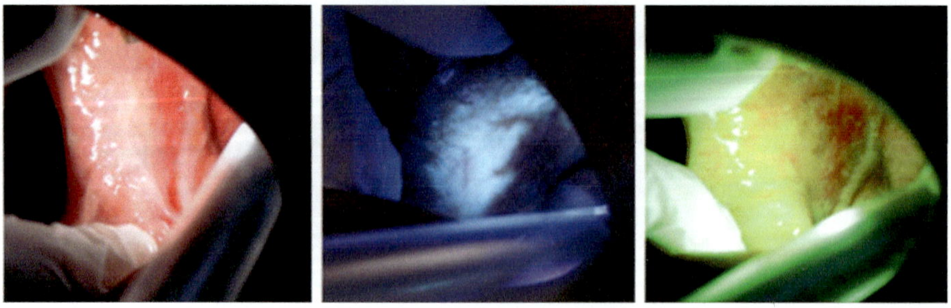

FIGURE 3.6 Oral lichen planus examined under (Left) white, (Centre) violet, and (Right) green illumination of identafi (adapted from Sharma et al., 2021).

has three layers of filters that exclusively isolate the fluorescence emitted by the FAD (Lajolo et al., 2022). Direct visualization of autofluorescence is made possible using filters present in the device, which occurs when a dental curing light is shone upon the mucosa (Moro et al., 2015). The light source is perpendicularly placed at a distance of 20 cm from the area under inspection (Lajolo et al., 2022). Out of 61 patients enrolled in a study, the naked eye examination identified 60 lesions whereas GOCCLES could identify 62 of the same, which were all grouped under dysplasia later after histopathological analysis (Moro et al., 2015).

The use of GOCCLES is encouraged in clinics equipped with well-trained experts and only in patients with suspected characteristic features (Moro et al., 2015). The application of the GOCCLES chip by attaching it to a smartphone camera gave a sensitivity of 66% and specificity of 48% (Lajolo et al., 2022). Even though this method cannot be relied upon for accurate differentiation of malignancy and normal mucosa, it can help locate lesions invisible to the naked eye and map the boundaries of the abnormalities before surgical intervention (Lajolo et al., 2022).

Autofluorescence Spectroscopy

Autofluorescence spectroscopy has demonstrated its biomedical applications in various fields (Nayak et al., 2006; Kamath et al., 2007; Kamath et al., 2009; Bhat et al., 2009; Chowdary et al., 2009; Kamath et al., 2011; Prabhu et al., 2012; Raghushaker et al., 2019; Mukunda et al., 2022b). It is another method that has gained popularity as a medium to ascertain the minute characteristic changes of premalignancy from the regular behaviour of the tissue. The mucosal properties will be subjected to alterations due to switches in the architecture and intrinsic reactions (Balasubramaniam et al., 2015). This principal aspect is considered under this method; that is, the modifications are assessed upon the illumination of the tissue under near-UV regions to instantaneously analyze the samples (Balasubramaniam et al., 2015; Veld et al., 2005). A thoroughly sanitized probe is used to analyze the emissions and a spectrograph is obtained, which is further subjected to computerized studies (Balasubramaniam et al., 2015). Conventionally, only one excitation wavelength was in use, but since the composition of tissues varies, so do their spectral characteristics. Multiple wavelengths can be imparted to increase the efficiency of the detection (Wang et al., 1999). The study conducted on 20 oral lichen planus patients, it was seen that there is a decline in the overall intensity due to the differences in the light attributes in abnormal conditions, making it an efficacious primary step in patient examination (Ramesh et al., 2021). This method has shown good sensitivity and specificity when used in correspondence with appropriate statistics (Paderni et al., 2011). This method facilitates appropriate disease management and reduces the number of additional biopsy tests (Paderni et al., 2011). With the provision of portability, non-invasiveness, and ease of operation, this becomes one of the most reliable methods for diagnosis and prognosis (Paderni et al., 2011).

The light source previously in use, including xenon lamps (under elevated pressure or pulse), has now been replaced with LED systems because of their inefficient performance in deep-UV wavelengths and excessive heat generation that exceeds their functional parameters and causes deleterious consequences on the whole setup (Mukunda et al., 2022a). LEDs are resourcefully systematized and come at reasonable prices and power stability (Mukunda et al., 2022a). The latest versions of LEDs have been developed to overcome the limitations of lower optical power and background emissions (Mukunda et al., 2022a).

During disease progression, the optical redox ratio of NADH to FAD values are replaced with respect to the metabolism of the cell (Duran-Sierra et al., 2020). The levels are seen to decline with a surge in the active functioning of the cell, along with which their lifetimes are also affected (Duran-Sierra et al., 2020) The structural rearrangement of the epithelia also contributes to the lowering of the connective tissue autofluorescence (Duran-Sierra et al., 2020). The shift from oxidative phosphorylation to glycolysis occurs as cancer progresses, which alters the cellular microenvironment and affects the fluorescence lifetime of NADH and FAD (Hinsdale et al., 2021). a two-channel light source was developed by Huang et al., featuring a handheld device with two excitation LEDs with

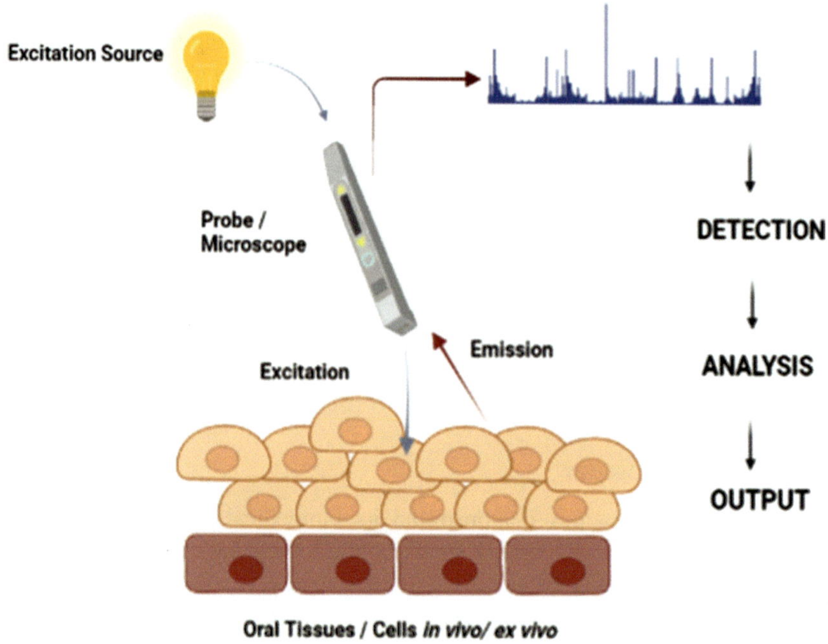

FIGURE 3.7 General workflow of optical spectroscopy in oral cancer.

low radiation hazards, a Charge-Coupled Device (CCD) camera with filters at 479 and 525 nm, and a mini motor to change filters (Huang et al., 2018). The device examines the image excited at 460 nm, and two images excited at both wavelengths are captured by pressing a button (Huang et al., 2018). The autofluorescence heterogeneity and intensities are evaluated to support the differentiation (Huang et al., 2018). Because of the fixed focus lens, unlike fibre, this method is not suitable for the least accessible areas (Huang et al., 2018). Only two images can be captured at a time, and proper ventilation is required to avoid fogging of lens (Huang et al., 2018; Figure 3.7).

NANOTECHNOLOGY

Widely used in various fields of life sciences, nanotechnology is an eloquent platform that is implemented in upgrading the standards of medical approaches in oral cancer (Nour et al., 2022). The advancement in nanofabrication is utilized for a better understanding of the condition (Nour et al., 2022). The safety of nanoparticles on cells is evaluated and then employed to characterize their spectral properties, which are important in precision detection (Nour et al., 2022). The small nanometre-sized fragments of metals have peculiar behaviours of absorption and emission of light (Manohar et al., 2011). A plasma is generated by the unbound electrons of these, which upon excitation resonate in visible regions, by the method of surface plasmon resonance (SPR) (Manohar et al., 2011). Gold nanoparticles were seen to express SPR, by which the absorptive rates are increased, making them easily detectable (Nour et al., 2022). The deeper understanding of nanoparticles has also opened avenues to improve screening techniques, as in the case of using metallic nanoparticles as contrast agents in capturing images using the principles of photoacoustics (Manohar et al., 2011).

Another study investigated how the matrix metalloproteinases that disrupt tissue architecture were differentially expressed in the case of carcinoma using a nanocomposite to function on fluorescence resonance energy transfer (FRET) and upconversion fluorescence and proposed the potential of this replacing other less specific methods (Chan et al., 2018). Apart from its extensive use in biosensing, this technology is also applied to perform immune reaction-based experiments like enzyme-linked immunoassay (ELISA), to sense the presence and abundance of biomarker molecules in the samples in sensitive manners (Chakraborty et al., 2018). Nanodiagnostics is the method of developing miniature devices using the principles of nanotechnology, where biomolecules are exploited microelectromechanically to conclude with a highly sensitive result, using minute volumes of test material, within a short span of time (Singhal et al., 2021).

Nanoparticles are highly stable with a longer half-life, can penetrate depths of tissues, and successfully detour the defencive barriers of the body; therefore, they need to be explored more (Perfézou et al., 2012). There is a requirement to enhance the sensitivity of the procedures performed using this technology (Perfézou et al., 2012).

PHOTOACOUSTIC SPECTROSCOPY

The molecules will absorb pulses of laser and generate heat, eventually creating a temporary pulse of pressure in the form of acoustics (Raghushaker et al., 2021). Such signals can retrieve information about the changes happening in the molecules (Raghushaker et al., 2021). This technique thereby gives an overview of the alterations that proteins undergo due to the influence of aromatic amino acids (Raghushaker et al., 2021). It combines the capabilities of investigating tissues in depth without using any ionization (Borg et al., 2018). Contrasting agents that can be present within the cells or outsourced can be used for obtaining images with exceptional contrasts (Rodrigues et al., 2021). Mostly, intrinsic contrast agents like haemoglobin are used, whereas the latest techniques make use of agents provided from external sources to enhance the process with better precision (Borg et al., 2018). Molecular contrast agents are preferred due to their size, which retain and enhance the permeation of the molecules into the cells (Borg et al., 2018; Table 3.3).

DNA ANALYSIS

The genome is subjected to analysis for chromosomal structural or numerical aberrations. The gene expression has associations with the nature of the disease and future consequences (Ziober et al., 2008).

DNA Microarray

The process of allowing the DNA fragments from the sample to hybridize with the probes etched onto the surface of a slide is called microarray. This works on the principle of complementary base pair hybridization of the genetic material in the sample to the spotted probes on the chip (Madhura et al., 2020b). The specimen is processed for fragmenting the DNA and allowed to hybridize with fluorescently labelled probes, and the signals are analyzed for gene expression (Madhura et al., 2020a). The effects of structural or chromosomal aberrations of genes also promote a better understanding of the changes at the molecular level, which can be used to classify the conditions accordingly (Chai et al., 2020).

Next-Generation Sequencing

Apart from the expense in conducting this procedure, next-generation sequencing (NGS) yields a wide range of information in a short duration, including the sequence arrangement, aberrations, and molecular networks involved by the genes (Madhura et al., 2020b). The abnormal expression of the oncogenes, tumour-suppressor genes, and miRNAs are understood by this method (Madhura et al., 2020). It can also elucidate the relationships of various factors that cause distinct characteristics of malignancy (Madhura et al., 2020).

TABLE 3.3

List of Contrast Agents for the Enhancement of Images in Photoacoustic Spectroscopy

S.No	Contrast Agents	Structural Characteristics	Properties	Drawbacks	Examples	References
1	Cyanine dyes	Two nitrogen-containing moieties joined with a linear polymethine chain; changes in chain length shift the spectral properties	-NIR absorption -Water stable and soluble -Tunable absorption and emission spectra -High molecular extinction coefficients	-Photo-isomerization changes the ground state absorption spectra -Dye aggregation causes changes in absorption spectra -Prone to irreversible oxidative photobleaching	Cy3, Cy5, Cy7-based indocyanine dyes	Mishra et al. (2000); Taniguchi and Lindsey (2018)
2	Xanthene dyes	Oxygen bridge fusing the benzene rings	-Restricted rotational freedom -High fluorescence quantum yield	-Relatively low response	Fluorescein	Rurack and Spieles. (2011) ; Magde et al. (2002)
3	Phthalein dyes	Lacks the fused O bridge unlike Xanthene	-Increased rotational freedom -Two times more signal than fluorescein	-Aggregates formed when in high concentration can cause thermal deactivation	Phenolphthalein dye	Borg and Rochford (2018); Boguta and Wróbel. (2001)
4	Rhodamine dyes	Structurally similar to Xanthine dyes, but their terminal hydroxyl and ketone moieties are replaced with electron-rich amino groups	-High quantum yields -Small stoke shifts	-Indirect quantification is required as the relative method yields incorrect quantum	Rhodamine B	Borg and Rochford (2018)
5	Squaraine and croconaine dyes	Central electron deficient four membrane ring derived from squaric acid with a quadrupolar resonance-stabilized donor-p-acceptor-p-donor zwitterionic structure	-Comer available ones have limited structural diversity	-Use of liposomes to prevent aggregation and increase monomer photostability and chemical stability to reduce photobleaching and cytotoxicity might be required	2,3-bis(4-(dibenzylamino) phenyl)cyclobuta-1,3-diene-1,3-bis(olate) squaraine	Borg and Rochford (2018)
6	Tetrapyrrole	Porphyrin-phospholipid conjugate	Porphysomes developed using porphyrin lipid-derived liposome are effective for tumour ablation	Heating above the transition temperature of the liposome lead to lowering the photoacoustic signal	Porphyrosome, porphyrin-phospholipid conjugate, porphyrin nanodroplets	Borg and Rochford (2018); Jin et al. (2013); Paproski et al. (2016); Huynh et al.(2012)
7	BODIPY dyes	Based on 4,4-difluoro-4-bora-3a,4a-diaza-s-indacene structure	Fluorescent probe – NIR absorbing ones have high photostability and greater PA response than ICG cyanine dyes	Highly dependent on the incident laser fluence	aza-BODIPY PA ratiometric probes, NIR absorbing BODIPY dyes	Borg and Rochford (2018); Li et al. (2015) ; Ni et al. (2016) ; Ni et al. (2017)
8	Curcumin dyes	Bright yellow pigment in the rhizomes of curcuma plant	Antibacterial, antifungal, antiviral, antitumour activity; enhances PA signals due to photon absorption	Limited to 532 nm laser excitation	Borondifluoride curcumin derivative; bis(phenyl) curcuminBF2 derivative	Bairwa et al. (2014); Frenette et al. (2014)

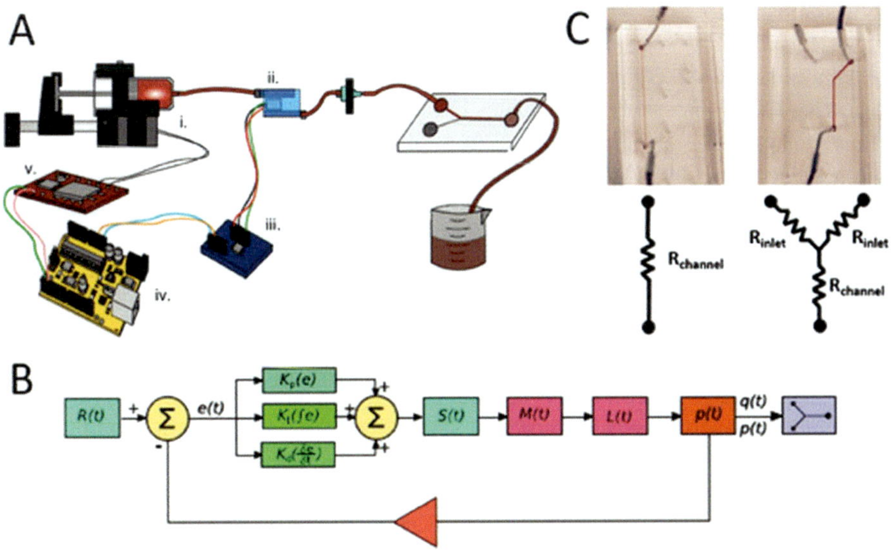

FIGURE 3.8 Arduino-controlled pressure in a microfluidic system (A), through PID or bang-bang algorithm (B) to control the desired flow pattern (C) (adapted from Nguyen et al., 2018).

MICROFLUIDICS

Microfluidics is the amalgamation of numerous strategies used to analyze the samples (Madhura et al., 2020b). The specimen is subjected to procedures like PCR, microarray, FISH, methylation pattern, and transcription regulation (Madhura et al., 2020b). The test specimen in minute concentrations can be processed in a relatively short timespan, using minimal reaction ingredients, also foreshortening the risks of hazards and contamination (Ziober et al., 2008).

The apparatus must be designed to tackle the challenges presented by the physical parameters such as fluid mixing, heat, and flow (Ziober et al., 2008). The samples are collected into a cassette containing antibodies corresponding to the protein of interest and the cells. Whatever is obtained is subjected to respective reagent treatments, or the components are isolated for further testing (Ziober et al., 2008; Figure 3.8).

DISCUSSION

The detection of oral cancer at preliminary stages can refine the treatment modalities and the quality of life after treatment (Fatakdawala et al., 2013). The five-year survival rates drastically reduce from 82% in the first stage to 35% in the fourth stage (Fatakdawala et al., 2013). On this account, methods that accurately describe the nature of the deformities are developed (Fatakdawala et al., 2013). The abnormalities are directly visualized, palpated, or subjected to radiologic analysis (Fatakdawala et al., 2013). The gold standard method for detection is a surgical biopsy followed by histopathology. However, it is error-prone throughout the procedure, causes patient irritation, and is invasive. COE practices fail to pinpoint early signs and miss most of the non-obvious lesions that would develop into cancer (Singh et al., 2016). The staining methods can non-specifically stain the tissue, causing more false-positive results (Singh et al., 2016). Resecting the affected region requires precision in outlining the margins (Wu et al., 2018). If the tests are not performed by experienced personnel, there are chances of errors and false-negative outcomes (Wu et al., 2018). Currently available imaging mechanisms depend on the tissue architecture and metabolism (Wu et al., 2018). The developing technology is aiding the images to be obtained in real time (Wu et al., 2018). Huff et al. performed a study by combining COE with VELscope and had the potential to detect more aberrations compared to COE alone (Huff et al., 2009). Another study conducted by Bhatia et al obtained 54.3% specificity with VELscope whereas combining the above two gave a specificity of 97.9% (Bhatia et al., 2014). By better understanding the disease's nature, altered metabolism, and factors affecting the progression, like changes in the tumour microenvironment, the discovery and development of new tools are made easy (Wu et al., 2018). Raman spectroscopy (RS) works on developing a spectrum from scattered light corresponding to the vibrational patterns of the bindings between molecules when they receive or liberate energy (Wu et al., 2018). The peaks obtained represent the cell's biochemical signature (Wu et al., 2018). It is advantageous over other methods as there is no interruption due to the absorbance of water, not involving fluorescence with less background noise using a narrow laser beam, facilitating the method to perform with very little analyte (Wu et al., 2018). Even though most of these have been theoretically or experimentally proven to be effective, many are not yet commercialized (Wu et al., 2018). The inconvenience associated with the portability of the large instruments is another challenge in a practical scenario (Wu et al., 2018). Diligently subjecting these to ensure thorough safety and efficiency trials is of paramount importance before applying them to the population (Wu et al., 2018). The amalgamation of extant methods with newer interdisciplinary findings will improve overall coherence (Wu et al., 2018). Multimodal strategies are thereby being employed that utilize each method's advantages. FLIm, UBM, and PAI techniques were brought together in a study to trace the cells' biological, chemical, structural, morphological, and vascular changes (Fatakdawala et al., 2013). Such activities will assist in compiling information that complements each other and meticulously verifying the results (Fatakdawala et al., 2013). The last ten years have witnessed revolutionary changes in the field of molecular sciences, and NGS has evolved to characterize the genomic signatures of this disease (Chai et al., 2020). Frameworks of the genome and transcriptome have been extensively studied, making it possible to target networks for developing therapeutics, predicting the nature of progression, drug repurposing, and understanding the immunological influences (Chai et al., 2020). The cooperative efforts of the active and experienced torchbearers in the field will be of service to expand and excel in diagnostics (Wu et al., 2018).

CONCLUSION

The mortality rate caused by OSCC has come down in recent years, but not to a satisfactory level, and it offers a five-year survival rate of only 50% (Singh et al., 2016). With a predilection higher in males and the predominance of various contributing factors, maximum prevalence is observed in Asia and Europe (Singh et al., 2016). A physician or dentist could diagnose oral carcinoma, but they tend to miss the earliest symptoms and signs (Smith, 2002; Chakraborty et al., 2019) especially among populations with a lower risk of having the disease. Despite having numerous disease management methods, the death rates caused by oral cancer remain high (Balasubramaniam et al., 2015). Conventional techniques like that of biopsy are more sensitive than studying the biology of the sample cells. Yet, it is invasive, causes pain and discomfort, and the results can be biased due to a lack of expertise (Goldoni et al., 2021). The latest advances in imaging have satisfactory potential but can result in erroneous outcomes (Chakraborty et al., 2019).

Some control could be established with chemotherapy, radiotherapy, and surgical interventions (Figueiredo et al., 2010). Yet, a routine check-up is important for the early detection of the condition (Reddy & Praveen, 2017). The chances of false reporting of a previously absent disease can occur by overestimating the signs through the conventional staining or operative method (Smith, 2002). Most importantly, there is a requirement for the expertise of the technicians to deduce the findings correctly (Smith, 2002). Correctly identifying the stage of the disease and its spread is predominantly important, which can be performed by the techniques available now. Still, these are neither completely optimized nor accurate in diagnosing the metastatic nature of the tumour. Because of this, better approaches are necessary to identify the symptoms at preliminary stages to avoid advancement and to administer timely treatment. Fine-tuning the existing methods by incorporating various factors that would be beneficial for speedy and efficient diagnosis is being done. Researchers are combining methods like VELscope and RS to improve the sensitivity of the detection process (Jeng et al., 2020).

The biomedical approach to the field of oral health has undergone massive changes from conventional oral examinations to multiplexed tests (Ziober et al., 2008).

The field of nanotechnology is advancing to relieve the burdens of the existing systems (Singhal et al., 2021). The nanosized materials have more surface area and facilitate the adherence of multi-ligands, optimizing the working of the instrument and generating signals of amplified intensities (Singhal et al., 2021). More calibrations and ethical clearances are required before commercializing this, limiting it to the preliminary stage (Singhal et al., 2021).

Synthetic biology is another emerging discipline to diagnose conditions, using bioengineering (Madhura et al., 2020b). This raises ethical and financial concerns due to the complexity involved in designing and implementing the tools in a biologically safe manner (Madhura et al., 2020).

The social and economic elements and the unfamiliarity of the populace are major reasons for progression of disease and advances to the final stages (Singhal et al., 2021). The extensive upsurge in the number of cases, spreading the right information regarding the importance of routine check-ups, and reducing the exposure and use of carcinogenic substances are essential steps (Singhal et al., 2021). Making people aware of the genetic predisposition to have this cancer at a later stage in life also encourages them to undergo tests that could prolong their lives and reduce the chances of progression and death (Singhal et al., 2021).

ACKNOWLEDGMENT

The authors would like to thank Manipal School of Life Sciences, Manipal Academy of Higher Education, Manipal, for infrastructure facilities. They would also like to thank DBT-BUILDER (BT/INF/22/SP43065/2021) for providing an infrastructure grant.

REFERENCES

Adeoye, J., Alade, A. A., Zhu, W. Y., Wang, W., Choi, S. W., & Thomson, P. (2022). Efficacy of hypermethylated DNA biomarkers in saliva and oral swabs for oral cancer diagnosis: Systematic review and meta-analysis. *Oral Diseases*, *28*(3), 541–558.

Almadori, G., Bussu, F., Galli, J., Limongelli, A., Persichilli, S., Zappacosta, B., … & Giardina, B. (2007). Salivary glutathione and uric acid levels in patients with head and neck squamous cell carcinoma. *Head & Neck: Journal for the Sciences and Specialties of the Head and Neck*, *29*(7), 648–654.

Arantes, L. M. R. B., De Carvalho, A. C., Melendez, M. E., & Lopes Carvalho, A. (2018). Serum, plasma and saliva biomarkers for head and neck cancer. *Expert Review of Molecular Diagnostics*, *18*(1), 85–112.

Arya, S., Chaukar, D., & Pai, P. (2012). Imaging in oral cancers. *Indian Journal of Radiology and Imaging*, *22*(3), 195–208.

Awais, M., Ghayvat, H., Krishnan Pandarathodiyil, A., Nabillah Ghani, W. M., Ramanathan, A., Pandya, S., ... & Faye, I. (2020). Healthcare professional in the loop (HPIL): Classification of standard and oral cancer-causing anomalous regions of oral cavity using textural analysis technique in autofluorescence imaging. *Sensors*, *20*(20), 5780.

Awan, K. H. (2015). Efficacy of autofluorescence imaging as an adjunctive technique for examination and detection of oral potentially malignant disorders: A systematic review. *The Journal of Contemporary Dental Practice*, *16*, 741–749.

Awan, K. H., Morgan, P. R., & Warnakulasuriya, S. (2011). Utility of chemiluminescence (ViziLite™) in the detection of oral potentially malignant disorders and benign keratoses. *Journal of Oral Pathology & Medicine*, *40*(7), 541–544.

Awan, K. H., Yang, Y. H., Morgan, P. R., & Warnakulasuriya, S. (2012). Utility of toluidine blue as a diagnostic adjunct in the detection of potentially malignant disorders of the oral cavity—A clinical and histological assessment. *Oral Diseases*, *18*(8), 728–733.

Baba, A., Ojiri, H., Ikeda, K., Yamauchi, H., Ogino, N., Seto, Y., ... & Mogami, T. (2018). Essentials on oncological imaging: Postoperative computed tomography and magnetic resonance imaging of oral tongue cancer. *Canadian Association of Radiologists Journal*, *69*(4), 458–467.

Bairwa, K., Grover, J., Kania, M., & Jachak, S. M. (2014). Recent developments in chemistry and biology of curcumin analogues. *RSC Advances*, *4*(27), 13946–13978.

Balasubramaniam, A. M., Sriraman, R., Sindhuja, P., Mohideen, K., Parameswar, R. A., & Haris, K. M. (2015). Autofluorescence based diagnostic techniques for oral cancer. *Journal of Pharmacy & Bioallied Sciences*, *7*(Suppl 2), S374.

Bhat, R. A., Kamath, S. D., Mahato, K. K., & Ray, S. (2009). Autofluorescence of normal, benign, and malignant ovarian tissues: A pilot study. *Photomedicine and Laser Surgery, 27*(2), 325.

Bhatia, N., Matias, M. A. T., & Farah, C. S. (2014). Assessment of a decision making protocol to improve the efficacy of VELscope™ in general dental practice: A prospective evaluation. *Oral Oncology, 50*(10), 1012–1019.

Boguta, A., & Wróbel, D. (2001). Fluorescein and phenolphthalein—Correlation of fluorescence and photoelectric properties. *Journal of Fluorescence*, *11*(2), 129–137.

Borg, R. E., & Rochford, J. (2018). Molecular photoacoustic contrast agents: Design principles & applications. *Photochemistry and Photobiology*, *94*(6), 1175–1209.

Borse, V., Konwar, A. N., & Buragohain, P. (2020). Oral cancer diagnosis and perspectives in India. *Sensors International*, *1*, 100046.

Calin, G. A., & Croce, C. M. (2006). MicroRNA signatures in human cancers. *Nature Reviews Cancer, 6*(11), 857–866.

Chai, A. W. Y., Lim, K. P., & Cheong, S. C. (2020). Translational genomics and recent advances in oral squamous cell carcinoma. *Seminars in Cancer Biology, 61,* 71–83.

Chai, Y. D., Zhang, L., Yang, Y., Su, T., Charugundla, P., Ai, J., ... & Hu, S. (2016). Discovery of potential serum protein biomarkers for lymph node metastasis in oral cancer. *Head & Neck*, *38*(1), 118–125.

Chakraborty, D., Ghosh, D., Kumar, S., Jenkins, D., Chandrasekaran, N., & Mukherjee, A. (2022). Nano-diagnostics as an emerging platform for oral cancer detection: Current and emerging trends. *Wiley Interdisciplinary Reviews: Nanomedicine and Nanobiotechnology, 15*, e1830.

Chakraborty, D., Natarajan, C., & Mukherjee, A. (2019). Advances in oral cancer detection. *Advances in Clinical Chemistry*, *91*, 181–200.

Chakraborty, D., Viveka, T. S., Arvind, K., Shyamsundar, V., Kanchan, M., Alex, S. A., ... & Mukherjee, A. (2018). A facile gold nanoparticle–based ELISA system for detection of osteopontin in saliva: Towards oral cancer diagnostics. *Clinica Chimica Acta*, *477*, 166–172.

Chan, Y. C., Chan, M. H., Chen, C. W., Liu, R. S., Hsiao, M., & Tsai, D. P. (2018). Near-infrared-activated fluorescence resonance energy transfer-based nanocomposite to sense MMP2-overexpressing oral cancer cells. *ACS Omega*, *3*(2), 1627–1634.

Chaudhari, A., Hegde-Shetiya, S., Shirahatti, R., & Agrawal, D. (2013). Comparison of different screening methods in estimating the prevalence of precancer and cancer amongst male inmates of a jail in Maharashtra, India. *Asian Pacific Journal of Cancer Prevention*, *14*(2), 859–864.

Chen, X. J., Zhang, X. Q., Liu, Q., Zhang, J., & Zhou, G. (2018). Nanotechnology: A promising method for oral cancer detection and diagnosis. *Journal of Nanobiotechnology*, *16*(1), 1–17.

Chen, Y. W., Lin, J. S., Wu, C. H., Lui, M. T., Kao, S. Y., & Fong, Y. (2007b). Application of in vivo stain of methylene blue as a diagnostic aid in the early detection and screening of oral squamous cell carcinoma and precancer lesions. *Journal of the Chinese Medical Association*, *70*(11), 497–503.

Cheng, Y. S. L., Rees, T., & Wright, J. (2014). A review of research on salivary biomarkers for oral cancer detection. *Clinical and Translational Medicine*, *3*(1), 1–10.

Cherry, K. D., Schwarz, R. A., Yang, E. C., Vohra, I. S., Badaoui, H., Williams, M. D., ... & Richards-Kortum, R. (2019). Autofluorescence imaging to monitor the progression of oral potentially malignant disordersImaging to monitor oral potentially malignant disorders. *Cancer Prevention Research*, *12*(11), 791–800.

Chowdary, M. V. P., Mahato, K. K., Kumar, K. K., Mathew, S., Rao, L., Krishna, C. M., & Kurien, J. (2009). Autofluorescence of breast tissues: Evaluation of discriminating algorithms for diagnosis of normal, benign, and malignant conditions. *Photomedicine and Laser Surgery, 27*(2), 241–252.

De Veld, D. C., Skurichina, M., Witjes, M. J., Duin, R. P., Sterenborg, H. J., & Roodenburg, J. L. (2005). Autofluorescence and diffuse reflectance spectroscopy for oral oncology. *Lasers in Surgery and Medicine: The Official Journal of the American Society for Laser Medicine and Surgery*, *36*(5), 356–364.

Dhanuthai, K., Rojanawatsirivej, S., Thosaporn, W., Kintarak, S., Subarnbhesaj, A., Darling, M., ... & Shakib, P. A. (2018). Oral cancer: A multicenter study. *Medicina oral, patologia oral y cirugia bucal*, *23*(1), e23.

Dodeigne, C., Thunus, L., & Lejeune, R. (2000). Chemiluminescence as diagnostic tool. A review. *Talanta*, *51*(3), 415–439.

Du, G. F., Li, C. Z., Chen, H. Z., Chen, X. M., Xiao, Q., Cao, Z. G., ... & Cai, X. (2007). Rose bengal staining in detection of oral precancerous and malignant lesions with colorimetric evaluation: A pilot study. *International Journal of Cancer*, *120*(9), 1958–1963.

Duran-Sierra, E., Cheng, S., Cuenca-Martinez, R., Malik, B., Maitland, K. C., Cheng, Y. L., ... & Jo, J. A. (2020). Clinical label-free biochemical and metabolic fluorescence lifetime endoscopic imaging of precancerous and cancerous oral lesions. *Oral Oncology*, *105*, 104635.

Ebihara, A., Krasieva, T. B., Liaw, L. H. L., Fago, S., Messadi, D., Osann, K., & Wilder-Smith, P. (2003). Detection and diagnosis of oral cancer by light-induced fluorescence. *Lasers in Surgery and Medicine: The Official Journal of the American Society for Laser Medicine and Surgery*, *32*(1), 17–24.

Elimairi, I., Altay, M. A., Abdoun, O., Elimairi, A., Tozoglu, S., Baur, D. A., & Quereshy, F. (2017). Clinical relevance of the utilization of vital Lugol's iodine staining in detection and diagnosis of oral cancer and dysplasia. *Clinical Oral Investigations*, *21*(2), 589–595.

Epstein, J. B., Scully, C., & Spinelli, J. (1992). Toluidine blue and Lugol's iodine application in the assessment of oral malignant disease and lesions at risk of malignancy. *Journal of Oral Pathology & Medicine*, *21*(4), 160–163.

Farah, C. S., Dost, F., & Do, L. (2019). Usefulness of optical fluorescence imaging in identification and triaging of oral potentially malignant disorders: A study of VELscope in the LESIONS programme. *Journal of Oral Pathology & Medicine*, *48*(7), 581–587.

Farah, C. S., McIntosh, L., Georgiou, A., & McCullough, M. J. (2012). Efficacy of tissue autofluorescence imaging (VELScope) in the visualization of oral mucosal lesions. *Head & Neck*, *34*(6), 856–862.

Fatakdawala, H., Poti, S., Zhou, F., Sun, Y., Bec, J., Liu, J., ... & Marcu, L. (2013). Multimodal in vivo imaging of oral cancer using fluorescence lifetime, photoacoustic and ultrasound techniques. *Biomedical Optics Express*, *4*(9), 1724–1741.

Figueiredo, P. T., Leite, A. F., Freitas, A. C., Nascimento, L. A., Cavalcanti, M. G., Melo, N. S., & Guerra, E. N. (2010). Comparison between computed tomography and clinical evaluation in tumour/node stage and follow-up of oral cavity and oropharyngeal cancer. *Dentomaxillofacial Radiology*, *39*(3), 140–148

Frenette, M., Hatamimoslehabadi, M., Bellinger-Buckley, S., Laoui, S., La, J., Bag, S., ... & Rochford, J. (2014). Shining light on the dark side of imaging: excited state absorption enhancement of a bis-styryl BODIPY photoacoustic contrast agent. *Journal of the American Chemical Society*, *136*(45), 15853–15856.

Gandolfo, S., Pentenero, M., Broccoletti, R., Pagano, M., Carrozzo, M., & Scully, C. (2006). Toluidine blue uptake in potentially malignant oral lesions in vivo: Clinical and histological assessment. *Oral Oncology*, *42*(1), 88–94.

Goldoni, R., Scolaro, A., Boccalari, E., Dolci, C., Scarano, A., Inchingolo, F., ... & Tartaglia, G. (2021). Malignancies and biosensors: A focus on oral cancer detection through salivary biomarkers. *Biosensors*, *11*(10), 396.

Guneri, P., Epstein, J. B., Ilhan, B., Kaya, A., & Boyacioglu, H. (2013). Clinical application of a digital method to improve the accuracy of color perception in toluidine blue stained oral mucosal lesions. *Quintessence International*, *44*(8), 619–627.

Helsper, J. T. (1972). Staining techniques: Screening tests for oral cancer. *CA: A Cancer Journal for Clinicians*, *22*(3), 172–175.

Hinsdale, T. A., Malik, B. H., Cheng, S., Benavides, O. R., Giger, M. L., Wright, J. M., ... & Maitland, K. C. (2021). Enhanced detection of oral dysplasia by structured illumination fluorescence lifetime imaging microscopy. *Scientific Reports*, *11*(1), 1–11.

Ho, A. S., Kim, S., Tighiouart, M., Gudino, C., Mita, A., Scher, K. S., ... & Zumsteg, Z. S. (2017). Metastatic lymph node burden and survival in oral cavity cancer. *Journal of Clinical Oncology*, *35*(31), 3601.

Huang, T. T., Chen, K. C., Wong, T. Y., Chen, C. Y., Chen, W. C., Chen, Y. C., ... & Huang, J. S. (2018). Two-channel autofluorescence analysis for oral cancer. *Journal of Biomedical Optics*, *24*(5), 051402.

Huff, K., Stark, P. C., & Solomon, L. W. (2009). Sensitivity of direct tissue fluorescence visualization in screening for oral premalignant lesions in general practice. *Gen Dent*, *57*(1), 34–38.

Huynh, E., Lovell, J. F., Helfield, B. L., Jeon, M., Kim, C., Goertz, D. E., ... & Zheng, G. (2012). Porphyrin shell microbubbles with intrinsic ultrasound and photoacoustic properties. *Journal of the American Chemical Society*, *134*(40), 16464–16467.

Ibrahim, S. S., Al-Attas, S. A., Darwish, Z. E., Amer, H. A., & Hassan, M. H. (2014). Effectiveness of the Microlux/DLTM chemiluminescence device in screening of potentially malignant and malignant oral lesions. *Asian Pacific Journal of Cancer Prevention*, *15*(15), 6081–6086

Jeng, M. J., Sharma, M., Sharma, L., Huang, S. F., Chang, L. B., Wu, S. L., & Chow, L. (2020). Novel quantitative analysis using optical imaging (VELscope) and spectroscopy (Raman) techniques for oral cancer detection. *Cancers*, *12*(11), 3364.

Jin, C. S., Lovell, J. F., Chen, J., & Zheng, G. (2013). Ablation of hypoxic tumors with dose-equivalent photothermal, but not photodynamic, therapy using a nanostructured porphyrin assembly. *ACS Nano*, *7*(3), 2541–2550.

Kamath, S. D., & Mahato, K. K. (2007). Optical pathology using oral tissue fluorescence spectra: Classification by principal component analysis and k-means nearest neighbor analysis. *Journal of Biomedical Optics*, *12*(1), 014028–014028.

Kamath, S. D., & Mahato, K. K. (2009). Principal component analysis (PCA)-based k-nearest neighbor (k-NN) analysis of colonic mucosal tissue fluorescence spectra. *Photomedicine and Laser Surgery*, *27*(4), 659–668.

Kamath, S. D., Ray, S., & Mahato, K. K. (2011). Photoacoustic spectroscopy of ovarian normal, benign, and malignant tissues: A pilot study. *Journal of Biomedical Optics*, *16*(6), 067001.

Katakura, A., Kamiyama, I., Takano, N., Shibahara, T., Muramatsu, T., Ishihara, K., … & Shouno, T. (2007). Comparison of salivary cytokine levels in oral cancer patients and healthy subjects. *The Bulletin of Tokyo Dental College*, *48*(4), 199–203.

Kaur, J., Srivastava, R., & Borse, V. (2021). Recent advances in point-of-care diagnostics for oral cancer. *Biosensors and Bioelectronics*, *178*, 112995.

Kaur, M., Handa, U., Mohan, H., & Dass, A. (2016). Evaluation of brush cytology and DNA image cytometry for the detection of cancer of the oral cavity. *Diagnostic Cytopathology*, *44*(3), 201–205.

Keur, J. J., & Macdonald, R. (1986). Dental panoramic radiography with intra-oral X-ray tubes. *Australian Dental Journal*, *31*(1), 40–43.

Kikuta, S., Iwanaga, J., Todoroki, K., Shinozaki, K., Tanoue, R., Nakamura, M., & Kusukawa, J. (2018). Clinical application of the IllumiScan fluorescence visualization device in detecting oral mucosal lesions. *Cureus*, *10*(8), e3111.

Koch, F. P., Kaemmerer, P. W., Biesterfeld, S., Kunkel, M., & Wagner, W. (2011). Effectiveness of autofluorescence to identify suspicious oral lesions—A prospective, blinded clinical trial. *Clinical Oral Investigations*, *15*(6), 975–982.

Kordbacheh, F., Bhatia, N., & Farah, C. S. (2016). Patterns of differentially expressed genes in oral mucosal lesions visualised under autofluorescence (VELscope™). *Oral Diseases*, *22*(4), 285–296.

Kurita, H., Kamata, T., Li, X., Nakanishi, Y., Shimane, T., & Koike, T. (2012). Effectiveness of vital staining with iodine solution in reducing local recurrence after resection of dysplastic or malignant oral mucosa. *British Journal of Oral and Maxillofacial Surgery*, *50*(2), 109–112.

Lajolo, C., Tranfa, M., Patini, R., Fiorino, A., Musarra, T., Boniello, R., & Moro, A. (2022). Clinical evaluation of the optical filter for autofluorescence glasses for oral cancer curing light exposed (GOCCLES®) in the management of potentially premalignant disorders: A retrospective study. *International Journal of Environmental Research and Public Health*, *19*(9), 5579.

Lalla, Y., Matias, M. A. T., & Farah, C. S. (2015). Oral mucosal disease in an Australian urban Indigenous community using autofluorescence imaging and reflectance spectroscopy. *Australian Dental Journal*, *60*(2), 216–224.

Lalla, Y., Matias, M. A. T., & Farah, C. S. (2016). Assessment of oral mucosal lesions with autofluorescence imaging and reflectance spectroscopy. *The Journal of the American Dental Association*, *147*(8), 650–660

Leonard, J. R., & Beck, W. L. (1971). Hematoporphyrin fluorescence: An aid in diagnosis of malignant neoplasms. *The Laryngoscope*, *81*(3), 365–372.

Li, H., Zhang, P., Smaga, L. P., Hoffman, R. A., & Chan, J. (2015). Photoacoustic probes for ratiometric imaging of copper (II). *Journal of the American Chemical Society*, *137*(50), 15628–15631.

Lingen, M. W., Kalmar, J. R., Karrison, T., & Speight, P. M. (2008). Critical evaluation of diagnostic aids for the detection of oral cancer. *Oral Oncology*, *44*(1), 10–22.

López-Jornet, P., & De la Mano-Espinosa, T. (2011). The efficacy of direct tissue fluorescence visualization in screening for oral premalignant lesions in general practice: An update. *International Journal of Dental Hygiene*, *9*(2), 97–100.

Madhura, M. G., Rao, R. S., Patil, S., Alhazmi, Y. A., Jafer, M., Habib, S. R., & Awan, K. H. (2020a). Minimally invasive procedures for the recognition and diagnosis of oral precancer and cancer. *Disease-a-Month*, *66*(12), 101033.

Madhura, M. G., Rao, R. S., Patil, S., Fageeh, H. N., Alhazmi, A., & Awan, K. H. (2020b). Advanced diagnostic aids for oral cancer. *Disease-a-Month*, *66*(12), 101034.

Magde, D., Wong, R., & Seybold, P. G. (2002). Fluorescence quantum yields and their relation to lifetimes of rhodamine 6G and fluorescein in nine solvents: Improved absolute standards for quantum yields¶. *Photochemistry and Photobiology, 75*(4), 327–334.

Manohar, S., Ungureanu, C., & Van Leeuwen, T. G. (2011). Gold nanorods as molecular contrast agents in photoacoustic imaging: The promises and the caveats. *Contrast Media & Molecular Imaging, 6*(5), 389–400.

Mazumder, S., Datta, S., Ray, J. G., Chaudhuri, K., & Chatterjee, R. (2019). Liquid biopsy: miRNA as a potential biomarker in oral cancer. *Cancer Epidemiology, 58*, 137–145.

Mishra, A., Behera, R. K., Behera, P. K., Mishra, B. K., & Behera, G. B. (2000). Cyanines during the 1990s: A review. *Chemical Reviews, 100*(6), 1973–2012.

Mittal, N., Palaskar, S., & Shankari, M. (2012). Rose Bengal staining-diagnostic aid for potentially malignant and malignant disorders: A pilot study. *Indian Journal of Dental Research, 23*(5), 561.

Moafa, I., Hoving, C., van den Borne, B., & Jafer, M. (2022). Dentists' perceptions and usability testing toward the implementation of the ISAC, a comprehensive oral cancer intervention in dental practices: A qualitative study in Jazan region, Saudi Arabia. *BMC Health Services Research, 22*(1), 1–12.

Montero, P. H., & Patel, S. G. (2015). Cancer of the oral cavity. *Surgical Oncology Clinics, 24*(3), 491–508.

Morikawa, T., Kosugi, A., & Shibahara, T. (2019). The utility of optical instrument "ORALOOK®" in the early detection of high-risk oral mucosal lesions. *Anticancer Research, 39*(5), 2519–2525.

Moro, A., De Waure, C., Di Nardo, F., Spadari, F., Mignogna, M. D., Giuliani, M., ... & Pelo, S. (2015). The GOCCLES® medical device is effective in detecting oral cancer and dysplasia in dental clinical setting. Results from a multicentre clinical trial. *Acta Otorhinolaryngologica Italica, 35*(6), 449.

Mukunda, D. C., Joshi, V. K., Chandra, S., Siddaramaiah, M., Rodrigues, J., Gadag, S., ... & Mahato, K. K. (2022a). Probing nonenzymatic glycation of proteins by deep ultraviolet light emitting diode induced autofluorescence. *International Journal of Biological Macromolecules, 213*, 279–296.

Mukunda, D. C., Joshi, V. K., & Mahato, K. K. (2022b). Light emitting diodes (LEDs) in fluorescence-based analytical applications: A review. *Applied Spectroscopy Reviews, 57*(1), 1–38.

Murube, J. (2014). Rose bengal: The second most commonly used surfocular vital stain. *The Ocular Surface, 12*(1), 14–22.

Nae, A., O'Leary, G., Feeley, L., Fives, C., Fitzgerald, B., Chiriac, E., & Sheahan, P. (2019). Utility of CT and MRI in assessment of mandibular involvement in oral cavity cancer. *World Journal of Otorhinolaryngology-Head and Neck Surgery, 5*(02), 71–75.

Nagaraju, K., Prasad, S., & Ashok, L. (2010). Diagnostic efficiency of toluidine blue with Lugol's iodine in oral premalignant and malignant lesions. *Indian Journal of Dental Research, 21*(2), 218.

Nayak, G. S., Kamath, S., Pai, K. M., Sarkar, A., Ray, S., Kurien, J., ... & Mahato, K. K. (2006). Principal component analysis and artificial neural network analysis of oral tissue fluorescence spectra: Classification of normal premalignant and malignant pathological conditions. *Biopolymers: Original Research on Biomolecules, 82*(2), 152–166.

Neville, B. W., Damm, D. D., Allen, C. M., & Bouquot, J. E. (2009). *Text book of oral and maxillofacial pathology* (p. 692). St. Louis, MO: Elsevier.

Nguyen, T., Andreasen, S. Z., Wolff, A., & Bang, D. D. (2018). From lab on a chip to point of care devices: The role of open source microcontrollers. *Micromachines, 9*(8), 403.

Ni, Y., Kannadorai, R. K., Peng, J., Yu, S. W. K., Chang, Y. T., & Wu, J. (2016). Naphthalene-fused BODIPY near-infrared dye as a stable contrast agent for in vivo photoacoustic imaging. *Chemical Communications, 52*(77), 11504–11507.

Ni, Y., Kannadorai, R. K., Yu, S. W. K., Chang, Y. T., & Wu, J. (2017). Push–pull type meso-ester substituted BODIPY near-infrared dyes as contrast agents for photoacoustic imaging. *Organic & Biomolecular Chemistry, 15*(21), 4531–4535.

Nosratzehi, T. (2018). Oral lichen planus: An overview of potential risk factors, biomarkers and treatments. *Asian Pacific Journal of Cancer Prevention: APJCP, 19*(5), 1161.

Nour, M., Hamdy, O., Faid, A. H., Eltayeb, E. A., & Zaky, A. A. (2022). Utilization of gold nanoparticles for the detection of squamous cell carcinoma of the tongue based on laser-induced fluorescence and diffuse reflectance characteristics: An in vitro study. *Lasers in Medical Science, 37*(9), 3551–3560.

Paderni, C., Compilato, D., Carinci, F., Nardi, G., Rodolico, V., Lo Muzio, L., ... & Campisi, G. (2011). Direct visualization of oral-cavity tissue fluorescence as novel aid for early oral cancer diagnosis and potentially malignant disorders monitoring. *International Journal of Immunopathology and Pharmacology, 24*(2_suppl), 121–128.

Pałasz, P., Adamski, Ł., Górska-Chrząstek, M., Starzyńska, A., & Studniarek, M. (2017). Contemporary diagnostic imaging of oral squamous cell carcinoma—A review of literature. *Polish Journal of Radiology*, *82*, 193–202.

Paproski, R. J., Forbrich, A., Huynh, E., Chen, J., Lewis, J. D., Zheng, G., & Zemp, R. J. (2016). Porphyrin nanodroplets: Sub-micrometer ultrasound and photoacoustic contrast imaging agents. *Small*, *12*(3), 371–380.

Pérez, M. G. S., Bagán, J. V., Jiménez, Y., Margaix, M., & Marzal, C. (2015). Utility of imaging techniques in the diagnosis of oral cancer. *Journal of Cranio-Maxillofacial Surgery*, *43*(9), 1880–1894.

Perfézou, M., Turner, A., & Merkoçi, A. (2012). Cancer detection using nanoparticle-based sensors. *Chemical Society Reviews*, *41*(7), 2606–2622.

Petruzzi, M., Lucchese, A., Baldoni, E., Grassi, F. R., & Serpico, R. (2010). Use of Lugol's iodine in oral cancer diagnosis: An overview. *Oral Oncology*, *46*(11), 811–813.

Prabhu, V., Rao, S. B., Chandra, S., Kumar, P., Rao, L., Guddattu, V., ... & Mahato, K. K. (2012). Spectroscopic and histological evaluation of wound healing progression following Low Level Laser Therapy (LLLT). *Journal of Biophotonics, 5*(2), 168–184.

Raghushaker, C. R., Chandra, S., Chakrabarty, S., Kabekkodu, S. P., Satyamoorthy, K., & Mahato, K. K. (2019). Detection of mitochondrial dysfunction in vitro by laser-induced autofluorescence. *Journal of Biophotonics, 12*(11), e201900056.

Raghushaker, C. R., D'Souza, M., Urala, A. S., Ray, S., & Mahato, K. K. (2020). An overview of conventional and fluorescence spectroscopy tools in oral cancer diagnosis. *Lasers in Dental Science*, *4*(4), 167–179.

Raghushaker, C. R., Rodrigues, J., Nayak, S. G., Ray, S., Urala, A. S., Satyamoorthy, K., & Mahato, K. K. (2021). Fluorescence and photoacoustic spectroscopy-based assessment of mitochondrial dysfunction in oral cancer together with machine learning: A pilot study. *Analytical Chemistry*, *93*(49), 16520–16527.

Rai, V., Mukherjee, R., Ghosh, A. K., Routray, A., & Chakraborty, C. (2018). "Omics" in oral cancer: New approaches for biomarker discovery. *Archives of Oral Biology*, *87*, 15–34.

Rajkumar, K., Ramya, R., Nandhini, G., Rajashree, P., Ramesh Kumar, A., & Nirmala Anandan, S. (2015). Salivary and serum level of CYFRA 21-1 in oral precancer and oral squamous cell carcinoma. *Oral Diseases*, *21*(1), 90–96

Ramesh, S., Nazeer, S. S., Thomas, S., Vivek, V., & Jayasree, R. S. (2021). Optical diagnosis of oral lichen planus: A clinical study on the use of autofluorescence spectroscopy combined with multivariate analysis. *Spectrochimica Acta Part A: Molecular and Biomolecular Spectroscopy*, *248*, 119240.

Reddy, R. S., & Praveen, K. N. S. (2017). Optical coherence tomography in oral cancer: A transpiring domain. *Journal of Cancer Research and Therapeutics*, *13*(6), 883.

Reddy, S. G., Kanala, S., Chigurupati, A., Kumar, S. R., Poosarla, C. S., & Ramana, R. B. V. (2012). The sensitivity and specificity of computerized brush biopsy and scalpel biopsy in diagnosing oral premalignant lesions: A comparative study. *Journal of Oral and Maxillofacial Pathology*, *16*(3), 349.

Riaz, A., Shreedhar, B., Kamboj, M., & Natarajan, S. (2013). Methylene blue as an early diagnostic marker for oral precancer and cancer. *SpringerPlus*, *2*(1), 1–7.

Rivera, C. (2015). Essentials of oral cancer. *International Journal of Clinical and Experimental Pathology*, *8*(9), 11884

Rodrigues, J., Amin, A., Raghushaker, C. R., Chandra, S., Joshi, M. B., Prasad, K., ... & Mahato, K. K. (2021). Exploring photoacoustic spectroscopy-based machine learning together with metabolomics to assess breast tumor progression in a xenograft model ex vivo. *Laboratory Investigation*, *101*(7), 952–965.

Rurack, K., & Spieles, M. (2011). Fluorescence quantum yields of a series of red and near-infrared dyes emitting at 600–1000 nm. *Analytical Chemistry*, *83*(4), 1232–1242.

Schiegnitz, E., Kämmerer, P. W., Schön, H., Blatt, S., Berres, M., Sagheb, K., & Al-Nawas, B. (2018). Proinflammatory cytokines as serum biomarker in oral carcinoma—A prospective multi-biomarker approach. *Journal of Oral Pathology & Medicine*, *47*(3), 268–274.

Schwarzenbach, H., Nishida, N., Calin, G. A., & Pantel, K. (2014). Clinical relevance of circulating cell-free microRNAs in cancer. *Nature Reviews Clinical Oncology*, *11*(3), 145–156.

Sharma, D., Rimal, J., Maharjan, I. K., Shrestha, A., & Regmee, P. (2021). Evaluation of oral potentially malignant disorders with autoflorescence, reflectance spectroscopy and vital staining and their correlation with histopathology—Hospital based prospective study. *Oral Oncology*, *118*, 105312.

Shintani, S., Yoshihama, Y., Ueyama, Y., Terakado, N., Kamei, S., Fijimoto, Y., ... & Matsumura, T. (2001). The usefulness of intraoral ultrasonography in the evaluation of oral cancer. *International Journal of Oral and Maxillofacial Surgery*, *30*(2), 139–143.

Singh, A., Thukral, C. L., Gupta, K., Sood, A. S., Singla, H., & Singh, K. (2017). Role of MRI in evaluation of malignant lesions of tongue and oral cavity. *Polish Journal of Radiology*, *82*, 92–99.

Singh, S. P., Ibrahim, O., Byrne, H. J., Mikkonen, J. W., Koistinen, A. P., Kullaa, A. M., & Lyng, F. M. (2016). Recent advances in optical diagnosis of oral cancers: Review and future perspectives. *Head & Neck*, *38*(S1), E2403–E2411.

Singhal, J., Verma, S., Kumar, S., & Mehrotra, D. (2021). Recent advances in nano-bio-sensing fabrication technology for the detection of oral cancer. *Molecular Biotechnology*, *63*(5), 339–362.

Smith, P. W. (2002). Fluorescence emission-based detection and diagnosis of malignancy. *Journal of Cellular Biochemistry*, *87*(S39), 54–59.

Stoffels, I., Morscher, S., Helfrich, I., Hillen, U., Leyh, J., Burton, N. C., ... & Klode, J. (2015). Metastatic status of sentinel lymph nodes in melanoma determined noninvasively with multispectral optoacoustic imaging. *Science Translational Medicine*, *7*(317), 317ra199.

Taguchi, Y., Toratani, S., Matsui, K., Hayashi, S., Eboshida, N., Hamada, A., ... & Yanamoto, S. (2022). Evaluation of oral mucosal lesions using the IllumiScan® fluorescence visualisation device: Distinguishing squamous cell carcinoma. *International Journal of Environmental Research and Public Health*, *19*(16), 10414.

Taniguchi, M., & Lindsey, J. S. (2018). Database of absorption and fluorescence spectra of> 300 common compounds for use in photochem CAD. *Photochemistry and Photobiology*, *94*(2), 290–327.

Thankappan, K., Subramanian, S., Balasubramanian, D., Kuriakose, M. A., Sankaranarayanan, R., & Iyer, S. (2021). Cost-effectiveness of oral cancer screening approaches by visual examination: Systematic review. *Head & Neck*, *43*(11), 3646–3661.

Tiwari, L., Kujan, O., & Farah, C. S. (2020). Optical fluorescence imaging in oral cancer and potentially malignant disorders: A systematic review. *Oral Diseases*, *26*(3), 491–510

Towle, R., Prisman, E., Guillaud, M., & Garnis, C. (2021). A circulating biomarker for the early detection of recurrent oral cancer. *Oral Oncology*, *118*, 7.

Vijayakumar, V., Reghunathan, D., Edacheriyan, B., & Mukundan, A. (2019). Role of toluidine blue staining in suspicious lesions of oral cavity and oropharynx. *Indian Journal of Otolaryngology and Head & Neck Surgery*, *71*(1), 142–146.

Vonk, J., de Wit, J. G., Voskuil, F. J., & Witjes, M. J. H. (2021). Improving oral cavity cancer diagnosis and treatment with fluorescence molecular imaging. *Oral Diseases*, *27*(1), 21–26.

Vonk, J., Kukačka, J., Steinkamp, P. J., de Wit, J. G., Voskuil, F. J., Hooghiemstra, W. T. R., ... & Witjes, M. J. H. (2022). Multispectral optoacoustic tomography for in vivo detection of lymph node metastases in oral cancer patients using an EGFR-targeted contrast agent and intrinsic tissue contrast: A proof-of-concept study. *Photoacoustics*, *26*, 100362.

Waech, T., Pazahr, S., Guarda, V., Rupp, N. J., Broglie, M. A., & Morand, G. B. (2021). Measurement variations of MRI and CT in the assessment of tumor depth of invasion in oral cancer: A retrospective study. *European Journal of Radiology*, *135*, 109480.

Walsh, T., Liu, J. L., Brocklehurst, P., Glenny, A. M., Lingen, M., Kerr, A. R., ... & Scully, C. (2013). Clinical assessment to screen for the detection of oral cavity cancer and potentially malignant disorders in apparently healthy adults. *Cochrane Database of Systematic Reviews*, *2013*(11), CD010173.

Wang, B., Wang, J. H., Liu, Q., Huang, H., Chen, M., Li, K., ... & Chu, P. K. (2014). Rose bengal-conjugated gold nanorods for in vivo photodynamic and photothermal oral cancer therapies. *Biomaterials*, *35*(6), 1954–1966.

Wang, C. Y., Chiang, H. K., Chen, C. T., Chiang, C. P., Kuo, Y. S., & Chow, S. N. (1999). Diagnosis of oral cancer by light-induced autofluorescence spectroscopy using double excitation wavelengths. *Oral Oncology*, *35*(2), 144–150.

Warnakulasuriya, S., & Kerr, A. R. (2021). Oral cancer screening: Past, present, and future. *Journal of Dental Research*, *100*(12), 1313–1320.

Warnakulasuriya, S., Johnson, N. W., & Van der Waal, I. (2007). Nomenclature and classification of potentially malignant disorders of the oral mucosa. *Journal of Oral Pathology & Medicine*, *36*(10), 575–580.

Wei, J., Xie, G., Zhou, Z., Shi, P., Qiu, Y., Zheng, X., ... & Jia, W. (2011). Salivary metabolite signatures of oral cancer and leukoplakia. *International Journal of Cancer*, *129*(9), 2207–2217.

Wetzel, S. L., & Wollenberg, J. (2020). Oral potentially malignant disorders. *Dental Clinics*, *64*(1), 25–37.

Wong, T. S. C., & Wiesenfeld, D. (2018). Oral cancer. *Australian Dental Journal*, *63*, S91–S99.

Wu, C., Gleysteen, J., Teraphongphom, N. T., Li, Y., & Rosenthal, E. (2018). In-vivo optical imaging in head and neck oncology: basic principles, clinical applications and future directions. *International Journal of Oral Science*, *10*(2), 1–13.

Yan, Y. J., Cheng, N. L., Jan, C. I., Tsai, M. H., Chiou, J. C., & Ou-Yang, M. (2021). Band-selection of a portal LED-Induced autofluorescence multispectral imager to improve oral cancer detection. *Sensors*, *21*(9), 3219.

Yan, Y. J., Huang, T. W., Cheng, N. L., Hsieh, Y. F., Tsai, M. H., Chiou, J. C., …& Ou-Yang, M. (2017). Portable LED-induced autofluorescence spectroscopy for oral cancer diagnosis. *Journal of Biomedical Optics, 22*(4), 045007.

Zhang, L., & Wang, X. (2022). Optical resolution photoacoustic microscopy imaging in the detection of early oral cancer under image reconstruction algorithm. *Computational and Mathematical Methods in Medicine, 2022*, 6077748.

Zhong, L. P., Chen, G. F., Xu, Z. F., Zhang, X., Ping, F. Y., & Zhao, S. F. (2005). Detection of telomerase activity in saliva from oral squamous cell carcinoma patients. *International Journal of Oral and Maxillofacial Surgery, 34*(5), 566–570.

Ziober, B. L., Mauk, M. G., Falls, E. M., Chen, Z., Ziober, A. F., & Bau, H. H. (2008). Lab-on-a-chip for oral cancer screening and diagnosis. *Head & Neck: Journal for the Sciences and Specialties of the Head and Neck, 30*(1), 111–121.

4 Current Updates on Photoacoustic-Based Techniques for Breast Cancer Diagnosis

*Devika, Darshan Chikkanayakanahalli Mukunda, Jackson Rodrigues, Shaik Basha, Shimul Biswas, Ameera Hanif, and Krishna Kishore Mahato**

Department of Biophysics, Manipal School of Life Sciences, Manipal Academy of Higher Education, Manipal, Karnataka, India
*Corresponding Author

INTRODUCTION

According to GLOBOCAN 2020 statistics, with nearly 2.3 million new cases, breast cancer has been considered the most commonly diagnosed cancer. These high incidences are observed chiefly in developing countries in Asia and in South Africa, where there have been dramatic changes in women's lifestyles, environment, and abundant choices (Sung et al., 2021). Thus, this has proven to be a strong barrier against the world's reproductivity and mortality rates. Breast cancer results from factors like gender and genetics, increasing in severity and mortality at each stage (Jordan et al., 2019). One of the effective ways to check the alarming rate of cases is by developing high-throughput and specific monitoring strategies for improved patient prognosis. Especially in India, even with sufficient knowledge and a positive attitude toward breast cancer, there is a lack of screening practices among the general population (Pal et al., 2021).

The current non-invasive screening methods are observed to have poor diagnostic value. The widely popularized self-breast and clinical breast examinations have little to no effect on mortality (Jordan et al., 2019). Regular annual mammography (MAM) in women after 40 years of age is known to reduce the mortality rate significantly, but difficulty in positioning the tissue and high false-positive results have been its drawbacks (Coleman, 2017; Duffy et al., 2020). The significant harm among them is the overdiagnosis of breast cancer, for which the relative amount is up to 31% in the United States (Løberg et al., 2015). Even with optimal sensitivity, magnetic resonance imaging (MRI) has higher false-positive results with increasing breast density (Partovi et al., 2020).

The newer techniques, including positron emission tomography (PET) coupled with computed tomography (PET/CT), have shown limited spatial resolution, leading to lower sensitivity than MRI (Groheux and Hindie, 2021). Thus, the demand for techniques that overcome this limitation is high. Novel imaging approaches like digital breast tomosynthesis and molecular breast imaging are in the field, followed by the enhancement provided by radiomics and artificial intelligence. In addition, non-imaging techniques like breathing tests and biopsies provide validation (Mann et al., 2020). Biphotonic techniques like photoacoustic spectroscopy, Raman spectroscopy, fluorescence spectroscopy, and imaging are gaining popularity, especially photoacoustic methods, for providing higher specificity in discriminating malignant tissues cost-effectively (Balasundaram et al., 2021).

DOI: 10.1201/9781003545712-4

The current screening guidelines are mostly issued by developed countries like those of the World Health Organization (WHO), China Guideline Clearinghouse (CGC), and Cancer Australia, Ministry of Health (MOH), which suggest yearly MAM for the average risk population, whereas high-risk populations are advised to undergo both annual MRI and MAM from the age of 40 to 74 years. Analysis of these guidelines suggests the design of tailored guidelines for particular geographic locations considering their economic status, well-developed follow-up procedures, and the need for improvement in risk-identifying models (Ren et al., 2022). As discussed, the suggested technique, like MAM previously, has considerable limitations in certain cases; thus, sufficient evidence on the latest improved screening techniques, like optoacoustic, will improve the overall detection and treatment plans to correspond to tumor assessment in breast cancer patients.

PHOTOACOUSTIC IMAGING

Photoacoustic imaging (PAI) provides high-resolution and contrasting images using non-invasive and non-ionizing radiation. It can be used as an in vivo imaging tool. The main components of the PAI system include a photoacoustic signal generator, usually a pulsed laser used for PA signal generation, and a detector that receives the PA signal from the irradiated sample with an appropriate central frequency and bandwidth (Laufer, 2018).

Contrast Agents

Natural endogenous contrast agents, like hemoglobin, can provide sufficient contrast, but in the case of deep tissue imaging, exogenous contrast agents like cyanine and xanthine dyes are used (Borg and Rochford, 2018). In recent times, nanoparticles have been commonly used and studied as contrast agents. Astaxanthin-reduced gold nanoparticles (Atx-AuNPs) treated MDA-MB-231 cells with high amplitude PA signals compared to control due to the light-scattering property of the NPs (Bharathiraja et al., 2016). NH_4HCO_3 gas-generating liposomal nanoparticle (ABN) is an inexpensive and safe contrasting agent that has proven effective in both in vitro and in vivo studies (Xia et al., 2017). Glycol-chitosan-coated gold nanoparticles (GC-AuNPs) are also known to enhance the PA signals when incubated with the cancerous cell line, with increasing signal intensity along with the extended incubation period, observed under a dark field microscope (Sun et al., 2019).

Image Quality

The performance standards for PAI are often analyzed using phantom-based image quality assessment test methods, which are still in their early development stage compared to the validated tests for mature modalities like MRI, ultrasound, and CT. Based on the established tests with other techniques, standardization of image-quality-test methods is being developed for PAI to quantify the spatial resolution, penetration, artifacts, and sensitivity obtained, thus providing feedback for better product design and image quality assessment (Vogt et al., 2017; Palma-Chavez et al., 2021).

Phantom-based techniques utilize tissue-mimicking materials (TMMs) that closely resemble human tissue and suitable phantoms with nanoparticles to reciprocate the imaging mechanism, thereby characterizing and optimizing the PAI devices (Hsu et al., 2022). Here we discuss recent improvements in PA modalities and their applications in breast cancer imaging (Figure 4.1).

PHOTOACOUSTIC TOMOGRAPHY

Photoacoustic tomography (PAT) is a hybrid imaging modality, using photoacoustic (PA) signal-based detection of the optical absorption either by natural absorbent molecules like hemoglobin and

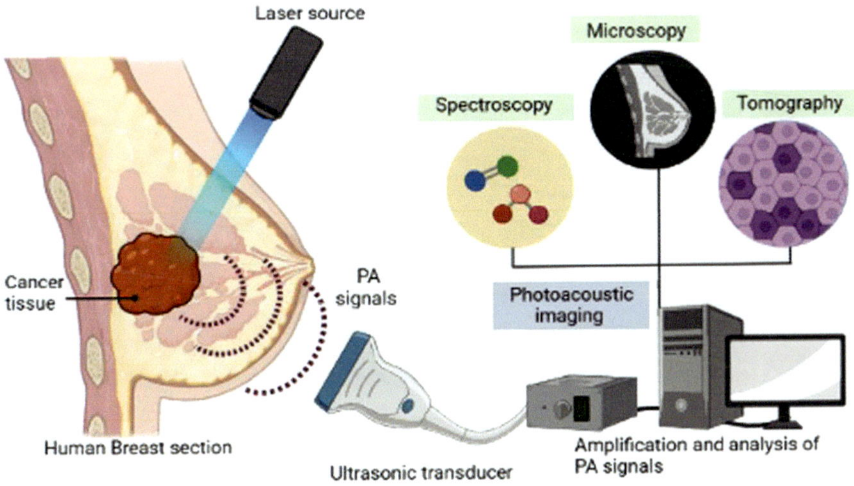

FIGURE 4.1 Photoacoustic effect-based techniques utilized in cancer detection and imaging (BioRender).

cytochrome or by exogenous contrast like nanoparticles, provides high resolution and sensitivity (Wang and Yao, 2016). One of the most promising modalities of PAT for early cancer detection is photoacoustic computed tomography (PACT). While widely used for advantageous features like good penetration, high spatial and temporal resolution, long scanning time, and limited apertures are known to hinder its efficiency (Zhang et al., 2020). Here to detect the ultrasound (US) waves from the sample, a wide variety of transducer arrays are utilized, serving specific purposes. For in vivo imaging, linear and planar arrays are better options as they can work within restricted regions, unlike cyclindrical, spherical or circular array (Rao et al., 2020).

Single-Breath-Hold Photoacoustic Computed Tomography

The challenges of PACT could be resolved by newer and more effective techniques like single-breath-hold photoacoustic computed tomography (SBH-PACT), which significantly reduces the long scanning period of PACT to a period of a single breath hold (i.e., approximately 15 s), thus considerably reducing the respiratory motion distortion. The system design includes a soft agar pillow that relieves discomfort while positioning the breast tissue, and a single illumination source at 1064 nm is illuminated from below. A full-ring ultrasonic 512-element transducer array is responsible for the observed high in-plane resolution. SBH-PACT is best known for its high spatiotemporal resolution, which allows the precise imaging of the high-density blood vessels characteristic of angiographic structures.

For validation purposes, a study on the blood vessel densities of tumor tissue confirmed eight tumors out of nine (Figure 4.2). These observations were not readily distinguishable in other conventional methods like mammograms. In continuation, these results were confirmed using methods like biopsies. In addition, SBH-PACT elastography was performed to identify the imaging deformations caused by breathing, which surprisingly led to the detection of a single tumor that had been missed during angiographic screening (Lin et al., 2018). These features of SBH-PACT can be applied to accurately assess the response to breast cancer treatments like neoadjuvant chemotherapy (NAC) by quantifying the relative blood vessel density and tumor size. The study design consisting of three patients with breast cancer undertaking NAC were imaged at three time points, i.e., before NAC, during NAC, and after NAC treatment, which revealed reliable results similar to the MRI reports (Lin et al., 2021).

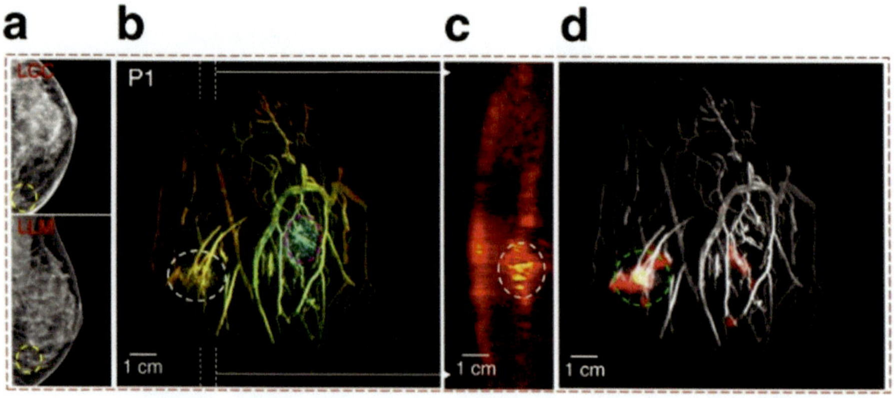

FIGURE 4.2 The images acquired from the SBH-PACT results of a 48-year-old patient suffering from invasive lobular carcinoma. (a) An X-ray mammogram of the breast in positions left cranial-caudal (LCC) and left lateral-medio (LLM). (b) Angiograms with depth indications, where the circles indicate the presence of tumor tissue. (c) Maximum amplitude projection images (MAP) of the breast tissue with marked circles at sagittal plane sections. (d) The vessel density map automatically indicates the tumor (green dashed circles) (Lin et al., 2018).

Multispectral Optoacoustic Tomography

Multispectral optoacoustic tomography (MSOT) provides a multi-wavelength (28 wavelengths ranging from 700 to 970 nm) option that is advantageous in imaging different absorbing agents with higher accuracy. The instrumentation includes a handheld system of MSOT scanner along with the half-arc shape arranged transducer array. The image and color map reconstruction were obtained by unmixing them mainly at four particular wavelengths corresponding to the absorption by endogenous agents like water, lipid, and oxyhemoglobin using MATLAB. The high-resolution insights on tumor patterns like oxygenation and deoxygenation of hemoglobin (often observed due to hypoxia in tumors) and total blood volume (TBV) (indicating the vascular heterogeneity in tumors compared to healthy tissue) provide increased confidence in differentiating the cancerous tissues in nonspecific breast cancer patients. Further advancements can be undertaken by reducing the depth and wavelength-dependent attenuation and considering the absorption by melanin in unmixing MSOT (Diot et al., 2017).

A study conducted by Stoffels et al. suggests MSOT as a better non-radioactive alternative in detecting sentinel lymph nodes (SLNs) than conventional lymphoscintigraphic imaging in melanoma cases. In this study, real-time analysis was conducted by precisely labeling the SLNs using Indocyanine Green (ICG), validated in 83 patients (Stoffels et al., 2019). Additionally, MSOT has been effectively implemented to assess the tumor margin in 55-year-old humans suffering from infiltrating ductal carcinoma during breast conservation surgery, proving its broad applicability (Goh et al., 2018).

Different modes of MSOT have also been developed with specificity for other tumor parameters, such as volumetric multispectral optoacoustic tomography (vMSOT), which can simultaneously assess multiple tumor parameters like mass, oxygenation, viability, and drug response in rea time, with its adaptable setup that can be operated in handheld mode as well (Ermolayev et al., 2016) and eigenspectra multispectral optoacoustic tomography (eMSOT), widely used in tracking tumor hemodynamics, mainly whole tumor oxygenation levels during treatments like chemotherapy and bevacizumab (Liapis et al., 2020; Liapis et al., 2021).

Photoacoustic Mammography

To address the limitation of standard mammography in dense breast tissue scanning, a PAT-based instrument explicitly dedicated to breast imaging called photoacoustic mammography (PAMMG) was introduced, providing 3D images by scanning tissue around 53 mm depth (Kruger et al., 2013). Further, a second-generation prototype, PAM-02, was developed with better resolution, higher contrast ratio, and sensitivity through modifications such as a smaller detector element size and higher sensitivity capacitive micromachined ultrasonic transducer (CMUT). Additionally, the images from this were obtained by fusing ultrasound and PA signals, which are more appropriate for understanding both morphological features and functional components of breast tissue, like vasculature (Asao et al., 2016). To compensate for the limited view issue of these two prototypes due to the flat-shaped detectors, PAM-03 has been developed, equipped with a hemispherical-shaped detector array (HDA). The modified HDA detectors slightly improve the view field as they focus more on reconstructing tumor vasculature than on morphological imaging of tumor mass. The instrument's efficiency was assessed in 22 malignant breast tumors by MR image deformation, accurate detection of centripetal vasculature signals suggesting the tumor boundary, and observation of changes in tumor vasculature in response to a chemotherapeutic drug (Toi et al., 2017; Figure 4.3.

Twente Photoacoustic Mammoscope

The Twente Photoacoustic Mammoscope 1 (PAM 1) uses a 2D ultrasound flat detector array to visualize malignancies in vivo. Here, slight compression on the breast tissue is applied by an ultrasound detector and a glass plate that functions as a laser illumination source at 1064 nm. Reconstruction using MATLAB is performed. PA lesions were identified in the region with high-density PA pixels in an organization similar to that of lesions, whose positions were further analyzed by comparisons between cranio-caudal PA images and X-ray images.

An observational study to validate the effectiveness was conducted among 31 patients, where 32 malignancies were detected with good contrast and resolution out of 33 malignancies of breast cancer. Even though these provide good contrast imaging that is independent of breast density,

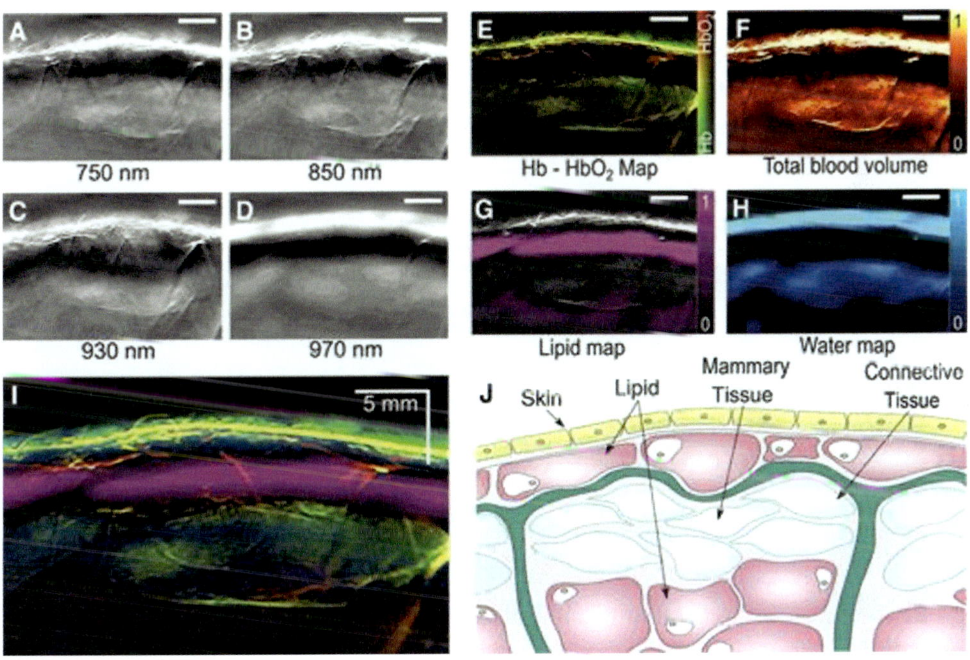

FIGURE 4.3 The maximum intensity projection (MIP) images produced from the PAM-03 at two different depths, a) 24 mm and b) 27 mm, from the skin surface. The different colors of the vessels correspond to their depth, such as the black arrow indicating the signal from the deepest region (Toi et al., 2017).

they accompany some challenges, such as the requirement for breast compression, which can be uncomfortable for the patient, the flat detector array used here is known to cause limited view, and most importantly, the image quality obtained has a 28% deviation from the histopathological study, which is considered the gold standard (Heijblom et al., 2016).

The advanced version of this, referred to as Twente Photoacoustic Mammoscope 2 (PAM 2), overcomes the limitations of PAM 1 by introducing the arc-shaped detector array that follows the breast contour with illumination, increasing the coverage. The patient, in a prone position places her breast inside the imaging tank, where illumination is provided using nine smaller beams at two wavelengths (755 nm for hemoglobin absorption and 1064 nm for deep tissue penetration) instead of a single beam so that it provides accurate 3D images with sufficient contrast depicting the breast vasculature. Further, the rotating detectors and the fiber bundles in the tank capture the breast tissue's tomographic scan at different positions, which are used for visualizing 3D volume reconstruction using maximum intensity projections (MIPs). A preliminary study to validate the efficiency was conducted on two healthy volunteers of 29 and 25 years of age and analyzed features like vessels, lactiferous ducts, nipple, and breast contour (Figure 4.4). This study exposed some limitations, like slight movements during scanning getting registered, which would require motion correction in later stages. To rectify this, breast immobilization, like a cup stabilizer, can be recommended (Schoustra et al., 2019a, 2019b).

Dual Scan Mammoscope

The most recent advancement in PAT-based instruments is the dual scan mammoscope (DSM) proposed by Nyayapathi et al., which, along with being a portable device, provides imaging in an upright position with an impressive image depth of nearly 7 cm (i.e., breast size of a D cup), thus ensuring its aid in early cancer detection for dense breast tissue. The scanning is performed on the patient in an upright position, using a foot-operated lift table, to which the breast tissue is manually positioned using the clamps. The imaging system includes a 3D-printed holder which positions the dichroic mirror at 45° towards the transducers for optimal PA signal detection. The acquired image quality is superior to MRI results and can simultaneously provide both ultrasonic and PA images, producing overlapping images for comprehensive analysis (Nyayapathi et al., 2020). The system's efficiency is further validated based on a report suggesting DSM's capacity to differentiate tumor tissue in 38 patient samples based on the vascular features with a confidence level of 95%. The

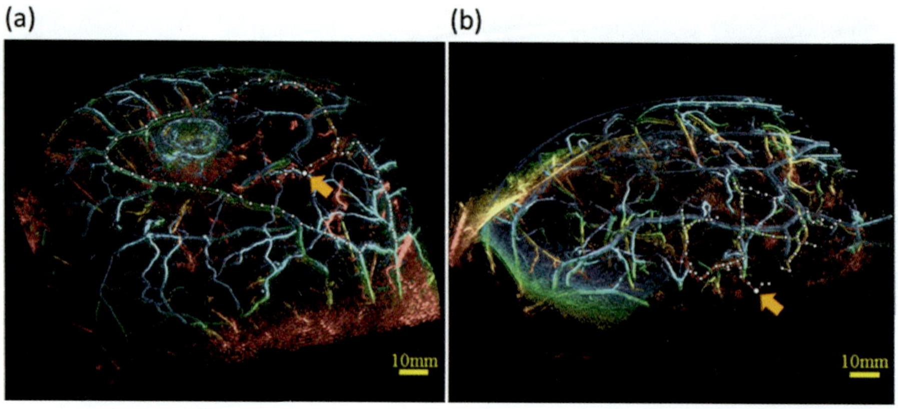

FIGURE 4.4 The MIP images obtained with illumination with 1064 nm on the right breast from a healthy volunteer 2. The illumination ratios are 67/33 in (a), displaying more structures, and 50/50 in (b), showing the localized signal intensity. The positions of the breast images include coronal in (a) and sagittal in (b). Yellow arrowheads in images (a, b) indicate the position of the nipple (Schoustra et al., 2019b).

detection has been observed to be dependent on the higher PA signal amplitude in regions of the tumor when compared to healthy breast tissue (Nyayapathi et al., 2021).

To further upgrade this instrument, the latest model, second-generation DSM (DSM-2), is designed with features like quasi-static elastography to characterize breast tissue based on the modified mechanical properties such as stiffness caused by collagen contraction often observed in tumor tissue. The improvement in case instrumentation includes a motorized compression system and the replacement of the 3D-printed holder with a precisely designed transducer and fiber combiner, which are proven to be advantageous in numerous ways, ensuring 45° placement of the mirror for optimized light passage and incorporation of long-lasting materials like polylactic acid, thus ensuring reproducible and calibrated results for each scan. In the case of imaging, the use of round-tip scanning increases the overall lateral coverage compared to regular scanning, thereby improving image quality and contrast (Zheng et al., 2021; Table 4.1).

PHOTOACOUSTIC MICROSCOPY

Photoacoustic microscopy (PAM) is a microscopic technique based on the PA effect, wherein acoustic signals generated from a laser-illuminated target region are detected using an ultrasonic transducer. The PAM images thus obtained have micron-scale resolution, providing both the 3D structure and function of the biological tissue sample without the need for optical sectioning (Liu and Yao, 2018; Wong et al., 2017). Based on the size of the laser foci, PAM could be classified as optical-resolution photoacoustic microscopy (OR-PAM) utilizing a focused illumination source used for flat cell sample imaging and acoustic-resolution photoacoustic microscopy (AR-PAM) with a weakly focused illumination source for whole body imaging (Jeon et al., 2019). In a recent study, AR-PAM has been used to track the responses, like tumor vasculature and hypoxia of breast tissue for antivascular treatment using bevacizumab, assuring its efficiency (J. Yang et al., 2019).

PAM is widely used to detect and differentiate cancer from normal breast tissue. UV-PAM is known to provide histologic-grade imaging from unprocessed breast cancer tissue, including information like the average nuclear size. This presents a high clinical significance as it can guide breast-conserving surgery to precisely remove the tumor, thereby reducing the rate of reoperations (Wong et al., 2017). This technique can also assist in predicting the metastasis stage along with the dual targeting nanoparticle-like status of the sentinel lymph node, whether it is inflamed or popliteal, thus aiding in choosing the effective therapeutic strategy (Dai et al., 2020). These also assist in tracking the deposition of nanodrugs like silver-coated gold nanorods with doxorubicin to ensure precise targeting against the molecules of triple-negative breast cancer (Jenkins et al., 2017).

Further, to improve the limited information on PAM, multimodal imaging techniques have emerged, like PAM integrated with US and confocal fluorescence microscopy (CFM) for multi-contrast imaging (Dadkhah and Jiao, 2021). Multimodal imaging techniques with photoacoustic, ultrasonic, and fluorescent imaging by targeted nanobubbles with AMD070 and Indocyanine Green have been shown to benefit both the detection and effective inhibition of breast cancer tissue (Shen et al., 2020). Additionally, to overcome the challenges of PAM, such as bulky transducers that limit its clinical applications, an emerging method called photoacoustic remote sensing (PARS) has been popularized. Still, more studies focused on breast cancer need to be analyzed (Bell et al., 2020).

PHOTOACOUSTIC SPECTROSCOPY

Photoacoustic spectroscopy (PAS) is a spectroscopic technique utilizing targeted laser-induced PA signals to measure the light spectrum of the molecules in the sample in either solid, gaseous, or liquid form. Features like low background interference and light scattering make it beneficial compared to conventional spectroscopy (Jin et al., 2022). Earlier studies have reported the effective differentiation of malignant and normal breast cancer tissue based on the PA spectra of the respective

TABLE 4.1
Comparative Analysis of the Configurations and Application of PAT-Based Instruments

Instrument	Optical Wavelength (nm)	Resolution	Penetration	Instrument Type	Detector System	Imaging Period	Best Suited Purpose
Twente Photoacoustic Mammoscope 2 (Schoustra et al., 2019b)	755, 1064	1 mm (spatial)	27 mm	Bench side type	12 arc-shaped arrays, each with 32 piezo composite elements, 1 MHz	4 minutes	Tumor vasculature
Photoacoustic mammography (PMMG) with hemispherical-shaped detector array (PAM-03) (Toi et al., 2017)	795, 755			Bench side type	Hemispherical-shaped detector array (HDA)	4 minutes	Tumor vasculature
Single-breath-hold photoacoustic computed tomography (SBH-PACT) (Lin et al., 2018)	1064	255 μm	4 cm	Bench side type	Full-ring 512-element ultrasonic transducer array, 2.25 MHz	~15 seconds	Angiogenic structures
Multispectral optoacoustic tomography (MSOT) (Diot et al., 2017)	28 wavelengths (700, 970)	12-bit digital resolution		Handheld type	Half arc shape arranged 256-element piezoelectric transducer array, 5 Hz	2–4 minutes	Tumor spatial heterogeneity
Dual Scan Mammoscope (DSM) (Nyayapathi et al., 2020)	1064	1 mm (spatial)	7 m	Bench side type	Linear, two 128-element ultrasonic transducer, 2.25 MHz	50 econds	High density breast imaging
Second generation DSM (DSM-2) (Zheng et al., 2021)	1064			Bench side type	Linear, two 128-element linear transducers with a bifurcated line-output fiber bundle	38 seconds	Tissue mechanical properties

serum samples with the highest accuracy (Priya et al., 2016). Recently, Rodrigues et al. have shown that PAS could aid in monitoring breast tumor progression based on metabolomics analysis, wherein significantly low serum tryptophan levels were observed in tumor tissue compared to normal tissue (Rodrigues et al., 2021). Even though it is effective for monitoring molecules like blood oxygen concentration (S. Yang et al., 2022) and glucose levels (Tanaka et al., 2020), they have been less explored in the context of breast cancer.

DEEP LEARNING IN PA IMAGING

Potential benefits of the PA devices can be further improved by overcoming the current limitations in image analysis, like the inverse problem and the interference of artifacts. Deep learning (DL) algorithms are a developing field widely popularized in medical imaging analysis, like intraoperative imaging, for their beneficial features such as shorter inference time, image reconstruction and quantification abilities, and applicability in real-time data processing. These usually train the network to develop the ability to generalize using the dataset. There are many architectures for DL, but the one most used in the context image processing of PAI is the convolutional neural network (CNN) (Deng et al., 2021; DiSpirito et al., 2021).

In particular, PA microscopy, along with DL, is known to assist in intraoperative real-time histology imaging by providing virtually stained histological images of the tissue in a label-free and slide-free form, thus allowing for rapid diagnosis (Cao et al., 2022). Further, MSOT has been improved by automated vascular assessment using a DL architecture known as Sparse-UNET (Chlis et al., 2020). Automated DL algorithms have been successfully applied in multispectral PA imaging for designing a system for cancer detection using CNN in ex vivo specimens (prostate and thyroid cancer). Another such model has been developed to recognize and classify cancer using a particle swarm with ant colony optimization recurrent neural network (PS-ACO-RNN) model (Jnawali et al., 2020; Sobhanan Warrier et al., 2022). In the context of the applicability of DL in breast cancer, few studies have been conducted. The Breast Numerical model has been automatically generated using DL for PA imaging, which are useful in developing open-source datasets with clinical data (Ma et al., 2020). Further, a newer DL model for PAI called social engineering optimization with deep transfer learning-based breast cancer detection and classification (SEODTL-BDC) has been designed to precisely detect and differentiate breast cancer into different classes (Althobaiti et al., 2022).

CHALLENGES

The PA-based instruments developed should be standardized for specific applications, and clinical trials must be conducted to validate the results. In order to ensure their position as a standard procedure in breast cancer screening, apart from efficiency and cost, the accessibility rate of false positives must also be improved (Ren et al., 2022). Further optimization of the detectors, transducers, and image quality assessment systems must be focused on.

The standard naming system for the techniques must be reviewed, as in the case of PAM or PAMMG, followed by numerals indicating each generation. On the other hand, the Twente Photoacoustic Mammoscope is denoted as PAM (Schoustra et al., 2019b). In addition, even photoacoustic microscopy is abbreviated as PAM, which leads to a lack of clarity when stated in the same document (Liu and Yao, 2018). Even though these three are based on the same PA principle, the system and utility of the instruments vastly differ.

CONCLUSION

The review of the current PA techniques developed for the screening and imaging of breast cancer proves significant improvement in ease of detection, shorter scanning time, and better resolution. The success of in vivo studies ensures that these can be effectively introduced in the clinical setup for screening breast cancer patients. The personalized recommendation of these techniques based on convenience, nature of the tumor, and other factors will increase the accuracy of tumor detection. Thus, further studies to overcome the current limitations, such as expensive instrumentation setup and lack of standardized protocol, can provide a better alternative to the current breast cancer screening techniques.

ACKNOWLEDGEMENT

The authors would like to thank Manipal School of Life Sciences, Manipal Academy of Higher Education, Manipal, for infrastructure facilities. They would also like to thank DBT-BUILDER (*BT/INF/22/SP43065/2021*) for providing an infrastructure grant.

REFERENCES

Althobaiti, M. M., Ashour, A. A., Alhindi, N. A., Althobaiti, A., Mansour, R. F., Gupta, D., & Khanna, A. (2022). Deep transfer learning-based breast cancer detection and classification model using photoacoustic multimodal images. *BioMed Research International, 2022*. https://doi.org/10.1155/2022/3714422

Asao, Y., Hashizume, Y., Suita, T., Nagae, K., Fukutani, K., Sudo, Y., Matsushita, T., Kobayashi, S., Tokiwa, M., Yamaga, I., Fakhrejahani, E., Torii, M., Kawashima, M., Takada, M., Kanao, S., Kataoka, M., Shiina, T., & Toi, M. (2016). Photoacoustic mammography capable of simultaneously acquiring photoacoustic and ultrasound images *Journal of Biomedical Optics*, *21*(11), 116009. https://doi.org/10.1117/1.JBO.21.11.116009

Balasundaram, G., Krafft, C., Zhang, R., Dev, K., Bi, R., Moothanchery, M., Popp, J., & Olivo, M. (2021). Biophotonic technologies for assessment of breast tumor surgical margins—A review. *Journal of Biophotonics*, *14*(1), e202000280. https://doi.org/10.1002/JBIO.202000280

Bell, K., Abbasi, S., Dinakaran, D., Taher, M., Bigras, G., van Landeghem, F. K. H., Mackey, J. R., & Haji Reza, P. (2020). Reflection-mode virtual histology using photoacoustic remote sensing microscopy. *Scientific Reports*, *10*(1). https://doi.org/10.1038/S41598-020-76155-6

Bharathiraja, S., Manivasagan, P., Bui, N. Q., Oh, Y. O., Lim, I. G., Park, S., & Oh, J. (2016). Cytotoxic induction and photoacoustic imaging of breast cancer cells using astaxanthin-reduced gold nanoparticles. *Nanomaterials*, *6*(4). https://doi.org/10.3390/NANO6040078

Borg, R. E., & Rochford, J. (2018). Molecular photoacoustic contrast agents: Design principles & applications. *Photochemistry and Photobiology*, *94*(6), 1175–1209. https://doi.org/10.1111/PHP.12967

Cao, R., Nelson, S. D., Davis, S., Liang, Y., Luo, Y., Zhang, Y., Crawford, B., & Wang, L. v. (2022). Label-free intraoperative histology of bone tissue via deep-learning-assisted ultraviolet photoacoustic microscopy. *Nature Biomedical Engineering*. https://doi.org/10.1038/S41551-022-00940-Z

Chlis, N. K., Karlas, A., Fasoula, N. A., Kallmayer, M., Eckstein, H. H., Theis, F. J., Ntziachristos, V., & Marr, C. (2020). A sparse deep learning approach for automatic segmentation of human vasculature in multispectral optoacoustic tomography. *Photoacoustics*, *20*. https://doi.org/10.1016/J.PACS.2020.100203

Coleman, C. (2017). Early detection and screening for breast cancer. *Seminars in Oncology Nursing*, *33*(2), 141–155. https://doi.org/10.1016/J.SONCN.2017.02.009

Dadkhah, A., & Jiao, S. (2021). Integrating photoacoustic microscopy with other imaging technologies for multimodal imaging. *Experimental Biology and Medicine*, *246*(7), 771. https://doi.org/10.1177/1535370220977176

Dai, Y., Yu, X., Wei, J., Zeng, F., Li, Y., Yang, X., Luo, Q., & Zhang, Z. (2020). Metastatic status of sentinel lymph nodes in breast cancer determined with photoacoustic microscopy via dual-targeting nanoparticles. *Light, Science & Applications*, *9*(1), 2047–7538. https://doi.org/10.1038/S41377-020-00399-0

Deng, H., Qiao, H., Dai, Q., & Ma, C. (2021). Deep learning in photoacoustic imaging: A review. *Journal of Biomedical Optics*, *26*(4). https://doi.org/10.1117/1.JBO.26.4.040901

Diot, G., Metz, S., Noske, A., Liapis, E., Schroeder, B., Ovsepian, S. v., Meier, R., Rummeny, E., & Ntziachristos, V. (2017). Multispectral Optoacoustic Tomography (MSOT) of human breast cancer. *Clinical Cancer Research*, *23*(22), 6912–6922. https://doi.org/10.1158/1078-0432.CCR-16-3200

DiSpirito, A., Vu, T., Pramanik, M., & Yao, J. (2021). Sounding out the hidden data: A concise review of deep learning in photoacoustic imaging. *Experimental Biology and Medicine*, *246*(12), 1355. https://doi.org/10.1177/15353702211000310

Duffy, S. W., Tabár, L., Yen, A. M., Dean, P. B., Smith, R. A., Jonsson, H., Törnberg, S., Chen, S. L., Chiu, S. Y., Fann, J. C., Ku, M. M., Wu, W. Y., Hsu, C. Y., Chen, Y. C., Svane, G., Azavedo, E., Grundström, H., Sundén, P., Leifland, K., Frodis, E., Ramos, J., Epstein, B., Åkerlund, A., Sundbom, A., Bordás, P., Wallin, H., Starck, L., Björkgren, A., Carlson, S., Fredriksson, I., Ahlgren, J., Öhman, D., Holmberg, L., & Chen, T. H. (2020). Mammography screening reduces rates of advanced and fatal breast cancers: Results in 549,091 women. *Cancer* Jul 1, *126*(13), 2971–2979. doi: 10.1002/cncr.32859. Epub 2020 May 11. PMID: 32390151; PMCID: PMC7318598.

Ermolayev, V., Dean-Ben, X. L., Mandal, S., Ntziachristos, V., & Razansky, D. (2016). Simultaneous visualization of tumour oxygenation, neovascularization and contrast agent perfusion by real-time three-dimensional optoacoustic tomography. *European Radiology*, *26*(6), 1843–1851. https://doi.org/10.1007/S00330-015-3980-0

Goh, Y., Balasundaram, G., Moothanchery, M., Attia, A., Li, X., Lim, H. Q., Burton, N., Qiu, Y., Putti, T. C., Chan, C. W., Iau, P., Tang, S. W., Ng, C. W. Q., Pool, F. J., Pillay, P., Chua, W., Sterling, E., Quek, S. T., & Olivo, M. (2018). Multispectral optoacoustic tomography in assessment of breast tumor margins during breast-conserving surgery: A first-in-human case study. *Clinical Breast Cancer*, *18*(6), e1247–e1250. https://doi.org/10.1016/j.clbc.2018.07.026

Groheux, D., & Hindie, E. (2021). Breast cancer: Initial workup and staging with FDG PET/CT. *Clinical and Translational Imaging*, *9*(3), 221–231. https://doi.org/10.1007/S40336-021-00426-Z/TABLES/2

Heijblom, M., Piras, D., van den Engh, F. M., van der Schaaf, M., Klaase, J. M., Steenbergen, W., & Manohar, S. (2016). The state of the art in breast imaging using the Twente Photoacoustic Mammoscope: Results from 31 measurements on malignancies. *European Radiology*, *26*(11), 3874–3887. https://doi.org/10.1007/S00330-016-4240-7

Hsu, H.-C., Wear, K. A., Joshua Pfefer, T., & Vogt, W. C. (2022). Tissue-mimicking phantoms for performance evaluation of photoacoustic microscopy systems. *Biomedical Optics Express*, *13*(3), 1357. https://doi.org/10.1364/BOE.445702

Jenkins, S. v., Nima, Z. A., Vang, K. B., Kannarpady, G., Nedosekin, D. A., Zharov, V. P., Griffin, R. J., Biris, A. S., & Dings, R. P. M. (2017). Triple-negative breast cancer targeting and killing by EpCAM-directed, plasmonically active nanodrug systems. *NPJ Precision Oncology*, *1*(1), 27. https://doi.org/10.1038/S41698-017-0030-1

Jeon, S., Kim, J., Lee, D., Baik, J. W., & Kim, C. (2019). Review on practical photoacoustic microscopy. *Photoacoustics*, *15*. https://doi.org/10.1016/J.PACS.2019.100141

Jin, Y., Yin, Y., Li, C., Liu, H., & Shi, J. (2022). Non-invasive monitoring of human health by photoacoustic spectroscopy. *Sensors*, *22*(3), 1155. https://doi.org/10.3390/S22031155

Jnawali, K., Chinni, B., Dogra, V., & Rao, N. (2020). Automatic cancer tissue detection using multispectral photoacoustic imaging. *International Journal of Computer Assisted Radiology and Surgery*, *15*(2), 309–320. https://doi.org/10.1007/S11548-019-02101-1

Jordan, V., Khan, M., & Prill, D. (2019). Breast cancer screening: Why can't everyone agree? *Primary Care*, *46*(1), 97–115. https://doi.org/10.1016/J.POP.2018.10.010

Kruger, R. A., Kuzmiak, C. M., Lam, R. B., Reinecke, D. R., del Rio, S. P., & Steed, D. (2013). Dedicated 3D photoacoustic breast imaging. *Medical Physics*, *40*(11). https://doi.org/10.1118/1.4824317

Laufer, J. (2018). Photoacoustic imaging: Principles and applications. In *Quantification of Biophysical Parameters in Medical Imaging*, pp. 303–324. https://doi.org/10.1007/978-3-319-65924-4_13/FIGURES/14

Liapis, E., Karlas, A., Klemm, U., & Ntziachristos, V. (2021). Chemotherapeutic effects on breast tumor hemodynamics revealed by eigenspectra multispectral optoacoustic tomography (eMSOT). *Theranostics*, *11*(16), 7813–7828. https://doi.org/10.7150/THNO.56173

Liapis, E., Klemm, U., Karlas, A., Reber, J., & Ntziachristos, V. (2020). Resolution of spatial and temporal heterogeneity in bevacizumab-treated breast tumors by eigenspectra multispectral optoacoustic tomography. *Cancer Research*, *80*(23), 5291–5304. https://doi.org/10.1158/0008-5472.CAN-20-1011

Lin, L., Hu, P., Shi, J., Appleton, C. M., Maslov, K., Li, L., Zhang, R., & Wang, L. v. (2018). Single-breath-hold photoacoustic computed tomography of the breast. *Nature Communications, 9*, 2352. https://doi.org/10.1038/s41467-018-04576-z

Lin, L., Tong, X., Hu, P., Invernizzi, M., Lai, L., & Wang, L. v. (2021). Photoacoustic computed tomography of breast cancer in response to neoadjuvant chemotherapy. *Advanced Science*, *8*(7). https://doi.org/10.1002/advs.202003396

Liu, W., & Yao, J. (2018). Photoacoustic microscopy: principles and biomedical applications. *Biomedical Engineering Letters*, *8*(2), 203. https://doi.org/10.1007/S13534-018-0067-2

Løberg, M., Lousdal, M. L., Bretthauer, M., & Kalager, M. (2015). Benefits and harms of mammography screening. *Breast Cancer Research: BCR*, *17*(1). https://doi.org/10.1186/S13058-015-0525-Z

Ma, Y., Yang, C., Zhang, J., Wang, Y., Gao, F., & Gao, F. (2020). Human breast numerical model generation based on deep learning for photoacoustic imaging. In *Annual International Conference of the IEEE Engineering in Medicine and Biology Society, 2020*, pp. 1919–1922. https://doi.org/10.1109/EMBC44109.2020.9176298

Mann, R. M., Hooley, R., Barr, R. G., & Moy, L. (2020). Novel approaches to screening for breast cancer. *Radiology, 297*(2), 266–285. https://doi.org/10.1148/RADIOL.2020200172/ASSET/IMAGES/LARGE/RADIOL.2020200172.FIG11H.JPEG

Nyayapathi, N., Lim, R., Zhang, H., Zheng, W., Wang, Y., Tiao, M., Oh, K. W., Fan, X. C., Bonaccio, E., Takabe, K., & Xia, J. (2020). Dual Scan Mammoscope (DSM) - A new portable photoacoustic breast imaging system with scanning in craniocaudal plane. *IEEE Transactions on Biomedical Engineering, 67*(5), 1321–1327. https://doi.org/10.1109/TBME.2019.2936088

Nyayapathi, N., Zhang, H., Zheng, E., Nagarajan, S., Bonaccio, E., Takabe, K., Fan, X. C., & Xia, J. (2021). Photoacoustic dual-scan mammoscope: Results from 38 patients. *Biomedical Optics Express, 12*(4), 2054. https://doi.org/10.1364/BOE.420679

Pal, A., Taneja, N., Malhotra, N., Shankar, R., Chawla, B., Awasthi, A., & Janardhanan, R. (2021). Knowledge, attitude, and practice towards breast cancer and its screening among women in India: A systematic review. *Journal of Cancer Research and Therapeutics, 17*(6), 1314–1321. https://doi.org/10.4103/JCRT.JCRT_922_20

Palma-Chavez, J., Pfefer, T. J., Agrawal, A., Jokerst, J. v., & Vogt, W. C. (2021). Review of consensus test methods in medical imaging and current practices in photoacoustic image quality assessment. *Journal of Biomedical Optics, 26*(9), 90901–90902. https://doi.org/10.1117/1.JBO.26.9.090901

Partovi, S., Sin, D., Lu, Z., Sieck, L., Marshall, H., Pham, R., & Plecha, D. (2020). Fast MRI breast cancer screening—Ready for prime time. *Clinical Imaging, 60*(2), 160–168. https://doi.org/10.1016/j.clinimag.2019.10.013

Priya, M., Rao, B. S. S., Chandra, S., Ray, S., M.D., S. M., Datta, A., Nayak, S. G., & Mahato, K. K. (2016). Photoacoustic spectroscopy based investigatory approach to discriminate breast cancer from normal: A pilot study. In *Photonic Therapeutics and Diagnostics XII, 9689*, pp. 416–420. https://doi.org/10.1117/12.2210900

Rao, A. P., Bokde, N., & Sinha, S. (2020). Photoacoustic imaging for management of breast cancer: A literature review and future perspectives. *Applied Sciences 2020, 10*(3), 767. https://doi.org/10.3390/APP10030767

Ren, W., Chen, M., Qiao, Y., & Zhao, F. (2022). Global guidelines for breast cancer screening: A systematic review. *The Breast : Official Journal of the European Society of Mastology, 64*, 85. https://doi.org/10.1016/J.BREAST.2022.04.003

Rodrigues, J., Amin, A., Raghushaker, C. R., Chandra, S., Joshi, M. B., Prasad, K., Rai, S., Nayak, S. G., Ray, S., & Mahato, K. K. (2021). Exploring photoacoustic spectroscopy-based machine learning together with metabolomics to assess breast tumor progression in a xenograft model ex vivo. *Laboratory Investigation, 101*(7), 952–965. https://doi.org/10.1038/s41374-021-00597-3

Schoustra, S. M., Huijink, R., Alink, L., op 't Root, T. J. P. M., Sprünken, D., Piras, D., Muller Kobold, W. F., Klazen, C. A. H., van der Schaaf, M. C., van den Engh, F. M., Steenbergen, W., & Manohar, S. (2019a). The Twente Photoacoustic Mammoscope 2: 3D vascular network visualization. In *Photons Plus Ultrasound: Imaging and Sensing 2019, 10878*, pp. 63–69. https://doi.org/10.1117/12.2507448

Schoustra, S. M., Piras, D., Huijink, R., Root, T. J. P. M. op 't, Alink, L., Kobold, W. M., Steenbergen, W., & Manohar, S. (2019b). Twente Photoacoustic Mammoscope 2: System overview and three-dimensional vascular network images in healthy breasts. *Journal of Biomedical Optics, 24*(12), 1. https://doi.org/10.1117/1.JBO.24.12.121909

Shen, D., Zhu, L., Liu, Y., Peng, Y., Lan, M., Fang, K., & Guo, Y. (2020). Efficacy evaluation and mechanism study on inhibition of breast cancer cell growth by multimodal targeted nanobubbles carrying AMD070 and ICG. *Nanotechnology, 31*(24). https://doi.org/10.1088/1361-6528/AB7E73

Sobhanan Warrier, G., Amirthalakshmi, T. M., Nimala, K., Thaj Mary Delsy, T., Stella Rose Malar, P., Ramkumar, G., & Raju, R. (2022). Automated recognition of cancer tissues through deep learning framework from the photoacoustic specimen. *Contrast Media & Molecular Imaging, 2022*. https://doi.org/10.1155/2022/4356744

Stoffels, I., Jansen, P., Petri, M., Goerdt, L., Brinker, T. J., Griewank, K. G., Poeppel, T. D., Schadendorf, D., & Klode, J. (2019). Assessment of nonradioactive multispectral optoacoustic tomographic imaging with conventional lymphoscintigraphic imaging for sentinel lymph node biopsy in melanoma. *JAMA Network Open, 2*(8). https://doi.org/10.1001/JAMANETWORKOPEN.2019.9020

Sun, I.-C., Ahn, C.-H., Kim, K., & Emelianov, S. (2019). Photoacoustic imaging of cancer cells with glycol-chitosan-coated gold nanoparticles as contrast agents. *Journal of Biomedical Optics*, *24*(12), 1. https://doi.org/10.1117/1.JBO.24.12.121903

Sung, H., Ferlay, J., Siegel, R. L., Laversanne, M., Soerjomataram, I., Jemal, A., & Bray, F. (2021). Global Cancer Statistics 2020: GLOBOCAN estimates of incidence and mortality worldwide for 36 cancers in 185 countries. *CA: A Cancer Journal for Clinicians*, *71*(3), 209–249. https://doi.org/10.3322/CAAC.21660

Tanaka, Y., Tajima, T., Seyama, M., & Waki, K. (2020). Differential continuous wave photoacoustic spectroscopy for non-invasive glucose monitoring. *IEEE Sensors Journal*, *20*(8), 4453–4458. https://doi.org/10.1109/JSEN.2019.2962251

Toi, M., Asao, Y., Matsumoto, Y., Sekiguchi, H., Yoshikawa, A., Takada, M., Kataoka, M., Endo, T., Kawaguchi-Sakita, N., Kawashima, M., Fakhrejahani, E., Kanao, S., Yamaga, I., Nakayama, Y., Tokiwa, M., Torii, M., Yagi, T., Sakurai, T., Togashi, K., & Shiina, T. (2017). Visualization of tumor-related blood vessels in human breast by photoacoustic imaging system with a hemispherical detector array. *Scientific Reports*, *7*. https://doi.org/10.1038/SREP41970

Vogt, W. C., Jia, C., Wear, K. A., M.D., B. S. G., & Pfefer, T. J. (2017). Phantom-based image quality test methods for photoacoustic imaging systems. *Journal of Biomedical Optics*, *22*(9), 095002. https://doi.org/10.1117/1.JBO.22.9.095002

Wang, L. v., & Yao, J. (2016). A practical guide to photoacoustic tomography in the life sciences. *Nature Methods*, *13*(8), 627. https://doi.org/10.1038/NMETH.3925

Wong, T. T. W., Zhang, R., Hai, P., Zhang, C., Pleitez, M. A., Aft, R. L., Novack, D. V., & Wang, L. V. (2017). Fast label-free multilayered histology-like imaging of human breast cancer by photoacoustic microscopy. *Science Advances*, *3*(5). https://doi.org/10.1126/SCIADV.1602168

Xia, J., Feng, G., Xia, X., Hao, L., & Wang, Z. (2017). NH4HCO3 gas-generating liposomal nanoparticle for photoacoustic imaging in breast cancer. *International Journal of Nanomedicine*, *12*, 1803–1813. https://doi.org/10.2147/IJN.S113366

Yang, J., Zhang, G., Li, Q., Liao, C., Huang, L., Ke, T., Jiang, H., & Han, D. (2019). Photoacoustic imaging for the evaluation of early tumor response to antivascular treatment. *Quantitative Imaging in Medicine and Surgery*, *9*(2), 160–170. https://doi.org/10.21037/qims.2018.11.06

Yang, S., Qiao, S., Liu, X., & Ma, Y. (2022). Highly sensitive measurement of oxygen concentration based on reflector-enhanced photoacoustic spectroscopy. *Sensors*, *22*(14), 5087. https://doi.org/10.3390/S22145087

Zhang, J., Duan, F., Liu, Y., & Nie, L. (2020). High-resolution photoacoustic tomography for early-stage cancer detection and its clinical translation. *Radiology: Imaging Cancer*, 2(3). https://doi.org/10.1148/rycan.2020190030

Zheng, E., Zhang, H., Goswami, S., Kabir, I. E., Doyley, M. M., & Xia, J. (2021). Second-generation dual scan mammoscope with photoacoustic, ultrasound, and elastographic imaging capabilities. *Frontiers in Oncology*, *11*, 779071. https://doi.org/10.3389/fonc.2021.779071

5 Characterization and Identification of Novel Bacterial Exopolysaccharides

*Nupura Manish Prabhune and Sudharshan Prabhu**

Department of Cell and Molecular Biology, Manipal School of Life Sciences, Manipal Academy of Higher Education, Manipal, Karnataka, India
*Corresponding Author

INTRODUCTION

The significance of bacterial metabolism lies in the synthesis of complex products from relatively simple substrates that enhance the survival, virality, distribution, and genetic diversity of bacteria (Amato et al., 2014). Polysaccharides represent an important class of bacterial metabolites, which are primarily classified into two types on the basis of their site of production – endopolysaccharides and exopolysaccharides (Nwodo et al., 2012). Polysaccharides that are produced intracellularly are known as endopolysaccharides, which mainly represent energy reserves in the form of glycogen and polyhydroxy butyrate. In contrast, exopolysaccharides are defined as extracellular polysaccharides produced by bacteria, which are well-known for their structural and functional diversity, such as cellulose, chitosan, xanthan, etc. While some exopolysaccharides form a uniform capsule called the glycocalyx around the bacterium to provide structural integrity (Nwodo et al., 2012), some others are secreted into the external environment. These secreted exopolysaccharides, such as alginate or poly-N-acetylglucosamine, play an important role in biofilm formation, along with additional functions such as bacterial adhesion and host defense (Roux et al., 2015; He et al., 2018; Ostapska et al., 2018).

The exopolysaccharides are produced through a biosynthetic process that is tightly regulated within the metabolic framework of the bacteria. While the entire mechanism is still not established, on a fundamental level, exopolysaccharide synthesis involves the stepwise assembly of monosaccharide units onto a lipid backbone from sugar nucleotide precursors, by the action of enzymes called glycosyltransferases and polymerases (Wzy protein), giving rise to an exopolysaccharide composed of either the same repeating monomer (homopolysaccharide) or different monomers (heteropolysaccharide). The exopolysaccharide is then translocated across the cytoplasm by an enzyme called flippase (Wzx protein) and is externally released by the OPX family of proteins (outer membrane polysaccharide transport proteins). This mechanism of exopolysaccharide synthesis is commonly known as the Wzx/Wzy-dependent pathway (Schmid et al., 2015), which represents the most integral pathway for producing structural exopolysaccharides such as xanthum gum, as well as important lipopolysaccharide markers across a range of pathogenic bacteria (Islam and Lam, 2014). Apart from this mechanism, some alternate methods of exopolysaccharide production include the ABC transporter-dependent pathway, which is responsible for producing capsular polysaccharides, and the synthase-dependent pathway, in which the polymerization and translocation of exopolysaccharides are performed without the requirement of flippases (Soumya et al., 2021).

The deep interest in the exploration of exopolysaccharides stems from their versatility and potential to offer a wide range of benefits across fields such as medicine, healthcare, and polymer design,

 DOI: 10.1201/9781003545712-5

to name a few. In the medical field in recent years, exopolysaccharides have found major applications as wound dressings, nano-formulations for drug delivery, as well as scaffolds for tissue engineering (Moscovici, 2015). Similarly, the pharmacological properties of exopolysaccharides (accounted for by their structural uniqueness) have also been investigated through multiple research studies, primarily in the form of anti-cancer, anti-bacterial, and immunomodulatory effects. In a study carried out by Wang et al. in 2019, the exopolysaccharide extract from a novel *Bifidobacterium* strain, composed of rhamnose, arabinose, and mannose, was observed to induce cell cycle arrest and apoptosis in a head and neck squamous carcinoma cell line. Further, in a study carried out in 2020 by Xu et al., the anti-bacterial and anti-biofilm potential of the exopolysaccharide extract from a novel *Lactobacillus casei* strain was demonstrated against *Staphylococcus aureus*. With respect to immunomodulation, in 2013 Guo et al. reported macrophage phagocytosis and complement system activation, induced by selenium exopolysaccharides isolated from a *Lactococcus lactis* strain in immunocompromised mice.

The studies highlighted above are merely representative of the comprehensive and extensive application-based research being devoted to the characterization of exopolysaccharides in the field of applied microbiology. The exopolysaccharide extracts under study are mainly sourced or isolated from novel strains identified in samples from seawater (Cho et al., 2004), soil (Singh and Haritash, 2021), mountainous regions (Uniyal et al., 2021), freshwater lakes (Cabello et al., 2018), and other environmental regions. During the process of investigation, multiple challenges and complexities come into play. To begin with, the process of sample collection is labor-intensive and cumbersome. Following this, to isolate the bacterial strains, culturing methods must be adopted (Bharti and Grimm, 2021). At this stage, if the selection of the growth media is not done appropriately, taking into account the diversity of bacterial species that may be present in the sample, we may lose crucial information. It is also necessary to consider that not all bacterial strains present in the sample may possess the ability to grow on the culture medium, due to which molecular identification techniques must be adopted beyond traditional culture methods (Ranjard et al., 2000). Moreover, the process of screening for unique exopolysaccharides, recognizing their functional importance, and scaling up their yield are also complicated aspects that require intensive analysis.

In such a scenario where there are major inconsistencies in the process of investigation, a standardized protocol for sample collection, bacterial culturing, molecular identification, and biological activity screening can help in streamlining the study of bacterial strains and their exopolysaccharide extracts, which will be elaborated and explained in the subsequent sections.

BACTERIAL CULTURE AND IDENTIFICATION

Exopolysaccharide-producing bacterial strains will be isolated from diverse environmental samples (soil, water, deep sea sediments, extremophiles). The isolated organisms will be cultured on different media (tryptic soy agar, yeast extract mannitol agar) at a temperature of 32°C, and the agar plates will be observed for the production of colonies (Figure 5.1) (Ali et al., 2020). Taxonomic identification will be done by sequencing of the 16S rRNA gene. The sequence data will be aligned with standard bacterial lineage sequences available on the EzTaxon e server, using the Basic Local Alignment Search Tool (BLAST).

EXTRACTION AND PURIFICATION OF EXOPOLYSACCHARIDES

To extract and purify the exopolysaccharides (EPS), the selected strains will be grown in yeast extract mannitol (YEM) broth at a temperature of 32°C, with 100 rpm agitation for 72 h. The cells will be separated by centrifugation (9000 pm, 10 min), and the supernatant will be mixed with chilled ethanol in a 1:3 ratio and kept for incubation overnight at 4°C to precipitate the EPS (Quesada et al., 1994). The precipitated EPS will be separated out by the process of centrifugation (10,000 rpm, 4°C, 20 min), and the protein content will be precipitated from the crude EPS by treating it with trichloroacetic acid (8% w/v), followed by centrifugation (10,000 rpm, 4°C, 15

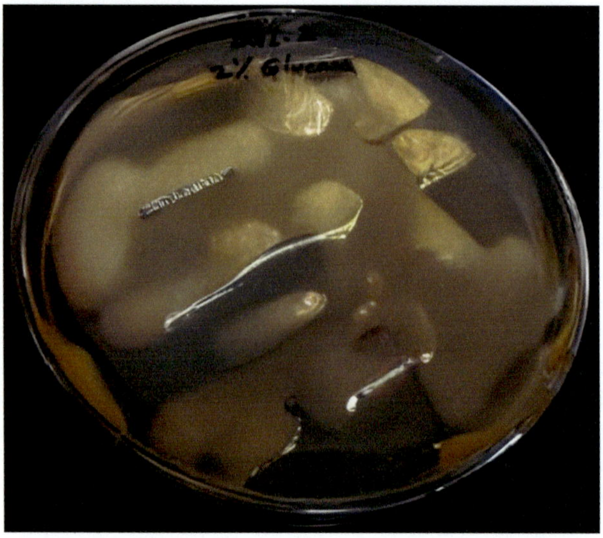

FIGURE 5.1 Production of exopolysaccharides on tryptic soy agar medium (Ali et al., 2020).

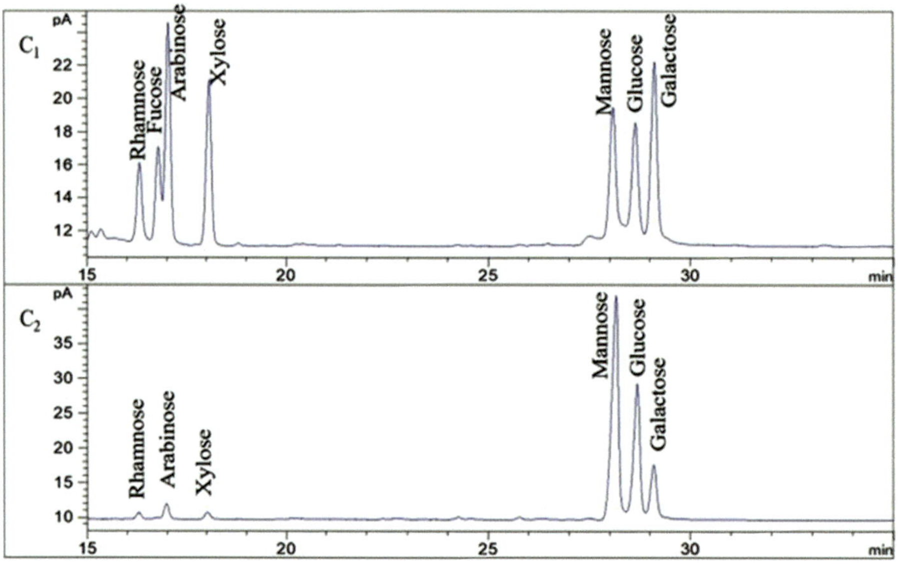

FIGURE 5.2 GC analysis of (C1) monosaccharide standards and (C2) exopolysaccharide sample (Huo et al., 2020).

min). The obtained EPS will then be subjected to dialysis with ultra-pure water using a dialysis membrane (12-kDa pore size). The dialyzed EPS will be freeze-dried, and the yield (dry weight) will be determined.

CHARACTERIZATION OF EXOPOLYSACCHARIDES

Biochemical Characterization

The isolated EPS will be evaluated for biochemical and structural characterization. The total carbohydrates will be evaluated by the phenol sulfuric acid method (DuBois et al., 1956), and

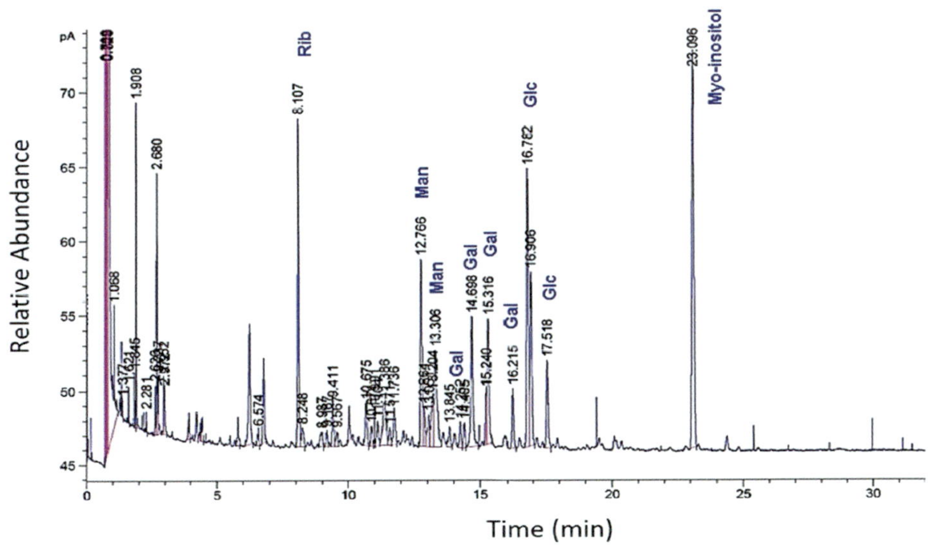

FIGURE 5.3 GC profiles of exopolysacharrides (Jivkova et al., 2022).

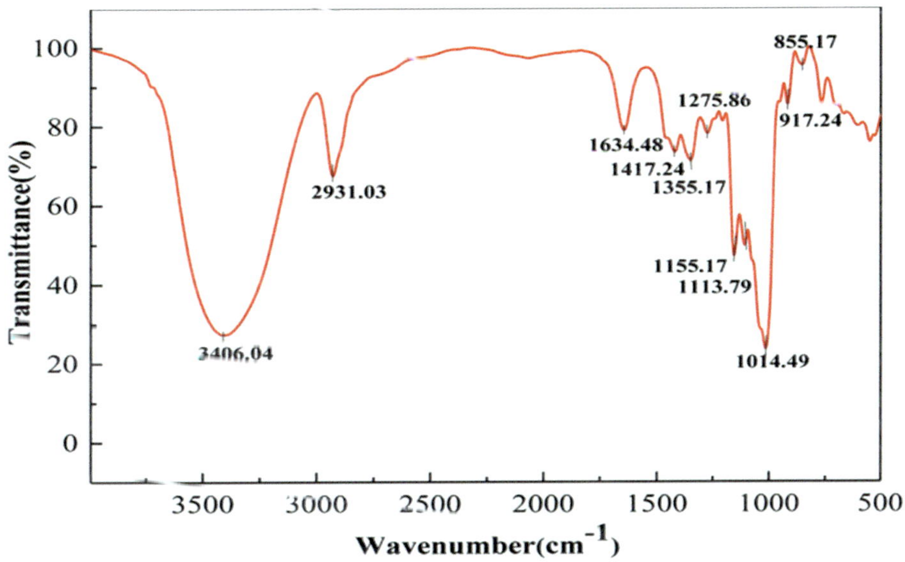

FIGURE 5.4 FTIR results of an exopolysaccharide sample (Hao et al., 2020).

proteins will be quantified using Bradford's method (Bradford, 1976). Standard spectrophotometric assays will be used to quantify phosphate content (Chen et al., 1956), sulfate (Terho and Hartiala, 1971), uronic acids (Blumenkrantz and Asboe-Hansen, 1973), and acetyl residues (McComb and McCready, 1957).

To determine the molecular weight of the isolated EPS, the purified EPS will be subjected to gel filtration chromatography using Sephacryl S-300 HR. The elution will be performed using ultrapure water at a flow rate of 1mL/min (Gan et al., 2020).

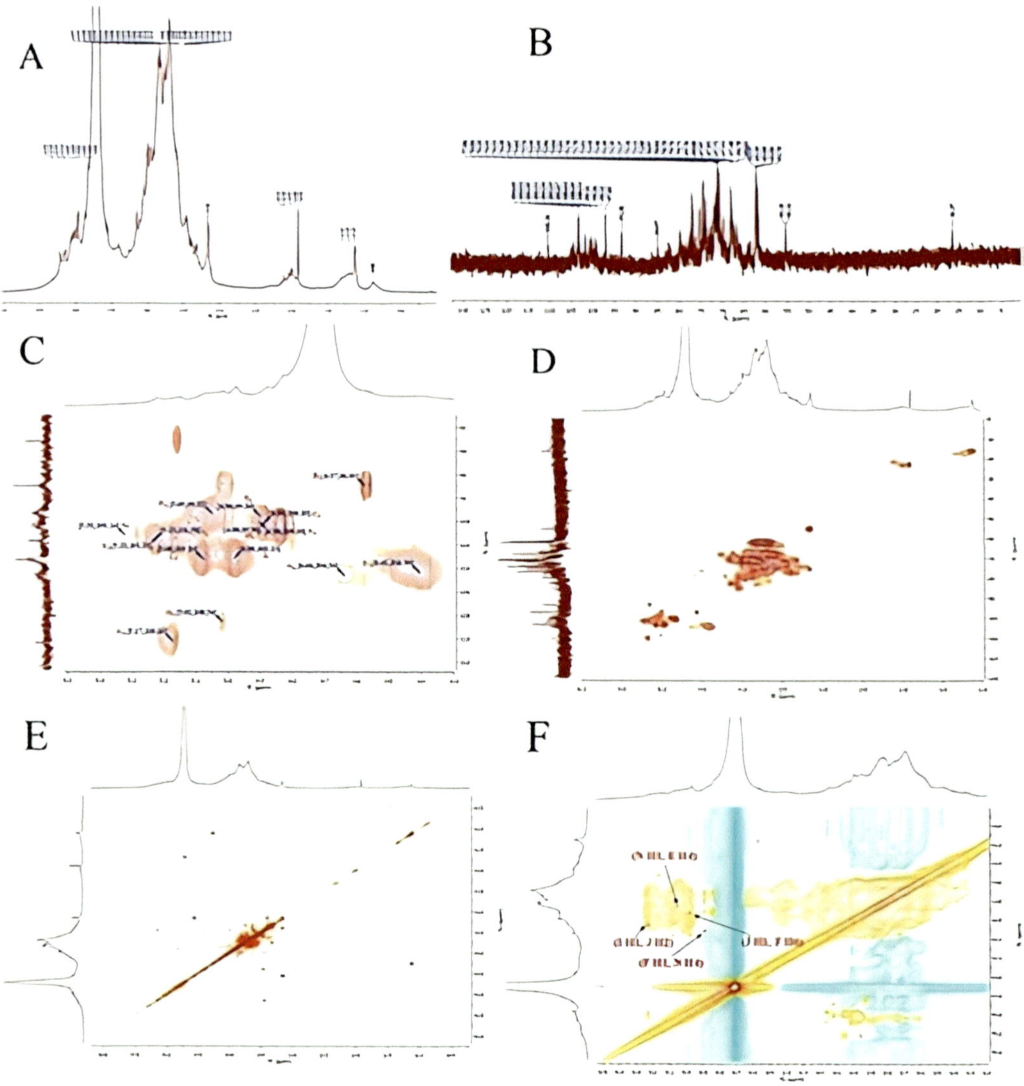

FIGURE 5.5 NMR spectra of the EPS sample. (A) ^{1}H, (B) ^{13}C, (C) HSQC-part of anomeric carbon/proton, (D) HSQC, (E) 1H–^{1}H COSY, and (F) NOESY (Huo et al., 2020).

Analysis of Monosaccharide Composition

Monosaccharide analysis of exopolysaccharides will be carried out according to the methodology outlined earlier (Blau and Halket, 1993). Hydrolysis of the EPS samples will be performed with 1 M methanolic HCl at a temperature of 80°C for a duration of 16 h. After this, the samples will be dried using nitrogen gas and then dissolved in dichloromethane and dimethyl acetamide. The samples will then be re-evaporated with nitrogen and further re-dissolved in *N,O*-bis(trimethylsilyl) trifluoroacetamide-trimethyl chlorosilane (BSTFA–TMCS). After keeping the samples at a temperature of 60°C for a duration of 1 h, analysis will be carried out using the GC-MS system (Figure 5.2; Figure 5.3), with DB-5ms column (30 × 0.25 mm). A 1 μL volume of the samples will be injected, and the injector temperature will be maintained at 260°C.

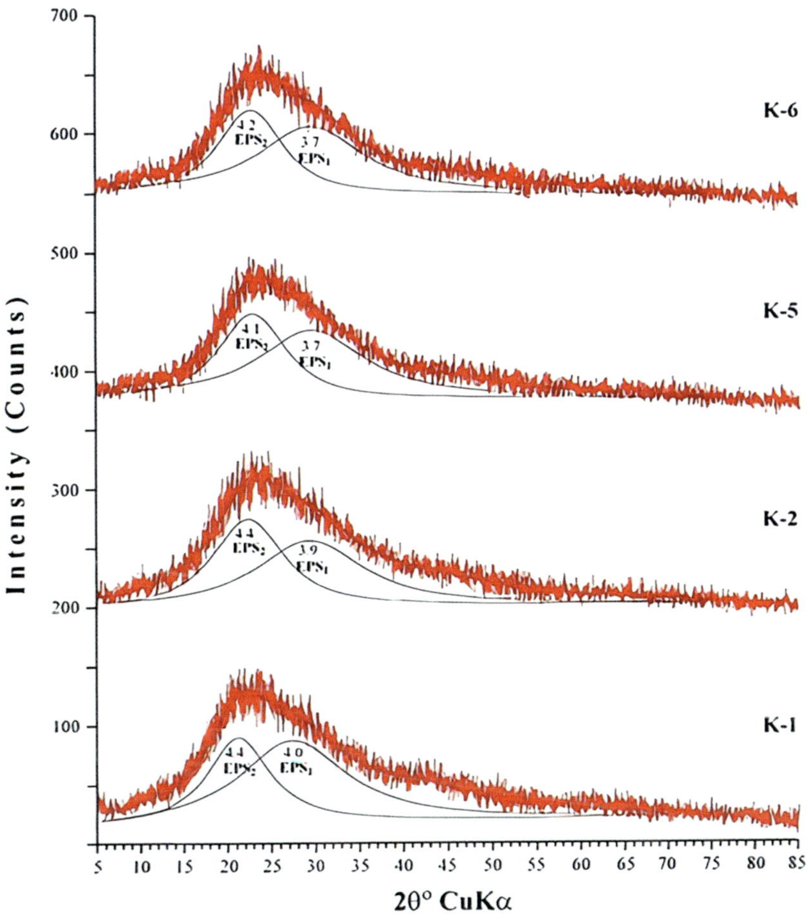

FIGURE 5.6 X-ray diffraction (XRD) patterns of EPS from different media. (K-1: dissolved EPS in LB medium, K-2: particulate EPS in LB medium, K-5: dissolved EPS in LB with whey, K-6: particulate EPS in LB with whey) (Solmaz et al., 2018).

The initial column temperature will be maintained at 50°C for 2 min and then increased to 310°C (10°C min^{-1} ramp), followed by 5 min of holding time. Monosaccharide identification will be done using a mass spectrometry (MS) detector, operated in the full-scan mode ranging from 35 to 700 *m/z* (Priyanka et al., 2015).

FOURIER TRANSFORM INFRARED SPECTROSCOPY ANALYSIS

The isolated EPS will be subjected to Fourier transform infrared (FTIR) spectroscopy analysis to investigate the chemical composition and structure (Figure 5.4). Briefly, the EPS will be pelleted with KBr, and the frequency will be recorded from 4000 to 500 cm^{-1} at a temperature of 25°C (room temperature) with 32 scans (Hao et al., 2020).

NUCLEAR MAGNETIC RESONANCE SPECTROSCOPY ANALYSIS

To determine the structure of EPS, nuclear magnetic resonance (NMR) spectroscopy studies will be performed according to the method outlined earlier (Huo et al., 2020). Approximately 25 mg of the lyophilized EPS will be dissolved in deuterium oxide (D_2O 9.9% D) and then further transferred to

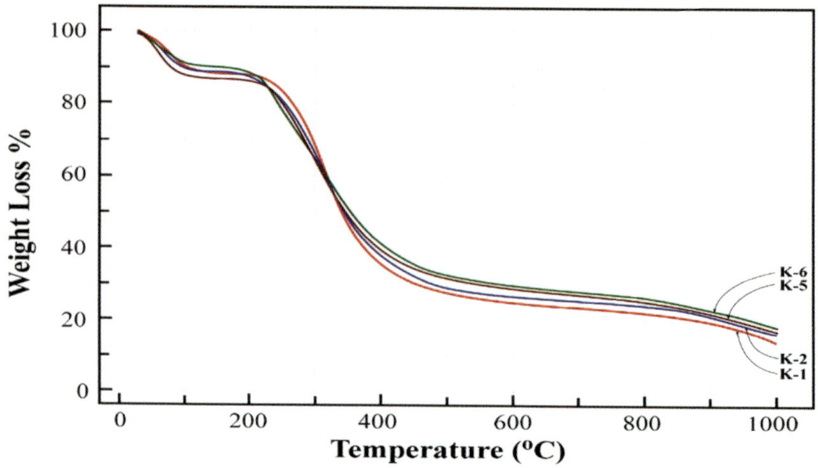

FIGURE 5.7 Thermogravimetric analysis (TGA) curves of EPS from different media (K-1: dissolved EPS in LB medium, K-2: particulate EPS in LB medium, K-5: dissolved EPS in LB with whey, K-6: particulate EPS in LB with whey) (Solmaz et al., 2018).

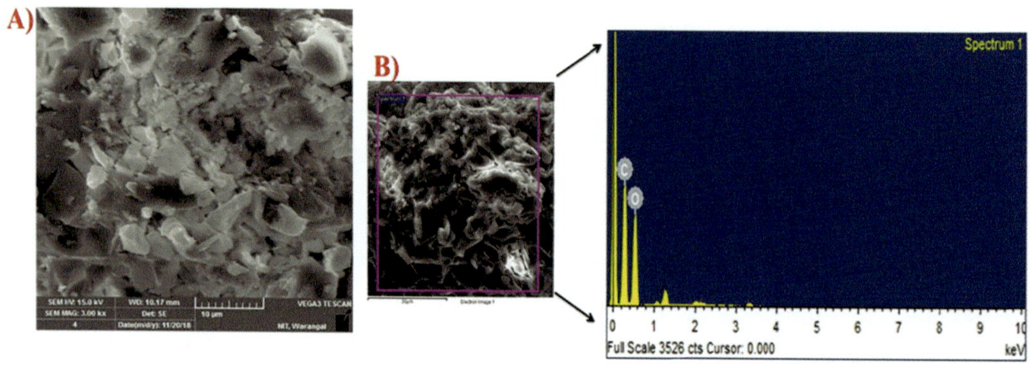

FIGURE 5.8 (A) SEM images showing the shape and surface morphology and (B) EDAX analysis of the EPS (Gangalla et al., 2021).

the NMR tube. 1D NMR, which includes ^{1}H and ^{13}C NMR, and 2D NMR, which includes COSY, HMBC, NOESY, HSQC, and TOCSY, will be analyzed using the NMR spectrometer (Figure 5.5). The data obtained will then be processed to determine the structure.

X-RAY DIFFRACTION ANALYSIS

To understand the crystalline structure, morphology, and properties of the EPS, X-ray diffraction (XRD) studies will be performed according to the method outlined previously (Figure 5.6) (Solmaz et al., 2018). The isolated pure EPS will be analyzed using an X-ray diffractometer that is equipped with a Cu-Kα radiation source ($\lambda = 1.54056$ Å). The equipment will be operated at 40 kV, 40 mA, with a scan speed of 0.005°, time/scan of 0.1 sec, and with a 0.2 mm slit. Scanning of the EPS will be done at an angle lying between 5° and 80°, and at a 0.02°/min step size. The diffracted X-rays will be calculated in accordance with Bragg's law ($d = \lambda/2 \sin\theta$), based on the method outlined earlier (Solmaz et al., 2018).

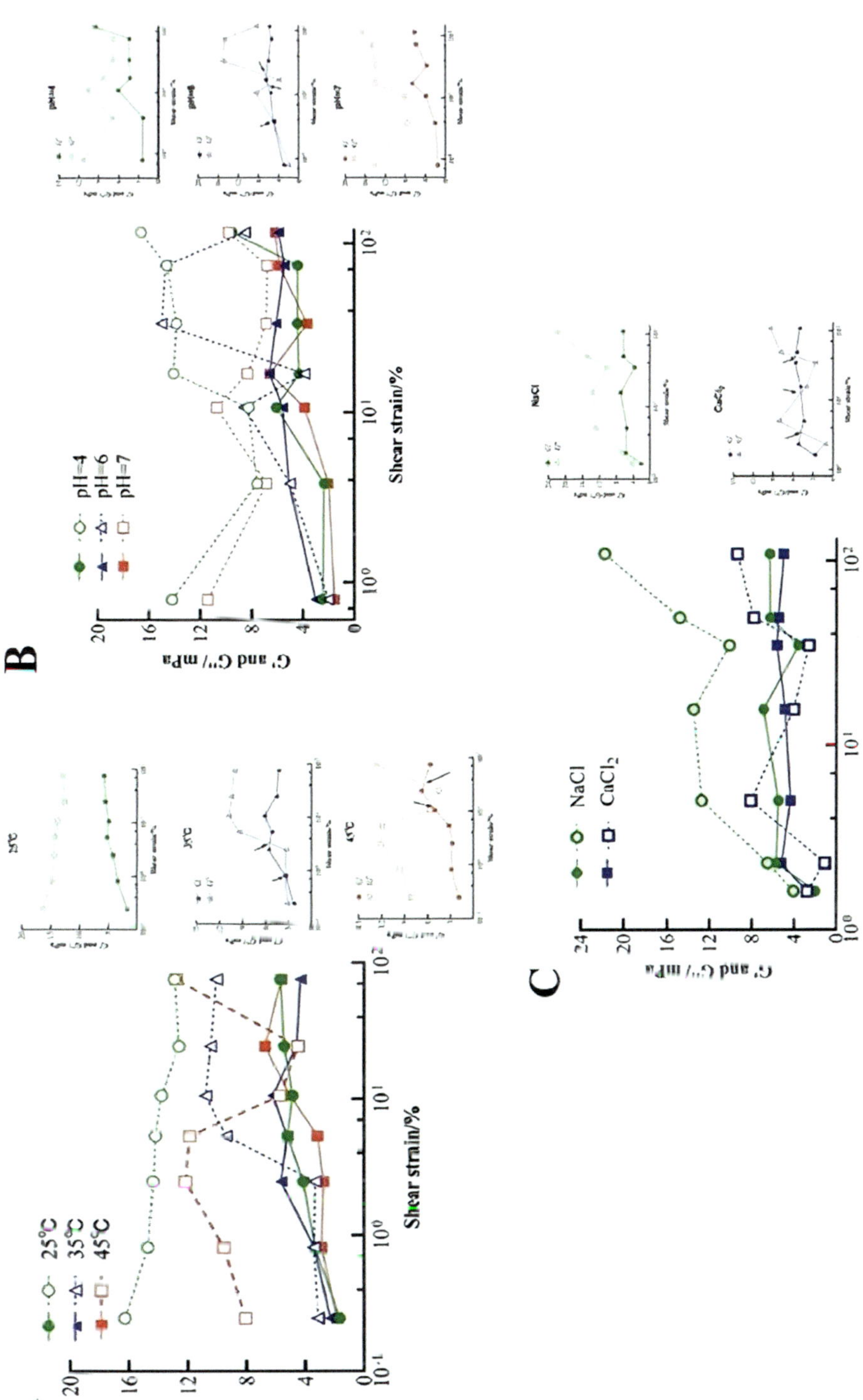

FIGURE 5.9 Rheological behavior at different temperatures (A), pH (B), and salt solution treatments (C).

THERMOGRAVIMETRIC ANALYSIS

Thermogravimetric analysis (TGA) is a thermal technique used to investigate changes in EPS characteristics in response to temperature change. TGA analysis will be performed in addition to differential thermal analysis (DTA), between 50 and 800°C on the thermogravimetric analyzer, according to the methodology outlined earlier (Figure 5.7) (Yang et al., 2021). Differential scanning calorimetry (DSC) studies of EPS will provide information about thermal changes in phase transition. The EPS sample will be heated from a temperature of 25°C (room temperature) to 260°C, with a ramp rate of 10°C/min (Jiang et al., 2021).

SCANNING ELECTRON MICROSCOPY AND ENERGY-DISPERSIVE X-RAY SPECTROSCOPY

Scanning electron microscopy (SEM) analysis of EPS will be performed to observe the structural morphology of the purified EPS, and energy-dispersive X-ray spectroscopy (EDX) studies will be carried out for the elemental analysis of EPS (Figure 5.8). The studies will be carried out according to the method outlined earlier (Gangalla et al., 2021).

RHEOLOGICAL CHARACTERIZATION

To understand the mechanical behavior of the EPS and to gain deeper insights regarding the viscoelastic properties of the EPS, rheological characterization will be conducted (Figure 5.9). Various treatment groups will be established to analyze different rheological behaviors of the EPS, such as viscosity and dynamic viscoelasticity, across different temperatures (25°C, 35°C, and 45°C), pH values (4, 6, and 7), and with different salt solutions (NaCl and $CaCl_2$), as per the methodology outlined previously (Li et al., 2023)

CONCLUSION

Identification of novel bacterial strains and the characterization of their exopolysaccharide extracts for biological importance is a complicated process, involving the analysis of multiple properties and parameters. This chapter provides a standardized workflow of the techniques used for the culture and molecular identification of bacterial strains, followed by the extraction of exopolysaccharides and their biochemical and structural characterization. The adoption of these techniques is crucial for recognizing the potential of novel bacterial exopolysaccharides and for further facilitating their applications across various fields, ranging from healthcare to polymer development and design.

REFERENCES

Ali P, Shah AA, Hasan F, Hertkorn N, Gonsior M, Sajjad W, Chen F. A glacier bacterium produces high yield of cryoprotective exopolysaccharide. *Frontiers in Microbiology*. 2020;10:1–16.

Amato SM, Fazen CH, Henry TC, Mok WW, Orman MA, Sandvik EL, Volzing KG, Brynildsen MP. The role of metabolism in bacterial persistence. *Frontiers in Microbiology*. 2014;5:1–9.

Bharti R, Grimm DG. Current challenges and best-practice protocols for microbiome analysis. *Briefings in Bioinformatics*. 2021;22:178–193.

Blau K, Halket J. *Handbook of derivatives for chromatography* (4th ed.) New York: John Wiley & Sons; 1993.

Blumenkrantz N, Asboe-Hansen G. New method for quantitative determination of uronic acids. *Analytical Biochemistry*. 1973;54:484–489.

Bradford MM. A rapid and sensitive method for the quantitation of microgram quantities of protein utilizing the principle of protein-dye binding. *Analytical Biochemistry*. 1976;72:248–254.

Cabello-Yeves PJ, Zemskaya TI, Rosselli R, Coutinho FH, Zakharenko AS, Blinov VV, Rodriguez-Valera F. Genomes of novel microbial lineages assembled from the sub-ice waters of Lake Baikal. *Applied and Environmental Microbiology*. 2018;84:1–21.

Chen PS, Toribara TT, Warner H. Microdetermination of phosphorus. *Analytical Chemistry.* 1956;28:1756–1758.
Cho JC, Vergin KL, Morris RM, Giovannoni SJ. Lentisphaera araneosa gen. nov., sp. nov, a transparent exopolymer producing marine bacterium, and the description of a novel bacterial phylum, Lentisphaerae. *Environmental Microbiology.* 2004;6:611–621.
DuBois M, Gilles KA, Hamilton JK, Rebers PT, Smith F. Colorimetric method for determination of sugars and related substances. *Analytical Chemistry.* 1956;28:350–356.
Gan L, Li X, Zhang H, Zhang R, Wang H, Xu Z, Peng B, Tian Y. Preparation, characterization and functional properties of a novel exopolysaccharide produced by the halophilic strain Halomonas saliphila LCB169T. *International Journal of Biological Macromolecules.* 2020;156:372–380.
Gangalla R, Sampath G, Beduru S, Sarika K, Govindarajan RK, Ameen F, Alwakeel S, Thampu RK. Optimization and characterization of exopolysaccharide produced by Bacillus aerophilus rk1 and its in vitro antioxidant activities. *Journal of King Saud University-Science.* 2021;33:1–9.
Guo Y, Pan D, Li H, Sun Y, Zeng X, Yan B. Antioxidant and immunomodulatory activity of selenium exopolysaccharide produced by Lactococcus lactis subsp. lactis. *Food Chemistry.* 2013;138:84–89.
Hao Y, Sun H, Zhang X, Wu L, Zhu Z. A novel polysaccharide from Pleurotus citrinopileatus mycelia: structural characterization, hypoglycemic activity and mechanism. *Food Bioscience.* 2020;37:1–13.
He X, Liu Y, Bai X, Yuan C, Li H. Alginate/albumin in incubation solution mediates the adhesion and biofilm formation of typical marine bacteria and algae. *Biochemical Engineering Journal.* 2018;139:25–32.
Huo J, Wu J, Huang M, Zhao M, Sun W, Sun X, Zheng F. Structural characterization and immuno-stimulating activities of a novel polysaccharide from Huangshui, a byproduct of Chinese Baijiu. *Food Research International.* 2020;136:1–12.
Islam ST, Lam JS. Synthesis of bacterial polysaccharides via the Wzx/Wzy-dependent pathway. *Canadian Journal of Microbiology.* 2014;60:697–716.
Jiang G, Gan L, Li X, He J, Zhang S, Chen J, Zhang R, Xu Z, Tian Y. Characterization of structural and physicochemical properties of an exopolysaccharide produced by Enterococcus sp. F2 from fermented soya beans. *Frontiers in Microbiology.* 2021:1–14.
Jivkova D, Sathiyanarayanan G, Harir M, Hertkorn N, Schmitt-Kopplin P, Sanhaji G, Fochesato S, Berthomieu C, Heyraud A, Achouak W, Santaella C. Production and characterization of a novel exopolysaccharide from ramlibacter tataouinensis. *Molecules.* 2022;27:1–17.
Li P, Bai Y, Yang Y, Li M. Structural characterization of exopolysaccharide produced by *Streptococcus thermophilus* and rheology of mixed protein-polysaccharide systems. *Food Bioscience.* 2023;55:1–10.
McComb EA, McCready RM. Determination of acetyl in pectin and in acetylated carbohydrate polymers. *Analytical Chemistry.* 1957;29:819–821.
Moscovici M. Present and future medical applications of microbial exopolysaccharides. *Frontiers in Microbiology.* 2015;6:1–11.
Nwodo UU, Green E, Okoh AI. Bacterial exopolysaccharides: functionality and prospects. *International Journal of Molecular Sciences.* 2012;13:14002–14015.
Ostapska H, Howell PL, Sheppard DC. Deacetylated microbial biofilm exopolysaccharides: it pays to be positive. *PLoS Pathogens.* 2018;14:1–8.
Priyanka P, Arun AB, Ashwini P, Rekha PD. Versatile properties of an exopolysaccharide R-PS18 produced by Rhizobium sp. PRIM-18. *Carbohydrate Polymers.* 2015;126:215–221.
Quesada E, Del Moral A, Béjar V. Comparative methods for isolation of Volcaniella eurihalina exopolysaccharide. *Biotechnology Techniques.* 1994;8:701–706.
Ranjard L, Poly F, Nazaret S. Monitoring complex bacterial communities using culture-independent molecular techniques: application to soil environment. *Research in Microbiology.* 2000;151:167–177.
Roux D, Cywes-Bentley C, Zhang YF, Pons S, Konkol M, Kearns DB, Little DJ, Howell PL, Skurnik D, Pier GB. Identification of poly-N-acetylglucosamine as a major polysaccharide component of the Bacillus subtilis biofilm matrix. *Journal of Biological Chemistry.* 2015;290:19261–19272.
Schmid J, Sieber V, Rehm B. Bacterial exopolysaccharides: biosynthesis pathways and engineering strategies. *Frontiers in Microbiology.* 2015;6:1–24.
Singh SK, Haritash AK. Catabolic enzyme activity and kinetics of pyrene degradation by novel bacterial strains isolated from contaminated soil. *Environmental Technology and Innovation.* 2021;23:1–10.
Solmaz KB, Ozcan Y, Mercan Dogan N, Bozkaya O, Ide S. Characterization and production of extracellular polysaccharides (EPS) by Bacillus pseudomycoides U10. *Environments.* 2018;5:1–16.
Soumya MP, Nampoothiri KM. An overview of functional genomics and relevance of glycosyltransferases in exopolysaccharide production by lactic acid bacteria. *International Journal of Biological Macromolecules.* 2021;184:1014–1025.

Terho TT, Hartiala KJ. Method for determination of the sulfate content of glycosaminoglycans. *Analytical Biochemistry*. 1971;41:471–476.

Uniyal S, Sharma RK, Kondakal V. New insights into the biodegradation of chlorpyrifos by a novel bacterial consortium: process optimization using general factorial experimental design. *Ecotoxicology and Environmental Safety*. 2021;209:1–10.

Wang L, Wang Y, Li Q, Tian K, Xu L, Liu G, Guo C. Exopolysaccharide, isolated from a novel strain Bifidobacterium breve lw01 possess an anticancer effect on head and neck cancer–genetic and biochemical evidences. *Frontiers in Microbiology*. 2019;10:1044.

Xu X, Peng Q, Zhang Y, Tian D, Zhang P, Huang Y, Ma L, Dia VP, Qiao Y, Shi B. Antibacterial potential of a novel Lactobacillus casei strain isolated from Chinese northeast sauerkraut and the antibiofilm activity of its exopolysaccharides. *Food & Function*. 2020;11:4697–4706.

Yang W, Wu J, Liu W, Ai Z, Cheng Y, Wei Z, Zhang H, Ma H, Cui F, Zhou C, Yang L. Structural characterization, antioxidant and hypolipidemic activity of Grifola frondosa polysaccharides in novel submerged cultivation. *Food Bioscience*. 2021;42:1–10.

6 Bioactive Glasses
Synthesis, Characterization, and Bioactivity

*Yash Goel, Swathi Sadashiva, and Bhisham Narayan Singh**

Department of Biotechnology, Manipal School of Life Sciences,
Manipal Academy of Higher Education, Manipal, Karnataka, India
*Corresponding Author

INTRODUCTION

Bone is a mineralized, living, and highly vascularized connective tissue composed of many types of cells. It has been established that through intercellular communication, the bone experiences a regular remodeling process throughout life [1]. This process is important to maintain calcium and phosphorus homeostasis and heal small injuries like hairline fractures. However, it is not possible to completely treat intense structural bone defects due to intense trauma and diseases like osteoporosis with just natural ability. The 1960s and 1970s saw the development of materials specifically for treating defects in bone and cartilage tissue. These innovations formed the basis of biomaterials. To evade toxicity, all initial biomaterials were designed to be biologically inert and not elicit any cellular reaction. These were called first-generation biomaterials. However, these materials did not adhere to bone tissue and were prone to eventual failure [2].

Later, biomaterials that can elicit a cellular response were developed. They are classified as second-generation biomaterials or third-generation biomaterials; for this property, they are applied in tissue engineering for the regeneration and repair of bone tissue [3]. Bioglass was developed by Larry L. Hench in 1969. It is a silica-based second-generation biomaterial that is inorganic, biodegradable, and biocompatible [4]. Hench defined bioglass as a substance that has the unique capacity to chemically attach to the mineral phase of bone tissue by forming its hydroxyapatite layer. Based on cumulative evidence, bioglass is said to have both osteoconductive and osteostimulative properties [5–7]. It was reported that bioglass exceeded materials such as hydroxyapatite and others, demonstrating its superior osteostimulating characteristics [8]. Additionally, in vitro and in vivo tests have linked it to angiogenesis, which indicates that bioglass helps in the vascularization of neoformed bone tissue [9]. Bioactive glass has been used in a wide range of dental [10] and orthopedic applications as a synthetic substitute for bone-grafting materials for the last couple of decades [11].

In 2005, the FDA approved the use of bioglass and bioglass-based products as bioactive materials with osteostimulating properties. Moreover, bioglass, which contains 45–50% silicon, can bind to soft tissues such as tympanic membranes (eardrums), muscles, nerve tissue, and skin. Due to this property, bioglass has also found tissue engineering applications for soft tissues in wound healing and radiotherapy [12, 13]. Bioglass has also found applications in the field of drug delivery systems with the ability to deliver therapeutic agents, antibiotics, and growth factors to aid tissue repair and other biomolecules [14–16]. In this chapter, we attempt to dive further into all the aspects of bioglass and discuss the mechanism of bioactivity and the resulting cellular response. The chapter also aims

DOI: 10.1201/9781003545712-6

to describe various ways in which the bioactivity of the bioglass can be regulated to tailor it for use in different physiological environments.

Bioglass as Bioactive Material

Bioactive materials are described as materials with the ability to develop a hydroxyapatite layer whereas bioactivity refers to a material's ability to form a chemical bond with bone tissue [17]. The formation of the HA layer on the material surface in dissolution media like SBF is a test for confirmation of the bioactivity of biomaterials [18]. Bioglass has been shown to develop a mineral analogous to the bone mineral phase when immersed in SBF, this allows the bioglass to bind to the bony apatite and collagen in the calcified bone tissue, imitating the interaction between the organic and inorganic phase of bone tissue. As a result, it is classed as a bioactive substance.

Bioactivity Mechanism

The mechanism of hydroxyapatite formation by the bioglass was described by Hench et al. [19] in their model of degradation of 45S5 bioglass. Submerging bioglass in SBF solution causes silica groups to hydrolyze through the rapid exchange of monovalent and divalent cations, such as Na^+ and Ca^{2+}, with H^+ or H_3O^+ in the solution to form silanols (Si-OH) at the glass-solution interface. As a result, the concentration of H^+ ions in the solution decreases, and the pH rises. The reaction continues, producing soluble silica in the form of silicic acid ($Si(OH)_4$) [20], which destroys the glass network and speeds up the reaction.

Released silica polymerizes and condenses on the bioglass surface, resulting in a silica-rich gel [21]. Calcium and phosphorus migrate to the gel's surface and condense into an amorphous HA layer, which crystallizes when it combines with hydroxyl or carbonate ions in the SBF solution, finally becoming hydroxy carbonate apatite (HCA) (Figure 6.1) [22]. At this stage, the bond and between the bioglass surface and bone tissue is possible. Following HCA formation, the biological mechanism involves adsorption and desorption of growth factors, followed by migration

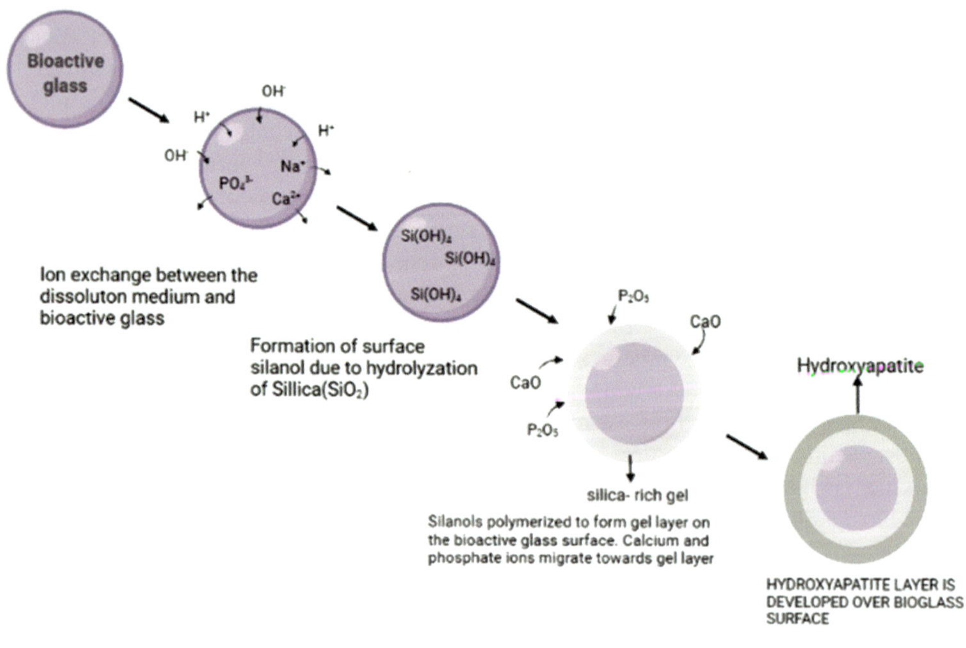

FIGURE 6.1 The mechanism of hydroxyapatite formation on a bioglass surface under physiological conditions (created using Biorender).

of macrophages and stem cells to the bioglass surface, their proliferation and differentiation into osteoblasts and osteocytes, and deposition of collagen-based extracellular matrix, which mineralizes to form new bone tissue while the glass dissolves over time [23].

Hench described different types of physicochemical interactions between the glass surface and bone tissue. The primary interaction included a strong covalent-ionic trans-ester bond formed by a phosphate group with organic and inorganic components of the bone matrix.

An electrostatic bond between the positively charged amino acids in collagen and the negatively charged glass surface. They also predicted other types of interactions, such as hydrogen bonding between the hydroxylated glass surface and carbonyl groups in amino acids present in collagen [24].

Cellular Response to Ionic Dissolution Product

As mentioned earlier, bone tissue undergoes a remodeling process throughout life; this property has been correlated with the bioavailability of the appropriate biomolecules, growth factors, and inorganic ions such as calcium, phosphorus, and silicon (all constituents of bioglass) along with other elements [25]. The bioavailability of these ions in the in vitro and in vivo systems increases due to the dissolution of glass, and therefore, it elicits a response from native cells. Calcium stimulates extracellular matrix mineralization and mitigates osteoblast proliferation and differentiation by regulating gene expression of growth factors like IGF-1 and IGF-2 [26, 27]. It has previously been suggested that the presence of phosphate ions is essential for osteoblasts to form calcium phosphate deposition and extracellular mineralized matrix [28]. By controlling the gene expression of matrix Gla protein (MGP), inorganic phosphorus controls bone formation in osteoblasts [29]. Silicon is involved in the osteoblast differentiation, bone calcification [30] production of collagen 1 [31], and osteoblast vacuole formation [32]. It is also thought to affect the biomineralization of bone tissue by affecting the incorporation of other trace elements [33].

Bioglass is associated with the upregulation of gene expression in osteoblasts, which regulates the maturation and proliferation of osteoblasts (Figure 6.2). It was reported that utilizing C-microarray analysis confirmed the role of ionic products of bioglass dissolution in the up-regulation of numerous genes expressed during osteoblast proliferation and differentiation [34]. Hench identified genes that were activated or upregulated by ionic dissolution products of bioglasses in primary human osteoblasts. This includes genes involved in DNA synthesis, repair, and recombination, such as HMG 1 and RFC 38, transcription factors, and cell cycle regulators. regulators, including cyclin D1, cyclin-dependent kinase inhibitor 1 (CDKN1A), and cyclin-K; apoptosis regulators such as calpain and DAD1; cell surface receptors, including CD44 marker protein (a specific phenotypic marker of osteocyte differentiation) and integrin β1; growth factors such as insulin-like growth factor II (IGF-II) and vascular endothelial growth factor (VEGF); signal transduction molecules like MAPKAP kinase-2; and extracellular matrix compounds including matrix metalloproteinases 2

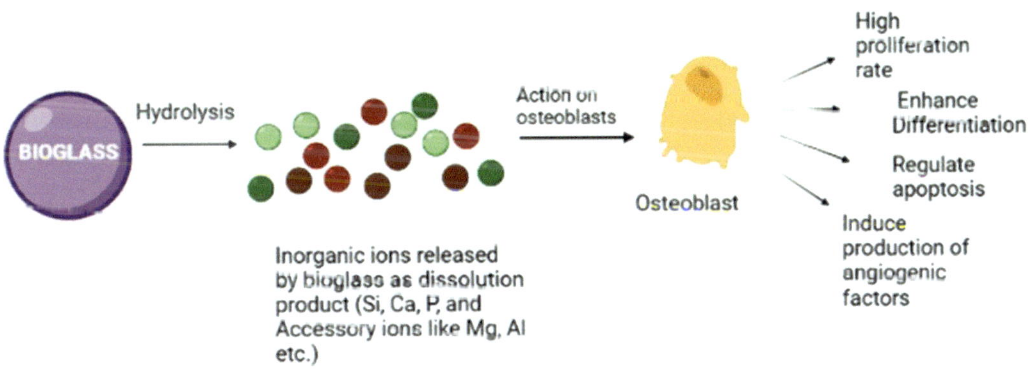

FIGURE 6.2 Cellular response to dissolution product of bioglass under physiological conditions (created using Biorender).

(MMP-2) and matrix metalloproteinase 14 precursor (MMP-14) and many more genes involved in osteoblast proliferation and differentiation [35]. The reported effect of bioglass-treated cell culture medium on the cell growth cycle of osteoblasts showed complete cell cycle growth in 2 days, suggesting a shorter S-G2 phase and faster cell cycle resulting in high osteoblast proliferation in culture medium [36]. Comparable results were seen in a study that reported increased alkaline phosphatase (ALP) activity until day 6 of cell culture with bioglass and then an increase in osteocalcin synthesis and increased number of osteoblasts in the growth phase of the cell cycle and reduced apoptotic activity by day 12 of culture [37].

Factors Regulating the Bioactivity of Bioglass

Biodegradable scaffolds are ideal for tissue engineering, and their degradation rate should correspond to that of tissue regeneration [38]. Therefore, it is necessary to know the degradation/dissolution kinetics of the scaffold to optimize it and achieve a proper response from the tissue. Bioglass attributes its bioactivity to many factors that are interconnected in a complex manner. Few properties of the bioglass and its surrounding environment have been established to play a key role in the dissolution of bioglass and can be broadly classified as chemical, morphological, or physical factors. The predominantly defining factors of bioactivity of bioglass have been discussed briefly.

Ionic Composition

This is perhaps the most important factor in determining the bioactive nature of bioglass. The ternary plot graph used for the design of potential bioglasses and glass-ceramics based on the percentage weight of each component in the system is an advantageous tool to predict the bioactive behavior of glass compositions [39]. According to the plot, only the bioglass with composition within the bioactivity boundary will develop an HA layer and be able to bind with bone tissue. Bioglass with bioactivity too low or too high will not bind to the bone tissue [40]. The essential compositional characteristics required to make the surface of bioglass bioactive are: SiO_2 concentration should be 60% mol or less in bioglass, and high CaO and Na_2O content [38].

Based on the ternary plot for the bioactive nature of bioglass, it is clear that the bioglass with 50–60% mol shows the optimum bioactivity. High calcium content ensures the formation of a large number of non-bonding oxygen (oxygen that does not participate in Si-O-Si polymer linkage) by bonding with silica and disrupting the silica network in bioglass, reducing network connectivity (cross-linking between silica monomers), which accelerates the dissolution of bioglass. The network connectivity of bioglass should be between 2.0 and 2.6 for an ideal response [41, 42]. It was reported that as the calcium-to-phosphate concentration ratio decreased (calcium is substituted for phosphorus), the glass network got denser (network connectivity increased), and pore size diminished, lowering bioactivity overall [23]. This could be explained by the formation of Si-O-P crosslinks in the glass network [43].

A high Ca/P concentration ratio is considered important for ensuring the formation of an HA layer with a Ca/P ratio of 1, which is analogous to hydroxyapatite. Araújo et al. reported the formation of a calcium-deficient (Ca/P is 1/1) layer on the bioglass surface with a Ca/P ratio of 2.3, which they attributed to the high phosphorus content of the bioglass [44]. Bioglass can be adjusted to obtain additional features that complement and enhance the bioglass's present properties. Bioglass is regularly doped with various other metals, for example, metals with antibiotic effects like silver, to prevent post-operation infection. Trace metals like copper, zinc, and boron have also shown improved angiogenesis and osteogenesis [45], and titanium that improves cell metabolism [46]. Some metal salts, such as iron oxides (magnetite), have also been used to create magnetically active bioglass for therapeutic responses, including targeted drug delivery [47]. The addition of these metal ions disrupts the silicon network, just like sodium and calcium. Thus affecting the thermal treatment temperature and release of silicon in the dissolution media [48]. The addition of elements like copper, zinc, aluminum, strontium, and magnesium enhances the bioactivity of bioglass [49]. However, boric acid is known to inhibit the dissolution rate of bioglass [50].

Morphological Factors

Morphological factors include the physical dimensions (specific surface area, particle size, pore size) surface chemistry (porosity/pore volume), and crystalline nature of the bioglass particles.

Specific Surface Area to Volume of Dissolution Media Ratio (SA/V Ratio)

Due to better diffusion mechanics, bioglass particles with a larger surface area dissolve faster than those with a smaller surface area. As the cations in the bioglass exchange with the ions in the media, a rise in calcium and phosphorus concentrations is observed in the media, resulting in co-precipitation of the hydroxyapatite. The larger surface area also enhanced the hydroxylation of the silicon network in bioglass particles and the formation of surface silanol. As a result, providing more nucleation sites for HCA deposition. It was previously reported that compared to lower surface area to volume of dissolution media (SA/V) ratios, greater SA/V ratios resulted in a faster rate of pH rise pH, indicating a high dissolution rate, and subsequently dropped rapidly, suggesting a significantly higher HA layer deposition rate [51].

Particle Size

Powders with spherical morphology and small size intrinsically have a large specific surface area, causing the process of material deterioration and resorption, as well as deposition of the Ca-P layer on the glass surface, to occur more quickly [52]. Sepulveda et al. used melt-derived 45S5 and sol–gel 58S bioglass to study the effect of the particle size on the bioactivity and found that particles with smaller sizes had better dissolution rates besides the differences in the specific surface area and composition of two bioglass samples. This indicates that variation in particle size is a defining factor in determining the dissolution rates of bioglass particles [53].

Porosity

The surface texture of bioglass plays a major role in determining its degradation rate. As bioglass is processed, it develops pores on its surface that give the bioglass a rough texture. Bigger pore size and higher porosity (pore volume) on the surface facilitate a higher rate of ion exchange and, therefore, increase the pH and Ca^{2+} concentration, which reduces the HCA induction time and increases the dissolution rate of the bioglass. Different gel glasses form HA layers at different rates. It was shown by Pereira et al. that for HA layer deposition on a bioglass surface, the particle with the largest pore diameter had the highest bioactivity, whereas silica gel with the smallest pore size had a slower degradation rate [54]. It was previously reported that results on the apatite-mineralization ability of Mesoporous bioglass, which has a pore diameter between 2–50 nm and high pore volume, was significantly higher and more dependent on its surface morphology than conventional melt-derived nano-bioglass [55].

Thermal Treatment and Crystallinity

Bioglass is subjected to calcination heated at an elevated temperature for phase stabilization and development of a polycrystalline microstructure. The sintering process to remove nitrates and alkoxides (and other by-products) from the bioglass requires a temperature above 500°C (subjected to precursors used); this is very close to phase-transition temperature (600°C) for glass material; above 600°C the amorphous glassy structure of the bioglass is lost, and the bioglass becomes crystalline. Such bioglass particles have higher mechanical strength than amorphous bioglass but lower bioactivity [56]. This is due to the surface morphology of the bioglass particle, such as surface area, pore size, and pore volume, decreasing with increasing heat-treatment temperature, as recorded by Pereira et al. [54] due to partial crystallization. These changes restrict the diffusion of ions into the solution, becoming a limiting factor, and the nucleation rate decreases. At even higher temperatures, ~1000°C completely crystallizes, which reduces bioactivity greatly. Therefore, the sintering temperature is advised to be kept low to avoid mass aggregation and fusion of particles as well.

Temperature and pH

Physical parameters of the local environment, like temperature and pH, have a positive effect on bioglass dissolution. Fagerlund et al. reported that the pH and average Si concentration in the solution increase with an increase in temperature, concluding that with the temperature rise, the rate of ion dissolution of glass also increased; this is because high temperature provided the activation energy to break Si–O bonds in the glass network [57]. Similarly, in the case of pH, the deposition of the HA layer is directly proportional to the local pH of the environment. The 45S5 bioglass dissolves faster at acidic pH, and the deposition rate of HA increases with an increase in pH due to the dissolution product [58]. Bingel et al. reported that at pH 5, the dissolution of bioglass was faster owing to the higher availability of protons (H^+ and H^{3O+}) in the medium, resulting in HA layer formation in a brief time as compared to at pH 7.3 and pH 9.0, where the apatite layer took longer time to form because of reduced ion dissolution rate [59].

Control of Bioactivity

Studies have repeatedly confirmed that due to the dissolution of bioglass, the pH of the dissolution medium rises sharply during the initial stages before then decreasing slowly and stabilizing over extended periods. Such a sudden and sharp increase in pH above the physiological pH of 7.3 must affect the biochemistry of the local environment, causing the denaturation of proteins, affecting cellular metabolism, and causing cell death [60]. Rismanchian et al. compared the cytotoxicity of micro-bioglass and nano-bioglass in different concentrations, and the data demonstrated that a higher concentration of bioglass caused higher cytotoxicity. Also, The nano-bioglass had higher cytotoxicity at the same concentration as micro-bioglass due to its high bioactivity, and the incubation period did not affect the results; this suggests that the cytotoxicity was dependent on the initial dissolution of bioglass [61].

It has also been reported that 58S bioglass showed slight cytotoxicity in earlier stages of in vitro cell study as compared to the 63S and 73S nano-bioglass [62]. Therefore, it can be seen how it is necessary to regulate and optimize the degradation of bioglass in a time-dependent manner to achieve high biocompatibility and proper bone adhesion. As pointed out earlier, the bioactivity mechanism is dependent on many interrelated factors, which makes it difficult to design and engineer bioglass particles with tailored dissolution kinetics to regulate bioactivity. It is achieved through modification of the surface morphology of the bioglass particles by several methods that can be classified into two – synthesis techniques (sol–gel processing and use of catalyst) and post-synthesis (encapsulation of bioglass) modifications. Below are some examples of how these routes are utilized to regulate bioactivity.

Sol–Gel Techniques and Nanoparticles

Bioglass is made in two main ways: melt-quench synthesis and sol–gel synthesis. Melt quench is a conventional technique used by hench to develop the first bioglass, where precursors are added together and heated at high temperatures to form bioglass [63]. Although it is still widely used today, the limitations of this method include the production of bulk material with micro-sized particles, which reduces the bioactivity potential of bioglass, as explained earlier. The method allows a limited range for silicon concentration in bioglass, and controlling the physical parameters like surface area, particle size, etc., of the particles is difficult. Maintaining partial crystallization of bioglass particles during sintering is also a challenge as melt quench requires a calcination temperature of around ~1400°C, which is high enough to cause crystallization in the bioglass. In the sol–gel method, the precursors to the individual components are dissolved to generate sol precursors like TEOS-a precursor for silica hydrolyses and condenses to form colloidal nanoparticles that later, under acidic conditions, form an inorganic silica network that incorporates the phosphate and calcium oxide produced by their respective precursors.

Thus, the final BG particles formed are also on the nanoscale. It also requires much lower processing temperatures (400 ~ 700°C) compared to the melt-quenching technique. Unlike dense melt-quenched glass, sol–gel nanoparticles contain inherent nanoporosity [64] and high surface area, and good biodegradability. The sol–gel method has received a lot of attention in the field of tissue engineering since then, and much research has been conducted to develop different methods of sol–gel processing, producing hosts of sol–gel-based techniques for the synthesis of bioglass that are available [65], tailored to develop bioglass with nanoparticles, nanofibers, nanofilms, and various other morphological structures [66]. Nanoparticles, unlike bulk materials, are highly dependent on their dimensions and specific surface area for their biological response. Sol–gel methods allow for regulating these physical parameters depending on the type of sol–gel method, the solvent to the molar concentration to precursor ratio (water/ethanol to TEOS), and the pH and temperature of the solution [67]. Improvements in sol–gel techniques have allowed for specific control of pore morphology using techniques like micro-emulsion [68] and template-assisted sol–gel synthesis, the morphology is modulated in a template species and concentration-dependent manner [69].

Bioglass size can be controlled by regulating the processing parameters such as the number of catalysts like ammonia, and the water-to-TEOS ratio used to form a sol [70]. It has been established that the water/alcohol to TEOS ratio regulates the hydration and condensation reaction during sol–gel synthesis, which regulates the morphology of the bioglass. A higher ethanol:TEOS ratio produces nanoparticles with a bigger pore size due to higher rates of hydrolyzation vs condensation [71]. The pH of sol–gel and the drying temperature are also determining factors in the development of the microstructure of the bioglass nanoparticle. When sol–gel with a pH of more than 2 is dried at 40°C or above, it maintains mesoporosity and higher pH and temperature increase the pore size [72].

Effect of Catalyst

As mentioned earlier, the sol–gel process involves the hydrolysis of TEOS using acid as a catalyst. Most protocols utilize strong inorganic acids like nitric acid, hydrochloric acid, and sulfuric acid to catalyze the sol–gel reactions. Though health and environmental concerns [73] related to the use of these chemicals have led to the modification of sol–gel methods that substitute the inorganic acids with organic acids like citric acid or lactic acid [74–76], proving that these can be utilized as catalysts too. They are biodegradable and are required in very low concentrations compared to inorganic acids [77]. It has been recorded that those organic acids modified the morphology of the bioglass particle depending on the acid species and its concentration. Lei et al. led a comparative study on the effects of different organic acids (acetic acid, lactic acid, and citric acid) on bioglass morphology and reported that the bioglass particles had a regular spherical shape at certain citric acid and lactic acid concentrations, whereas at lower concentrations lactic acid produced irregular morphology. A similar effect was observed with lower and higher concentrations of citric acid. Acetic acid, on the other hand, produced irregular morphology at all concentrations.

In the same study, the effect of the concentration of acids was also studied, and the results suggested that it affected the particle size distribution of bioglass particles. Increasing acetic acid or citric acid concentration produced bioglass particles with a narrow particle size distribution, whereas increasing lactic acid concentration showed a wider particle size range [78]. Chakraborty et al conducted another study involving acetic acid, formic acid, and nitric acid as catalysts and concluded that acetic acid produced bioglass particles with superior porosity compared to others [79]. Dang et al. utilized a modified sol–gel technique and successfully synthesized mesoporous bioglass with the composition of $70SiO_2$-30CaO (mol.%) without using acid catalysts. The bioglass showed similar bioactive behavior to regular mesoporous bioglass. Although this approach is innovative, it requires more research to develop methods to synthesize bioglass with lower silicon concentrations [80].

Preconditioning of Bioglass

Preconditioning is a method used to combat the sudden pH rise in the culture medium. In this method, the bioglass particles are dissolved in a culture medium (cell culture medium, SBF, DMEM, etc.) for a definite period of time before incubating the cells. This way, the cells do not experience the cytotoxic effects due to the initial rise in the pH. Several studies have been conducted to precondition different types of bioglasses using various mediums, which are summarized elsewhere [81]. A study regarding the cytotoxic effect of 45S5-bioglass, preconditioned for different time periods, on bone marrow stem cells (BMSCs) showed that the cell survivability enhanced as the preconditioning time increased, and bioglass granules preconditioned for 72 h showed the highest cell density apart from bioglass-free culture media as shown in Figure 6.3 [82]. El-Ghannam et al. reported that osteoblasts showed a better response to bioglass preconditioned in SBF. Longer preconditioning time enhanced the response further [83]. The time required for preconditioning depends on the dissolution behavior of a particular bioglass as it will decide the final pH change in the solution; dense bulk particles dissolve slower than porous nanoscale particles and require a lower amount of time. Similarly, bioglass with high phosphorus content requires a lower amount of time to precondition due to a slow degradation rate than bioglass with higher silicon or calcium content [84].

Encapsulation with Polymer Matrix

Bioceramics (bioglass and other glass-ceramics), in general, is not ideal for load-bearing applications. When bioglass and glass ceramics are combined with biopolymers to develop composite materials, the mechanical characteristics and surface area of the bioglass and glass ceramics are increased [85]. It enhances the apatite formation on the bioglass because it provides additional nucleation sites for HA deposition. The polymers also decrease the risk of particle aggregation by creating a barrier around the particle. Natural polymers like collagen, gelatin, chitosan, and alginate and synthetic polymers such as co-poly(lactic-glycolic acid) (PLGA), polyethylene glycol

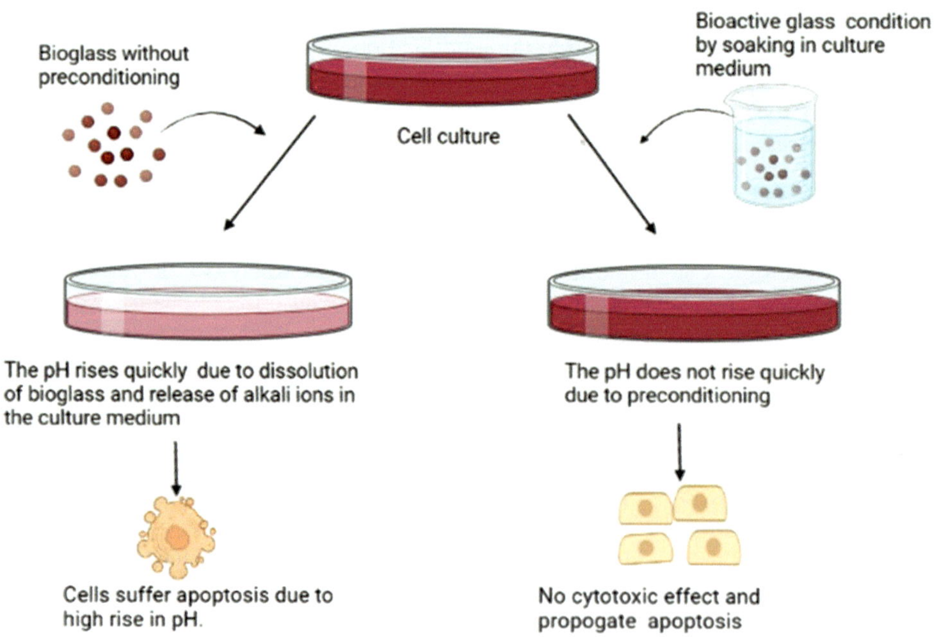

FIGURE 6.3 Bioglass without preconditioning raises pH quickly and causes cytotoxicity. Preconditioned bioglass does not raise pH quickly, and the cytotoxic effect is negated (created using Biorender).

(PEG), polycaprolactone (PCL) have been employed for the development of the bioglass/polymer composites [86].

Chitosan is a biopolymer with unique biological characteristics. Chitosan forms interconnected porous structures, allowing for the migration of cells through the scaffold [87]. Chitosan has been used for a broad number of different applications in biomedicine [88] and in the field of tissue engineering as scaffolds for hard and soft tissues [89–91]. In vitro evaluation of bioactive behavior of chitosan-bioglass composites revealed that 40% by volume of bioglass containing chitosan composites had uniform bioactivity [92]. Porous chitosan/nano-bioglass composite formed by the freeze cast method, on analysis, revealed the different nano-bioglass concentrations in the scaffold (10%, 30%, 50%) did not affect the rate of HA formation [93]. They observed that bioglass content in the scaffold produced a rough surface as the bioglass was partly exposed, negating the effect of polymer on the rate of HA layer formation on the scaffold.

Biomineralization studies conducted by Peter et al. showed a higher amount of mineral deposits on a mesoporous chitosan–gelatin nanocomposite scaffold incorporated with sol–gel NBG, which increased with increasing time of incubation attributed to the accelerated overall dissolution rate of bioglass [94]. Shalumon et al. reported the polycaprolactone (PCL)/chitosan composite fibrous scaffold incorporated with nano-bioglass particles had higher protein absorption, which can increase cell adhesion and biocompatibility of the scaffold [95]. Alginate is an anionic, hydrophilic polysaccharide made up of blocks of (1–4)-linked-D-mannuronic acid (M) and -L-guluronic acid (G) monomer. It offers a wide range of biomedical and tissue engineering uses due to its exceptional biocompatibility and biodegradability [96]. Shi et al. showed the influence of the additional amount of sodium alginate (SA) on the microstructure, porosity, pore size, swelling ratio, degradation profile, mechanical properties, and mineralization ability of bioglass-chitosan-alginate (BCA) composite, the BCA composite scaffolds developed a three-dimensional interconnected network structure with high porosity and pore size, which tends to facilitate the degradation of the scaffold and improve biomineralization. However, with the increase of SA concentration in the scaffold, the porosity and pore size of BCA gradually reduced. This significantly affected the polymer degradation [97]. Polyethylene glycol (PEG) is a highly biocompatible and biodegradable synthetic polymer. It possesses a straight polyether diol chain that has hydroxyl groups that allow PEG to bind to proteins covalently. This property enhances the protein adsorption on the surface of the polymer [98]. Kumar et al. conducted a study regarding the coating of nano-bioglass and nanohydroxyapatite particles with PEG and concluded that PEG increased the overall porosity of the scaffold, and it can regulate the rate of dissolution of bioglass in a concentration-dependent manner, i.e., with increasing PEG content the scaffold showed slower degradation [99].

The application of PEG as a surfactant in the preparation of calcium-silicate ceramics enhanced the surface morphology of the scaffold. The degradation of PEG during thermal treatment inhibited the densification of the particle. As a result, the surface area, pore volume, and porosity increased as the concentration of PEG increased until the increment of PEG concentration was more than 10% w/v. After that, the density of materials started increasing [100].

Jadidi et al. developed a PLGA/bioglass composite with different concentrations of PLGA, and analysis revealed that based on the Archimedes porosimetry method. The density of the PLGA coatings deposited was proportionate to the increasing polymer concentrations, which enhanced the total particle size, but the mean porosity levels of scaffold declined. The data showed that PLGA effectively decreased the biodegradation of the NBG-PEG scaffold, which could have also accounted for the high biocompatibility of the scaffold [101].

CONCLUSION

In this chapter, we discussed the mechanism of biomineralization of bioglass, cellular response to the ionic dissolution product of bioglass, and major factors regulating the bioactivity of bioglass. We also discussed how some methods like sol–gel synthesis, use of catalyst, and polymeric encapsulation of bioglass can be of importance in regulating various factors that influence overall bioactivities. The use of a particular catalyst has shown morphological modifications, but it has limited effect compared to that of using the right sol–gel technique to design and fabricate the bioglass nanoparticles with precision parameter control and polymeric encapsulation. The versatility of the sol–gel technique allows for the configuration of the technique to better adapt to compositional challenges in the synthesis of bioglass nanoparticles. The polymers can be compared for their cellular responses and mechanical properties to enhance the range of applications of bioglass better. The aforementioned morphological and in vitro biomineralization studies of these scaffolds suggest that depending on the polymeric species and concentration of the polymer in the composite, the morphological properties of the scaffold can vary, which can be used to regulate the biomineralization process.

GRAPHICAL ABSTRACT

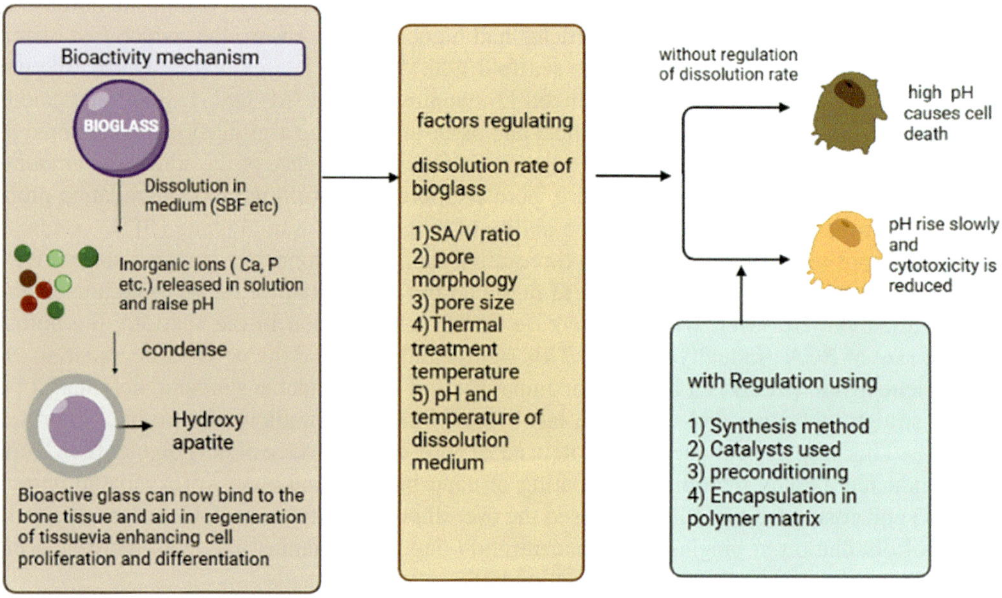

ABBREVIATIONS

ALP	Alkaline phosphatase
BCA	Bioglass chitosan alginate
BMSC	Bone marrow stem cells
Ca/P	Calcium/phosphorus
CaO	Calcium oxide
CDKN1A	Cyclin-dependent kinase inhibitor 1
DMEM	Dulbecco's modified eagle medium
DAD1	Defender against cell death 1
FDA	Food and Drug Administration
HA	Hydroxyapatite
HCA	Hydroxycarbonate apatite
HMG1	High mobility group box 1 protein

IGF1	Insulin-like growth factor 1
IGF 2	Insulin-like growth factor 2
MAPKAP kinase 2	Mitogen-activated protein kinase-activated protein kinase 2
MBG	Mesoporous bioglass
MGP	Matrix Gla protein
MMP-2	Matrix metalloproteinase 2
MMP-14	Matrix metalloproteinase 14
Na_2O	Sodium oxide
NBG	Nano-bioglass
PCL	Polycaprolactone
PLGA	Co-poly(lactic-glycolic acid)
PEG	Polyethylene glycol
RFC	Replication factor C
SA	Sodium alginate
SBF	Simulated biological fluid
S-G2 phase	Synthesis-growth 2 phase
Si	Silicon
Si-O-Si bonds	Silicon–oxygen–silicon bonds
TCP	Tricalcium phosphate
TEOS	Tetraethyl orthosilicate
VEGF	Vascular endothelial growth factor

REFERENCES

1. R. Florencio-Silva, G.R.D.S. Sasso, E. Sasso-Cerri, M.J. Simões, P.S. Cerri, Biology of bone tissue: Structure, function, and factors that influence bone cells, *Biomed Res. Int. 2015* (2015). https://doi.org/10.1155/2015/421746.
2. L.L. Hench, I. Thompson, Twenty-first century challenges for biomaterials, *J. R. Soc. Interface. 7* (2010) S379. https://doi.org/10.1098/RSIF.2010.0151.FOCUS.
3. H. Qu, H. Fu, Z. Han, Y. Sun, Biomaterials for bone tissue engineering scaffolds: A review, *RSC Adv. 9* (2019) 26252–26262. https://doi.org/10.1039/C9RA05214C.
4. L.L. Hench, The story of bioglass, *J. Mater. Sci. Mater. Med. 17* (2006) 967–978. https://doi.org/10.1007/S10856-006-0432-Z.
5. P. Ducheyne, A. El-Ghannam, I. Shapiro, Effect of bioactive glass templates on osteoblast proliferation and in vitro synthesis of bone-like tissue, *J. Cell. Biochem. 56* (1994) 162–167. https://doi.org/10.1002/JCB.240560207.
6. S.A. Ferreira, G. Young, J.R. Jones, S. Rankin, Bioglass/carbonate apatite/collagen composite scaffold dissolution products promote human osteoblast differentiation, *Mater. Sci. Eng. C. 118* (2021) 111393. https://doi.org/10.1016/J.MSEC.2020.111393.
7. N.W. Marion, W. Liang, G.C. Reilly, D.E. Day, M.N. Rahaman, J.J. Mao, Borate glass supports the in vitro osteogenic differentiation of human mesenchymal stem cells, *Mech. Adv. Mater. Struct. 12* (2005) 239–246. https://doi.org/10.1080/15376490590928615.
8. W.C.A. Vrouwenvelder, C.G. Groot, K. de Groot, Histological and biochemical evaluation of osteoblasts cultured on bioactive glass, hydroxylapatite, titanium alloy, and stainless steel, *J. Biomed. Mater. Res. 27* (1993) 465–475. https://doi.org/10.1002/JBM.820270407.
9. A.A. Gorustovich, J.A. Roether, A.R. Boccaccini, Effect of bioactive glasses on angiogenesis: A review of in vitro and in vivo evidences, *Tissue Eng. Part B Rev. 16* (2010) 199–207. https://doi.org/10.1089/TEN.TEB.2009.0416.
10. H.E. Skallevold, D. Rokaya, Z. Khurshid, M.S. Zafar, Bioactive glass applications in dentistry, *Int. J. Appl. Glass Sci. 20* (2019) 5960. https://doi.org/10.3390/IJMS20235960.
11. J.R. Jones, D.S. Brauer, L. Hupa, D.C. Greenspan, Bioglass and bioactive glasses and their impact on healthcare, *Int. J. Appl. Glass Sci. 7* (2016) 423–434. https://doi.org/10.1111/IJAG.12252.
12. F. Baino, G. Novajra, V. Miguez-Pacheco, A.R. Boccaccini, C. Vitale-Brovarone, Bioactive glasses: Special applications outside the skeletal system, *J. Non-Cryst. Solids. 432* (2016) 15–30. https://doi.org/10.1016/J.JNONCRYSOL.2015.02.015.

13. J. Wilson, A.E. Clark, E. Douek, J. Krieger, W.K. Smith, J.S. Zamet, Clinical applications of bioglass implants, *Bioceramics.* (1994) 415–422. https://doi.org/10.1016/B978-0-08-042144-5.50069-8.
14. A.M. El-Kady, M.M. Farag, Bioactive glass nanoparticles as a new delivery system for sustained 5-fluorouracil release: Characterization and evaluation of drug release mechanism, *J. Nanomater. 2015* (2015). https://doi.org/10.1155/2015/839207.
15. W. Li, P. Nooeaid, J.A. Roether, D.W. Schubert, A.R. Boccaccini, Preparation and characterization of vancomycin releasing PHBV coated 45S5 Bioglass®-based glass–ceramic scaffolds for bone tissue engineering, *J. Eur. Ceram. Soc. 34* (2014) 505–514. https://doi.org/10.1016/J.JEURCERAMSOC.2013.08.032.
16. N. Subhapradha, M. Abudhahir, A. Aathira, N. Srinivasan, A. Moorthi, Polymer coated mesoporous ceramic for drug delivery in bone tissue engineering, *Int. J. Biol. Macromol. 110* (2018) 65–73. https://doi.org/10.1016/J.IJBIOMAC.2017.11.146.
17. J. Sartuqui, A.N. Gravina, R. Rial, L.A. Benedini, L. Yahia, J.M. Ruso, P.V. Messina, Biomimetic fiber mesh scaffolds based on gelatin and hydroxyapatite nano-rods: Designing intrinsic skills to attain bone reparation abilities, *Colloids Surf. B Biointerfaces. 145* (2016) 382–391. https://doi.org/10.1016/J.COLSURFB.2016.05.019.
18. T. Kokubo, H. Takadama, How useful is SBF in predicting in vivo bone bioactivity?, *Biomaterials. 27* (2006) 2907–2915. https://doi.org/10.1016/J.BIOMATERIALS.2006.01.017.
19. L.L. Hench, D.L. Wheeler, D.C. Greenspan, Molecular control of bioactivity in sol-gel glasses, *J. Sol-Gel Sci. Technol. 13.* 1 (1998) 245–250. https://doi.org/10.1023/A:1008643303888.
20. M.N. Rahaman, D.E. Day, B. Sonny Bal, Q. Fu, S.B. Jung, L.F. Bonewald, A.P. Tomsia, Bioactive glass in tissue engineering, *Acta Biomater. 7* (2011) 2355–2373. https://doi.org/10.1016/J.ACTBIO.2011.03.016.
21. H. Ylänen, K.H. Karlsson, A. Itälä, H.T. Aro, Effect of immersion in SBF on porous bioactive bodies made by sintering bioactive glass microspheres, *J. Non-Cryst. Solids. 275* (2000) 107–115. https://doi.org/10.1016/S0022-3093(00)00245-3.
22. R. Borges, J. March, & A.C. Silva, Bioglass dissolution: a comparison between SBF solution and hydrolytic etching. 56 Brazilian Congress on Ceramics, 2012, Brazil.
23. H.K. Ting, S.J. Page, G. Poologasundarampillai, S. Chen, B. Yu, J.V. Hanna, J.R. Jones, Phosphate content affects structure and bioactivity of sol-gel silicate bioactive glasses, *Int. J. Appl. Glass Sci. 8* (2017) 372–382. https://doi.org/10.1111/IJAG.12322.
24. L.L. Hench, R.J. Splinter, W.C. Allen, T.K. Greenlee, Bonding mechanisms at the interface of ceramic prosthetic materials, *J. Biomed. Mater. Res. 5* (1971) 117–141. https://doi.org/10.1002/JBM.820050611.
25. N.J. Lakhkar, I.H. Lee, H.W. Kim, V. Salih, I.B. Wall, J.C. Knowles, Bone formation controlled by biologically relevant inorganic ions: Role and controlled delivery from phosphate-based glasses, *Adv. Drug Deliv. Rev. 65* (2013) 405–420. https://doi.org/10.1016/J.ADDR.2012.05.015.
26. S. Maeno, Y. Niki, H. Matsumoto, H. Morioka, T. Yatabe, A. Funayama, Y. Toyama, T. Taguchi, J. Tanaka, The effect of calcium ion concentration on osteoblast viability, proliferation and differentiation in monolayer and 3D culture, *Biomaterials. 26* (2005) 4847–4855. https://doi.org/10.1016/J.BIOMATERIALS.2005.01.006.
27. P.J. Marie, The calcium-sensing receptor in bone cells: A potential therapeutic target in osteoporosis, *Bone. 46* (2010) 571–576. https://doi.org/10.1016/J.BONE.2009.07.082.
28. M.G.M.G. Penido, U.S. Alon, Phosphate homeostasis and its role in bone health, *Pediatr. Nephrol. 27* (2012) 2039. https://doi.org/10.1007/S00467-012-2175-Z.
29. E.A. Abou Neel, W. Chrzanowski, J.C. Knowles, Biological performance of titania containing phosphate-based glasses for bone tissue engineering applications, *Mater. Sci. Eng. C. 35* (2014) 307–313. https://doi.org/10.1016/J.MSEC.2013.10.029.
30. E.M. Carlisle, Silicon: A possible factor in bone calcification, *Science. 167* (1970) 279–280. https://doi.org/10.1126/SCIENCE.167.3916.279.
31. D.M. Reffitt, N. Ogston, R. Jugdaohsingh, H.F.J. Cheung, B.A.J. Evans, R.P.H. Thompson, J.J. Powell, G.N. Hampson, Orthosilicic acid stimulates collagen type 1 synthesis and osteoblastic differentiation in human osteoblast-like cells in vitro, *Bone. 32* (2003) 127–135. https://doi.org/10.1016/S8756-3282(02)00950-X.
32. P. Valerio, M.M. Pereira, A.M. Goes, M.F. Leite, The effect of ionic products from bioactive glass dissolution on osteoblast proliferation and collagen production, *Biomaterials. 25* (2004) 2941–2948. https://doi.org/10.1016/J.BIOMATERIALS.2003.09.086.

33. W. Götz, E. Tobiasch, S. Witzleben, M. Schulze, Effects of silicon compounds on biomineralization, osteogenesis, and hard tissue formation, *Pharmaceutics. 11* (2019). https://doi.org/10.3390/PHARMACEUTICS11030117.
34. I.D. Xynos, A.J. Edgar, L.D.K. Buttery, L.L. Hench, J.M. Polak, Ionic products of bioactive glass dissolution increase proliferation of human osteoblasts and induce insulin-like growth factor II mRNA expression and protein synthesis, *Biochem. Biophys. Res. Commun. 276* (2000) 461–465. https://doi.org/10.1006/bbrc.2000.3503.
35. L.L. Hench, Genetic design of bioactive glass, *J. Eur. Ceram. Soc. 29* (2009) 1257–1265. https://doi.org/10.1016/J.JEURCERAMSOC.2008.08.002.
36. J.Y. Sun, Y.S. Yang, J. Zhong, D.C. Greenspan, The effect of the ionic products of Bioglass® Dissolution on human osteoblasts growth cycle in vitro, *J. Tissue Eng. Regen. Med. 1* (2007) 281–286. https://doi.org/10.1002/TERM.34.
37. I.D. Xynos, M.V.J. Hukkanen, J.J. Batten, L.D. Buttery, L.L. Hench, J.M. Polak, Bioglass 45S5 stimulates osteoblast turnover and enhances bone formation In vitro: implications and applications for bone tissue engineering, *Calcif. Tissue Int. 67* (2000) 321–329. https://doi.org/10.1007/S002230001134.
38. S. Srinivasan, R. Jayasree, K.P. Chennazhi, S.V. Nair, R. Jayakumar, Biocompatible alginate/nano bioactive glass ceramic composite scaffolds for periodontal tissue regeneration, *Carbohydr. Polym. 87* (2012) 274–283. https://doi.org/10.1016/J.CARBPOL.2011.07.058.
39. J.-A.N. Oliver, O. Akande, M. Ecker, Incorporation of novel elements in bioactive glass compositions to enhance implant performance, Current Concepts in Dental Implantology - From Science to Clinical Research [Working Title]. (2021). https://doi.org/10.5772/INTECHOPEN.99430.
40. J.R. Jones, Bioactive glasses, Bioceram. Clin. Appl. (2008) 266–283. https://doi.org/10.1533/9781845694227.2.266.
41. R. Hill, An alternative view of the degradation of bioglass, *J. Mater. Sci. Lett. 15* (1996) 1122–1125. https://doi.org/10.1007/BF00539955.
42. R.G. Hill, D.S. Brauer, Predicting the bioactivity of glasses using the network connectivity or split network models, *J. Non-Cryst. Solids. 357* (2011) 3884–3887. https://doi.org/10.1016/J.JNONCRYSOL.2011.07.025.
43. A. Tilocca, A.N. Cormack, Structural effects of phosphorus inclusion in bioactive silicate glasses, *J. Phys. Chem. B. 111* (2007) 14256–14264. https://doi.org/10.1021/JP075677O.
44. M. Araújo, M. Miola, G. Baldi, J. Perez, E. Verné, Bioactive glasses with low Ca/P ratio and enhanced bioactivity, *Materials. 9* (2016). https://doi.org/10.3390/MA9040226.
45. A. Hoppe, N.S. Güldal, A.R. Boccaccini, A review of the biological response to ionic dissolution products from bioactive glasses and glass-ceramics, *Biomaterials. 32* (2011) 2757–2774. https://doi.org/10.1016/J.BIOMATERIALS.2011.01.004.
46. W.C.A. Vrouwenvelder, C.G. Groot, K. de Groot, Better histology and biochemistry for osteoblasts cultured on titanium-doped bioactive glass: Bioglass 45S5 compared with iron-, titanium-, fluorine- and boron-containing bioactive glasses, *Biomaterials. 15* (1994) 97–106. https://doi.org/10.1016/0142-9612(94)90257-7.
47. N. Shankhwar, A. Srinivasan, Evaluation of sol-gel based magnetic 45S5 bioglass and bioglass-ceramics containing iron oxide, *Mater. Sci. Eng. C. 62* (2016) 190–196. https://doi.org/10.1016/J.MSEC.2016.01.054.
48. G.S. Lázaro, S.C. Santos, C.X. Resende, E.A. Dos Santos, Individual and combined effects of the elements Zn, Mg and Si on the surface reactivity of a SiO2·CaO·Na2O·P2O5 bioglass system, *J. Non-Cryst. Solids. 386* (2014) 19–28. https://doi.org/10.1016/J.JNONCRYSOL.2013.11.038.
49. B. Karakuzu-Ikizler, P. Terzioğlu, Y. Basaran-Elalmis, B.S. Tekerek, S. Yücel, Role of magnesium and aluminum substitution on the structural properties and bioactivity of bioglasses synthesized from biogenic silica, *Bioact. Mater.* 5 (2020) 66–73. https://doi.org/10.1016/J.BIOACTMAT.2019.12.007.
50. X. Lu, J. Kolzow, R.R. Chen, J. Du, Effect of solution condition on hydroxyapatite formation in evaluating bioactivity of B2O3 containing 45S5 bioactive glasses, *Bioact. Mater. 4* (2019) 207–214. https://doi.org/10.1016/J.BIOACTMAT.2019.05.002.
51. D.C. Greenspan, J.P. Zhong, G.P. LaTorre, Effect of surface area to volume ratio on in vitro surface reactions of bioactive glass particulates, *Bioceramics. 7* (1994) 55–60. https://doi.org/10.1016/B978-0-08-042144-5.50012-1.
52. C. Vichery, J.M. Nedelec, Bioactive glass nanoparticles: From synthesis to materials design for biomedical applications, *Materials. 9* (2016) 288. https://doi.org/10.3390/MA9040288.
53. P. Sepulveda, J.R. Jones, L.L. Hench, Characterization of melt-derived 45S5 and sol-gel-derived 58S bioactive glasses, *J. Biomed. Mater. Res. 58* (2001) 734–740. https://doi.org/10.1002/JBM.10026.

54. M.M. Pereira, A.E. Clark, L.L. Hench, Effect of texture on the rate of hydroxyapatite formation on gel-silica surface, *J. Am. Ceram. Soc. 78* (1995) 2463–2468. https://doi.org/10.1111/J.1151-2916.1995.TB08686.X.
55. C. Wu, J. Chang, Mesoporous bioactive glasses: Structure characteristics, drug/growth factor delivery and bone regeneration application, *Interface Focus. 2* (2012) 292–306. https://doi.org/10.1098/RSFS.2011.0121.
56. E. García-Gareta, M.J. Coathup, G.W. Blunn, Osteoinduction of bone grafting materials for bone repair and regeneration, *Bone. 81* (2015) 112–121. https://doi.org/10.1016/J.BONE.2015.07.007.
57. S. Fagerlund, P. Ek, L. Hupa, M. Hupa, Dissolution kinetics of a bioactive glass by continuous measurement, *J. Am. Ceram. Soc. 95* (2012) 3130–3137. https://doi.org/10.1111/J.1551-2916.2012.05374.X.
58. J. Cihlář, Hydrolysis and polycondensation of ethyl silicates. 1. Effect of pH and catalyst on the hydrolysis and polycondensation of tetraethoxysilane (TEOS), *Colloids Surf. A Physicochem. Eng. Asp. 70* (1993) 239–251. https://doi.org/10.1016/0927-7757(93)80298-S.
59. L. Bingel, D. Groh, N. Karpukhina, D.S. Brauer, Influence of dissolution medium pH on ion release and apatite formation of Bioglass® 45S5, *Mater. Lett. 143* (2015) 279–282. https://doi.org/10.1016/J.MATLET.2014.12.124.
60. A. Li, Y. Lv, H. Ren, Y. Cui, C. Wang, R.A. Martin, D. Qiu, In vitro evaluation of a novel pH neutral calcium phosphosilicate bioactive glass that does not require preconditioning prior to use, *Int. J. Appl. Glass Sci. 8* (2017) 403–411. https://doi.org/10.1111/IJAG.12321.
61. M. Rismanchian, N. Khodaeian, L. Bahramian, M. Fathi, H. Sadeghi-Aliabadi, In-vitro comparison of cytotoxicity of two bioactive glasses in micropowder and nanopowder forms, *Iran J. Pharm. Res. 12* (2013) 437. https://www.ncbi.nlm.nih.gov/pmc/articles/PMC3813266/.
62. V. Mortazavi, M. Mehdikhani Nahrkhalaji, M.H. Fathi, S.B. Mousavi, B. Nasr Esfahani, Antibacterial effects of sol-gel-derived bioactive glass nanoparticle on aerobic bacteria, *J. Biomed. Mater. Res. A. 94* (2010) 160–168. https://doi.org/10.1002/JBM.A.32678.
63. G. Kaur, O.P. Pandey, K. Singh, D. Homa, B. Scott, G. Pickrell, A review of bioactive glasses: Their structure, properties, fabrication and apatite formation, *J. Biomed. Mater. Res. A. 102* (2014) 254–274. https://doi.org/10.1002/JBM.A.34690.
64. S. Lin, C. Ionescu, K.J. Pike, M.E. Smith, J.R. Jones, Nanostructure evolution and calcium distribution in sol–gel derived bioactive glass, *J. Mater. Chem. 19* (2009) 1276–1282. https://doi.org/10.1039/B814292K.
65. K. Deshmukh, T. Kovářík, T. Křenek, D. Docheva, T. Stich, J. Pola, Recent advances and future perspectives of sol–gel derived porous bioactive glasses: A review, *RSC Adv. 10* (2020) 33782–33835. https://doi.org/10.1039/D0RA04287K.
66. S.W. et al., Review: Synthesis of nano bioactive glass or bioglass, *Rhazes: Green Appl. Chem. 10* (2020) 01–32. https://doi.org/10.48419/IMIST.PRSM/RHAZES-V10.23804.
67. J.R. Jones, Review of bioactive glass: From Hench to hybrids, *Acta Biomater. 9* (2013) 4457–4486. https://doi.org/10.1016/J.ACTBIO.2012.08.023.
68. X. Li, X. Chen, G. Miao, H. Liu, C. Mao, G. Yuan, Q. Liang, X. Shen, C. Ning, X. Fu, Synthesis of radial mesoporous bioactive glass particles to deliver osteoactivin gene, *J. Mater. Chem. B Mater. Biol. Med. 2* (2014) 7045–7054. https://doi.org/10.1039/C4TB00883A.
69. Q. Liang, Q. Hu, G. Miao, B. Yuan, X. Chen, A facile synthesis of novel mesoporous bioactive glass nanoparticles with various morphologies and tunable mesostructure by sacrificial liquid template method, *Mater. Lett. 148* (2015) 45–49. https://doi.org/10.1016/J.MATLET.2015.01.122.
70. K. Zheng, A.R. Boccaccini, Sol-gel processing of bioactive glass nanoparticles: A review, *Adv. Colloid Interface Sci. 249* (2017) 363–373. https://doi.org/10.1016/J.CIS.2017.03.008.
71. C.A. Milea, C. Bogatu, A. Duţă, The influence of parameters in silica sol-gel process , *Bull. Transilvania Univ. Braşov Series I: Eng. Sci. 4 (53)* (2011) 59–66 .
72. E.M. Valliant, C.A. Turdean-Ionescu, J.V. Hanna, M.E. Smith, J.R. Jones, Role of pH and temperature on silica network formation and calcium incorporation into sol–gel derived bioactive glasses, *J. Mater. Chem. 22* (2011) 1613–1619. https://doi.org/10.1039/C1JM13225C.
73. M. Poliakoff, J.M. Fitzpatrick, T.R. Farren, P.T. Anastas, Green chemistry: Science and politics of change, *Science. 297* (2002) 807–810. https://doi.org/10.1126/SCIENCE.297.5582.807.
74. J. Faure, R. Drevet, A. Lemelle, N. Ben Jaber, A. Tara, H. El Btaouri, H. Benhayoune, A new sol-gel synthesis of 45S5 bioactive glass using an organic acid as catalyst, *Mater. Sci. Eng. C. 47* (2015) 407–412. https://doi.org/10.1016/J.MSEC.2014.11.045.
75. Z. Hong, A. Liu, L. Chen, X. Chen, X. Jing, Preparation of bioactive glass ceramic nanoparticles by combination of sol-gel and coprecipitation method, *J. Non-Cryst. Solids. 355* (2009) 368–372. https://doi.org/10.1016/J.JNONCRYSOL.2008.12.003.

76. G.M. Luz, J.F. Mano, Preparation and characterization of bioactive glass nanoparticles prepared by sol-gel for biomedical applications, *Nanotechnology. 22* (2011). https://doi.org/10.1088/0957-4484/22/49/494014.
77. B.A.E. Ben-Arfa, H.R. Fernandes, I.M.M. Salvado, J.M.F. Ferreira, R.C. Pullar, Effects of catalysts on polymerization and microstructure of sol-gel derived bioglasses, *J. Am. Ceram. Soc. 101* (2018) 2831–2839. https://doi.org/10.1111/JACE.15450.
78. B. Lei, X. Chen, Y.H. Koh, Effects of acidic catalysts on the microstructure and biological property of sol–gel bioactive glass microspheres, *J. Sol-Gel Sci. Technol. 58*.3 (2011) 656–663. https://doi.org/10.1007/S10971-011-2441-8.
79. P.K. Chakraborty, J. Adhikari, P. Saha, Variation of the properties of sol–gel synthesized bioactive glass 45S5 in organic and inorganic acid catalysts, *Mater Adv. 2* (2021) 413–425. https://doi.org/10.1039/D0MA00628A.
80. T.H. Dang, T.H. Bui, E.V. Guseva, A.T. Ta, A.T. Nguyen, T.T.H. Hoang, X.V. Bui, Characterization of bioactive glass synthesized by sol-gel process in hot water, *Crystals. 10* (2020) 1–10. https://doi.org/10.3390/CRYST10060529.
81. F.E. Ciraldo, E. Boccardi, V. Melli, F. Westhauser, A.R. Boccaccini, Tackling bioactive glass excessive in vitro bioreactivity: Preconditioning approaches for cell culture tests, *Acta Biomater. 75* (2018) 3–10. https://doi.org/10.1016/J.ACTBIO.2018.05.019.
82. F. Hohenbild, M. Arango-Ospina, A. Moghaddam, A.R. Boccaccini, F. Westhauser, Preconditioning of bioactive glasses before introduction to static cell culture: What is really necessary?, *Methods Protoc. 3* (2020) 1–12. https://doi.org/10.3390/MPS3020038.
83. A. El-Ghannam, P. Ducheyne, I.M. Shapiro, Bioactive material template for in vitro, synthesis of bone, *J. Biomed. Mater. Res. 29* (1995) 359–370. https://doi.org/10.1002/JBM.820290311.
84. G.C. Reilly, S. Radin, A.T. Chen, P. Ducheyne, Differential alkaline phosphatase responses of rat and human bone marrow derived mesenchymal stem cells to 45S5 bioactive glass, *Biomaterials. 28* (2007) 4091–4097. https://doi.org/10.1016/J.BIOMATERIALS.2007.05.038.
85. D. Mohamad Yunos, O. Bretcanu, A.R. Boccaccini, Polymer bioceramic composites for tissue engineering scaffolds, *J. Mater. Sci. 43* (2008) 4433–4442. https://doi.org/10.1007/S10853-008-2552-Y.
86. R. Sergi, D. Bellucci, V. Cannillo, A review of bioactive glass/natural polymer composites: State of the art, *Materials. 13* (2020) 5560. https://doi.org/10.3390/MA13235560.
87. S. Vasiliu, M. Popa, M. Rinaudo, Polyelectrolyte capsules made of two biocompatible natural polymers, *Eur. Polym. J. 41* (2005) 923–932. https://doi.org/10.1016/J.EURPOLYMJ.2004.11.017.
88. V. Dodane, V.D. Vilivalam, Pharmaceutical applications of chitosan, *Pharm. Sci. Technolo. Today. 1* (1998) 246–253. https://doi.org/10.1016/S1461-5347(98)00059-5.
89. S. Ahmed, Annu, A. Ali, J. Sheikh, A review on chitosan centred scaffolds and their applications in tissue engineering, *Int. J. Biol. Macromol. 116* (2018) 849–862. https://doi.org/10.1016/J.IJBIOMAC.2018.04.176.
90. F. Croisier, C. Jérôme, Chitosan-based biomaterials for tissue engineering, *Eur. Polym. J. 49* (2013) 780–792. https://doi.org/10.1016/J.EURPOLYMJ.2012.12.009.
91. J.K. Francis Suh, H.W.T. Matthew, Application of chitosan-based polysaccharide biomaterials in cartilage tissue engineering: A review, *Biomaterials. 21* (2000) 2589–2598. https://doi.org/10.1016/S0142-9612(00)00126-5.
92. R. Ravarian, M. Craft, F. Dehghani, Enhancing the biological activity of chitosan and controlling the degradation by nanoscale interaction with bioglass, *J. Biomed. Mater. Res. A 103* (2015) 2898–2908. https://doi.org/10.1002/JBM.A.35423.
93. M. Pourhaghgouy, A. Zamanian, M. Shahrezaee, M.P. Masouleh, Physicochemical properties and bioactivity of freeze-cast chitosan nanocomposite scaffolds reinforced with bioactive glass, *Mater. Sci. Eng. C. 58* (2016) 180–186. https://doi.org/10.1016/J.MSEC.2015.07.065.
94. M. Peter, N.S. Binulal, S.V. Nair, N. Selvamurugan, H. Tamura, R. Jayakumar, Novel biodegradable chitosan-gelatin/nano-bioactive glass ceramic composite scaffolds for alveolar bone tissue engineering, *Chem. Eng. J. 158* (2010) 353–361. https://doi.org/10.1016/J.CEJ.2010.02.003.
95. K.T. Shalumon, S. Sowmya, D. Sathish, K.P. Chennazhi, S.V. Nair, R. Jayakumar, Effect of incorporation of nanoscale bioactive glass and hydroxyapatite in PCL/chitosan nanofibers for bone and periodontal tissue engineering, *J. Biomed. Nanotechnol. 9* (2013) 430–440. https://doi.org/10.1166/JBN.2013.1559.
96. J. Sun, H. Tan, Alginate-based biomaterials for regenerative medicine applications, *Materials 6* (2013) 1285–1309. https://doi.org/10.3390/MA6041285.

97. C. Shi, X. Hou, D. Zhao, H. Wang, R. Guo, Y. Zhou, Preparation of the bioglass/chitosan-alginate composite scaffolds with high bioactivity and mechanical properties as bone graft materials, *J. Mech. Behav. Biomed. Mater. 126* (2022) 105062. https://doi.org/10.1016/J.JMBBM.2021.105062.
98. V. Dave, A. Gupta, P. Singh, C. Gupta, V. Sadhu, K.R. Reddy, Synthesis and characterization of celecoxib loaded PEGylated liposome nanoparticles for biomedical applications, *Nano-Struct. Nano-Objects. 18* (2019). https://doi.org/10.1016/J.NANOSO.2019.100288.
99. P. Kumar, B.S. Dehiya, A. Sindhu, Synthesis and characterization of nHA-PEG and nBG-PEG scaffolds for hard tissue engineering applications, *Ceram. Int. 45* (2019) 8370–8379. https://doi.org/10.1016/J.CERAMINT.2019.01.145.
100. P. Kumar, M. Saini, V. Kumar, B.S. Dehiya, A. Sindhu, H. Fouad, N. Ahmad, A. Mahmood, M. Hashem, Polyethylene glycol (PEG) modified porous Ca5(PO4)2SiO4 bioceramics: Structural, morphologic and bioactivity analysis, *Coatings. 538*. 10 (2020) 538. https://doi.org/10.3390/COATINGS10060538.
101. A. Jadidi, E. Salahinejad, E. Sharifi, L. Tayebi, Drug-delivery Ca-Mg silicate scaffolds encapsulated in PLGA, *Int. J. Pharm. 589* (2020). https://doi.org/10.1016/J.IJPHARM.2020.119855.

7 Eggshell Membrane as Biomaterial for Tissue Engineering

Aman Dugar, Parichita Mishra, and Bhisham Narayan Singh

Department of Biotechnology, Manipal School of Life Sciences,
Manipal Academy of Higher Education, Manipal, Karnataka, India
*Corresponding Author

INTRODUCTION

The extracellular matrix (ECM) is a cellular secretion made of sugars, proteins, and other constituents present in all tissues and organs. It provides mechanical support or shape and is involved in homeostasis, differentiation, migration, and the healing process. The ECM is dynamic in nature and composition due to its regular remodeling via different mechanisms. Each tissue has its own specific ECM composition, and it varies significantly from tissue to tissue in terms of topology, constituents, and composition. This difference is what differentiates tissues in terms of their function. Figure 7.1 shows the composition difference in ECM components in different areas (McKee et al., 2019).

It is composed of fibrous proteins (elastins, laminin, fibronectin, and collagen) and proteoglycans in hydrated gel form. Proteoglycans are composed of glycosaminoglycan and specific core proteins. The core proteins are further subdivided into small leucine-rich proteoglycans, modular proteoglycans, and cell surface proteoglycans. The glycosaminoglycan chains are subdivided into sulfated (keratan sulfate, heparan sulfate, and chondroitin sulfate) and non-sulfated (hyaluronic acid). Small leucine-rich proteoglycans are involved in various signaling pathways involving EGFR, LRP1, and IGF1R. Modular proteoglycans help in the regulation of attachment, proliferation, and migration. Cell surface proteoglycans (glypicans and syndecans) act as co-receptors that aid in encountering ligands with signaling receptors. All these components are arranged into a 3D structure composed of the basement membrane and interstitial matrix. The basement membrane is present in blood vessels, epithelial, and endothelial tissues. Its major components are fibronectin, laminins, and collagen type IV. It is less porous but denser than the interstitial matrix. A vital HSPG of the basement membrane, perlecan binds to VEGF and FGF and thus involved in angiogenic processes. Unlike the basement membrane, the interstitial matrix is also found between connective tissue cells like tenocytes. Its major elements are elastin, collagen, and fibronectin, which aids in organizing the matrix structure (Frantz et al., 2010).

Collagen is the predominant fibrous protein within the ECM, constituting the main structural element of the ECM for providing tensile strength, support, migration, chemotaxis, and tissue development. Thirty different collagens have been classified, though not all of them are isolated to the ECM. It is produced by epithelial, endothelial, and fibroblast cells. Collagen type I is found in all tissues, but other forms are tissue-specific. Collagen types I, II, III, V, and XI are arranged to form fibrils that provide tensile strength. Fibroblasts exert pressure to organize these fibrils into sheets. These help in withstanding various stresses like shear, tension, and pressure. Molecules such as fibromodulin, elastin, and decorin regulate them. Elastin helps in recoiling in tissues like the dermis of the skin that undergo stretching. Tropoelastin assembles and gets cross-linked using its lysine residues and forms

DOI: 10.1201/9781003545712-7

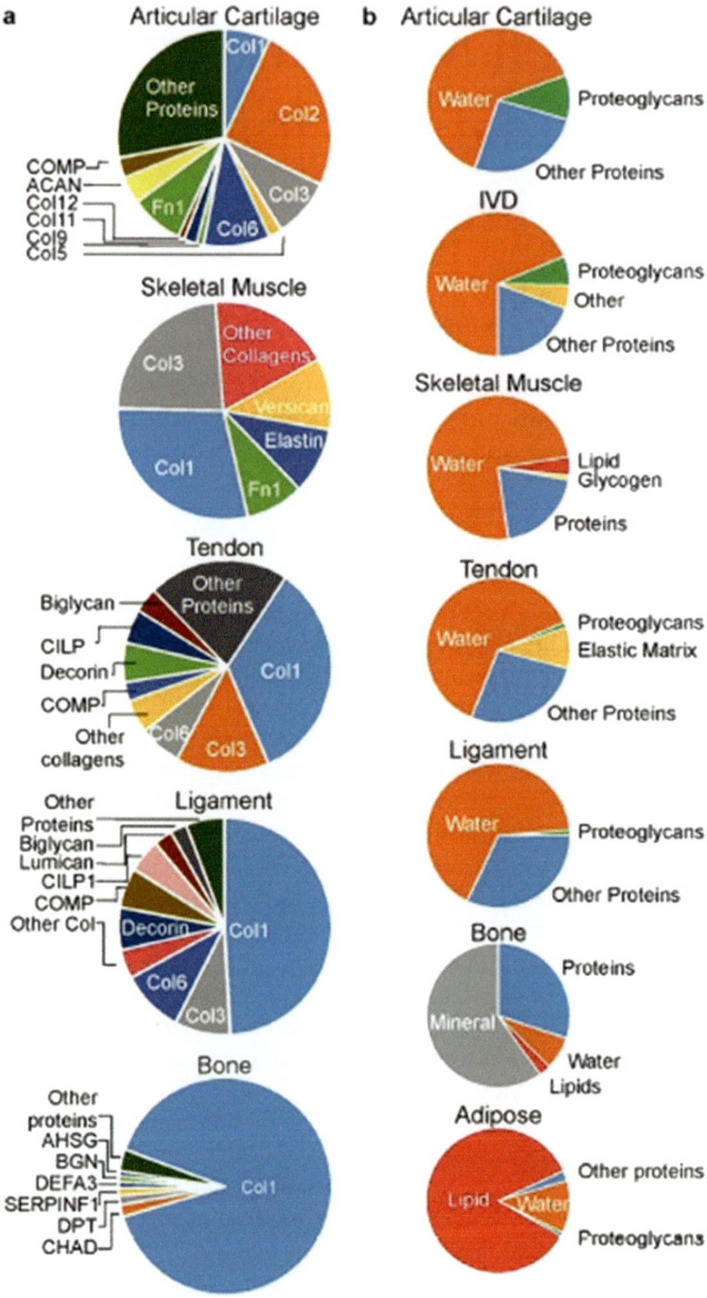

FIGURE 7.1 Comparative proteomic and overall composition of different connective tissues. (a) Relative abundance of ECM proteins derived from studies that investigated large numbers of proteins or complete proteomes in articular cartilage, skeletal muscle, tendon, ligament, and bone. For articular cartilage, skeletal muscle, ligament, and bone, the data are from a single study; for tendon, the relative abundance of the top ten proteins identified by both studies was averaged. (b) Relative quantity of whole tissue constituents in articular cartilage, skeletal muscle, ligament, tendon, bone, adipose tissue, and intervertebral disc (IVD) tissue (McKee et al., 2019).

fibers, which are then covered with fibrillins to form an elastic fiber. Fibronectin is present in the basement membrane of the ECM and has functions involving embryonic development, wound healing, and cell attachment. Like collagen, fibronectin is arranged in a mesh structure and is linked to integrins (cell surface receptors). The actin cytoskeleton is associated with the fibronectin matrix via integrin activity. Fibronectin has the characteristic of stretching many times while it is at rest. This unfolding of fibronectin helps expose integrin binding sites that result in pleiotropic changes in cellular behavior and hence acts as an extracellular mechano-regulator. Integrin α5β1 plays an important role in the fibronectin process assembly. This, with Arg-Gly-Asp (RGD), aids in the binding of protein to the cells. Laminins are present in the basement membrane of the ECM and perform vital roles regarding cell attachment, differentiation and proliferation via their integrins (Kular et al., 2014).

Tissue ECM homeostatic condition is maintained with the help of fibroblasts secreting metalloproteinases (MMPs), which again is balanced by tissue inhibitors of metalloproteinases (TIMPs) and other enzymes like transglutaminases and LOX. During processes like bone remodeling or neovascularization, MMPs (triggered by high cytokine or growth factors) help disintegrate ECM, which aids in the remodeling of ECM. Heparan or heparan sulfate keeps growth factors like VEGF, FGF, and TGF-β intact. Growth factors are involved in developmental processes and healing, especially tendon repair in adults. Cells and the matrix interact via focal adhesions. These adhesions could be of two types: regulatory proteins (paxillin and kinase) and cytoskeleton proteins (filamin and talin). These adhesions occur in cells attached to stiff substrates, while dynamic adhesions occur with cells attached to softer substrates. When the cell migrates, dynamic adhesion complexes help in the formation and disassembly as lamellipodia help propel the cell forward. When the cell reaches its destination, stable focal adhesions start to form. Integrins then signal cells to either increase or decrease ECM production, reorder forces they exert, or alter cytoskeleton structure. Integrins are transmembrane heterodimers that bind to ECM proteins like fibronectin and help in cell signaling. The cytoskeleton is made of microtubules and microfilaments. It is regulated depending on the mechanical forces exerted by the ECM. This is very important, as this elastic nature aids in stem cell differentiation towards particular lineages.

SUITABILITY OF EGGSHELL MEMBRANE AS A BIOMATERIAL

The eggshell membrane (ESM) contains collagenous fibers, structural proteins, glycoproteins, and the structural organization can be seen in Figure 7.2. Table 7.1 gives an idea of its composition, and Table 7.2 gives an idea of the function of each component. The inner ESM is 0.1–3 μm thick. The membrane possesses abundant amino acids too. Glutamic acid, glycine, and proline constitute more than 10% in both membranes (Nakano et al., 2003). Hydroxylysine constitutes about 0.19% and supports the evidence for the presence of collagen (Li et al., 2011). One research showed the presence of cysteine at about 10%, which is thought to be due to the presence of cysteine-rich proteins (CREMPs). These, due to its molecular interaction between disulfide linkage in cysteine-rich domains and collagen, provide strength to ESM (Yang et al., 2014). Even RGD (Arginine Glycine-Aspartate) tripeptide motif is present in ESM, which aids in cell attachment and reduces any immune reaction as ECM proteins contain RGD-attaching sites (Bellis, 2011). It even causes intracellular release of calcium ions that promote osteogenesis. Various ESM biological constituents could incorporate desired functionality into the artificial biosubstitute for tissue engineering applications. Collagen, on the other hand, is a triple helical structure that possesses tensile and compressive strength, which helps in providing anchorage. It helps chondrocytes to proliferate and mature, and during endochondral ossification, it also aids in hypertrophic deposition of ECM proteins (Rose & Hincke, 2009).

Hyaluronic acid, on the other hand, has high water retention capacity and provides antibacterial activity. It assists in hydrogen binding between adjacent N-acetyl and carboxyl groups, which helps in maintaining hydrophilicity and stiffness. It is also a shock absorber and provides lubrication in synovial joints (Laurent et al., 1992). Fibronectin promotes cell attachment and wound healing by

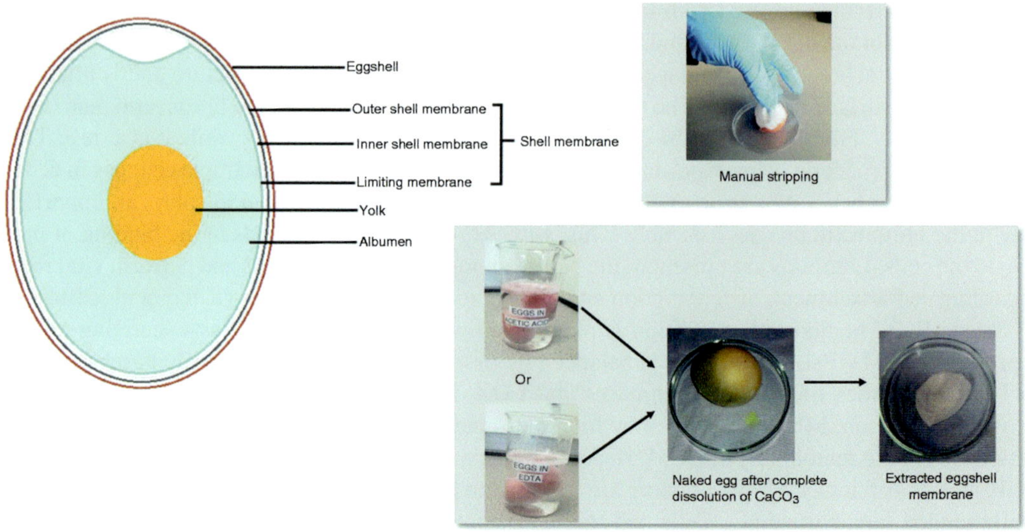

FIGURE 7.2 A schematic diagram representing the structural organization of a chicken egg alongside the most commonly used method for isolating the eggshell membrane (Mensah et al., 2021).

TABLE 7.1
Composition of Eggshell Membrane

Eggshell Membrane Composition		
Organic (80–85%)		**Inorganic (15–20%)**
Primary constituents (10%)	Secondary constituents (70–75%)	$CaCO_3$
Collagen type I	Osteopontin	
Collagen type V	Fibronectin	
Collagen type X	Other glycoproteins	
	Keratin	

promoting contraction (Valenick et al., 2005). A phosphorylated glycoprotein, osteopontin, is a key regulator of biomineralization, inflammation, and vascular calcification. It aids in cell proliferation and attachment, even as a soluble cytokine (O'Regan et al., 2000).

ECM composition and its properties change with aging and thus the way ECM aids in the healing process changes. Until now, we discussed ECM and the ESM, but now we need to be specific about how the composition of ECM changes with age so that only the desired scaffold can be made that replicates ECM. In newborns, injuries to the dermis and epidermis are fully restored in terms of architecture and function. Newborn skin contains a high amount of collagen type III, between 30% and 60%, whereas adult skin only holds 10% to 20% of collagen III. In newborn or young skin wounds, hyaluronic acid is present at high levels but for a shorter period of time. Hyaluronic acid enables cellular migration and collagen synthesis/deposition. The difference in proteoglycans like decorin and fibromodulin is also observed in the scarred phenotype in healed adult skin and the scarless phenotype in repaired newborn skin. In adult skin, a low level of decorin is observed with scar formation, whereas higher levels of fibromodulin are observed in fetal wounds than in adult wounds. In newborn skin wounds, the ratio of MMPs to TIMPs is very high, but during adult wound healing, high MMP-2 and low TIMPs are expressed (Lo et al., 2012; Lovvorn et al., 1999; Sawai et al., 1997; Sayani et al., 2000; Soo et al., 2000).

TABLE 7.2
Function of Each Component Present in Eggshell Membrane

Components of Eggshell Membrane	Functions	Reference
Collagen type I	An abundant protein that forms the mechanical and structural matrix of skin, cornea, bone, tendons, blood vessels, and other connective tissue.	Morello and Esposito (2012)
Collagen type V	A fibrillar collagen used for the fibrillation of type 1 and type 3 collagen. It even constitutes corneal stroma, bone matrix, and the interstitial matrix of lungs, placenta, muscles, and liver.	Leeming and Karsdal (2019)
Collagen type X	Regulates matrix mineralization and compartmentalizes matrix components, thus facilitating endochondral ossification.	He et al. (2019)
Chondroitin sulfate	Located in cartilage and plays an essential role in the elasticity and function of articular cartilage. It is also involved in processes like cell adhesion, growth, and proliferation	N. Maeda et al. (2011)
Glucosamine	It improves cartilage matrix components production like collagen type 2 and aggrecan and also hyaluronic acid in synovium explants	Jerosch (2011)
Hyaluronic acid	Involved in intracellular signaling, immune activation, cell proliferation, and migration	Nusgens (2010)
Desmosine and isodesmosine	Both cross links to give elasticity to connective tissues like skin and lungs	Mecham and Foster (1978)
Sialic acid	Glomerular filtration, neural plasticity, adhesion molecules involved in regulation of the immune response	Varki (2008)
Ovotransferrin	Influence nucleation and enhances the growth of calcite crystal	Gautron et al. (2001)
Osteopontin	Inhibits calcification of blood vessels, immune cell activation, bone resorption, and ECM remodeling	Singh et al. (2010)
Sialoprotein	Biomineralization	Malaval et al. (2008)
Agrin	Development of central nervous system (CNS) (specifically in axonal tracts), aggregation of acetylcholine receptors for synapsis	Nitkin and Rothschild (1990)
Keratin	Regulation of protein synthesis, cell size, apico-basal polarization and cell signaling	Karantza (2011)
Laminin	Cell attachment, differentiation and proliferation	Paulsson (1992)
Amino acids	Aids in protein synthesis	
Lysyl oxidase	To cross-link elastin and collagen to shape and provide strength to ECM	Tenti and Vannucci (2020)
ß-N-acetylglucosaminidase	ECM remodeling	Konopka (2012)
Fibronectin	Promote wound contraction, which is essential for wound healing. It is involved in attachment, differentiation and proliferation	Valenick et al. (2005)

For biomaterial application of ESM, it is important to harvest it successfully from eggshells. The separation step should be carried out in a way that makes industrial-scale production efficient and provides a good quality product. Manually peeling off the membrane from eggshells has received a lot of attention, though the outer membrane is strongly attached to the shell. Acidic treatment helps dissolve calcite and loosen or break interfacial bonds between the membrane and eggshell (Lunge et al., 2012; Xu et al., 2010). A few other separation techniques of ESM from raw shells are given in Table 7.3. The first method was used to analyse the ESM's capability to absorb gold ions from solutions. This was shown to be pH-dependent, as at pH 3.0 the absorption was observed to be optimum. The second method was used to study the absorption ability of radioactive elements like

TABLE 7.3
Different Approaches for the Extraction of Eggshell Membrane

Serial No.	Technique	Drawback	Reference
1	Treatment with 0.5 M HCl, then 1 hour immersion in 0.5 M NaOH, then peeling from eggshells	Collagen gets denatured strong alkali and acid treatment leads to loss of collagen that is acid soluble	Ishikawa et al. (2002)
2	Overnight treatment in 0.5 M HCl and then manual peeling off membrane	Collagen gets denatured strong alkali and acid treatment leads to loss of collagen that is acid soluble	Ishikawa et al. (1999)
3	1 hour 5% EDTA treatment and the manual peeling of the membrane off	Membrane needs to be manually separated	Soledad Fernandez et al. (2001)
4	24-hour treatment of EDTA then 3 min hexane treatment	None for now	Sachindra et al. (2014)
5	Crush the egg shells and remove the membrane in a valve	Efficiency of extraction of the membrane decreases as many membranes remain attached to shells and many gets contaminated	Thoroski (2003)
6	Boil the eggs and then expose pressurized gas through a needle at the base of shell	Membrane still attached to shells	Thoroski (2003)
7	Expose shells in water with turbulent force after settling	Only partial separation possible	Macneil, (1998)
8	Treatment at 100°C with alkali or acids for softening or dissolving the shells	Collagen solubilizes or denatures	Zeidler et al. (1991)

uranium, thorium, and actinides. In the third technique, treatment with Ethylenediaminetetraacetic acid (EDTA) caused partial demineralization that helped in the removal of the membrane using forceps. In the fourth method, EDTA again helped in the detachment of the membrane from shells, and hexane further aids in loosening the adhesiveness between the shell and membrane and even forming an emulsion consisting of the membrane, which further eases the process of separation of the membrane from the shell. One advantage of this method is that it doesn't contaminate the proteins in the membrane. In the fifth method, the eggshells are dried, which helps remove moisture and causes minute separation of the membrane and shell. The majority of membranes that are not separated undergo a series of steps. Centrifugation first helps in the removal of adhering liquid. The shells are then washed to remove residual liquid. Again, centrifugation helps in removing excess wash water. A tumble dryer is used to reduce the moisture content of eggshells to 1–3%. The vibratory screen then helps in removing the membrane. Using a hammer mill, the shells are crushed, and the membrane is removed pneumatically. In the sixth method, the pressurized gas used via a needle under the base of a boiled egg promotes loosening of the membrane, but the membrane still remains attached to the shell. In the seventh method, first, some membranes are separated as and when the shells are processed, disrupting the links between the membrane and shells. Then they are put in a tank with water where shells settle and turbulent forces then cause the membrane and shells to separate. In the eighth method, acid or alkali is used to either soften or dissolve the shells before mechanically peeling the membrane.

The reason for the lack of solubility of the membrane in water is the cross-linking between cystine, desmosines, and hydroxylysine or lysine, and also due to many disulfide bonds. Though using non-toxic solvents, it can be made soluble. There are different approaches, and one of the researches showed not to use very high temperature during the preparatory process of soluble ESM as ESM thermal stability is not that high and from 55°C, collagen fibers start degrading (Sah & Pramanik,

TABL 7.4

Approaches for Preparation of Soluble Eggshell Membrane

Serial No.	Chemicals	Temperature (°C)	Duration (hours)	pH	Conditions	Reference
1	5N β-thiopropionic acid aqueous solution	60–80	5	–	Dialyzed	Takigaura et al. (2009)
2	Performic acid oxidation followed by pepsin digestion	4 25	24 24–72	–	Dialyzed against water and lyophilized for storage	Takahashi et al. (1996a)
3	1.25 N 3-mercaptopropionic acid in 10% acetic acid	90	12	5	Continuous stirring	Yi et al. (2003)

2014). RGD should also remain intact to the soluble eggshell membrane as this is what integrins would identify; otherwise, it would lead to protein denaturation (Liddington & Ginsberg, 2002). Table 7.4 gives an idea of different approaches that could be used for preparing soluble ESM. With the performic acid approach, only small quantities of basic amino acids remained in the structure, whereas in the mercaptopropionic approach, collagen was denatured. In the performic acid approach, the yield of Soluble Egg Shell Membrane (SEP) was in the range of 16–39%. Cystine was converted to cysteic acid with the disruption of the disulphide bond. SEP with this approach even helped in promoting the reorganization of the collagen matrix with aligned molecular rearrangement. It even reduced the dissolution of the collagen matrix and increased its denaturation temperature. In the mercaptopropionic acid method, the acid acts as a reducing agent. This helps in cleaving the disulfide bonds between cysteine molecules which improves solubility. The highest yield was observed when 1.25N 3-mercaptopropionic acid was used with 10% acetic acid for 20.5 hours. The yield was 62%. Even the bioactivity of the prepared SEP via this method was checked using NIH3T3 cells. The findings were very insightful as when compared with collagen, the adhesion and proliferation times were the same; interestingly, the cell viability was slightly greater in SEP.

USE OF SCAFFOLDS IN TISSUE ENGINEERING

The spotlight for the research in the field of tissue engineering (TE) is particularly on the modification and development of biomaterials that must be cost-effective and could serve for either one or both functions: to promote the healing cascade or to provide a scaffold for tissue formation (Fisher & Mauck, 2013).

SCAFFOLDS

Scaffolds are supportive 3D structures developed for cellular interaction, adhesion, and migration. They are divided into three categories: ceramic scaffolds (hydroxyapatite, TCP), synthetic biopolymer-based scaffolds (PGA, PLGA, PLLA), and natural biomaterials (collagen (protein origin), chitosan (polysaccharide origin)) Furthermore, scaffolds must be able to replicate or mimic the ECM for targeted tissues like soft tissues (skin, cartilage) and hard tissues (bone, teeth) (Deb et al., 2018). Their purpose is to replicate the ECM or external environment to provide different signals to the cells grown on it. The interaction of the biomaterial must also be with tissues around the implantation site of the host.

Scaffolds provide a temporary template for cellular growth and allow the deposition of the cell's own ECM with the gradual resorption of the scaffolding biomaterial, which should aid or guide cell proliferation. Inertness is another important quality that needs to be present in a biomaterial. The bioconstruct needs to be free of any fibrotic or inflammatory reaction post-implantation; rather, it should provide immunoprotection to prevent the chance of rejection by the body. Post-implantation, the blood contact should not cause coagulation when the bioconstruct undergoes vascularization. Moreover, scaffolds must possess optimal porosity and suitable physicochemical properties to enable the diffusion of vital nutrients and maintain a suitable structure during tissue regeneration (Lee, 2007). These properties differ based on where these bioartificial constructs are implanted, such as in vasculature, interstitial space, or directly onto organs. For example, oxygen partial pressure and blood pressure at different sites vary, and the biochemical environment too, depending on the site of implantation. The difference in shear stress leads to changes in cell shape and cytoskeleton structure (O'Brien, 2011). The flow would influence cell proliferation. Functionalizing the polymer surface using different derivatives, like providing RGD motifs, triggers cell adhesion (Chan & Leong, 2008). It also helps allow specific cell adhesion due to differences in compatibility with different motifs. This helps in growing only the desired cell type.

Various techniques for preparing scaffolds include 3D printing, porogen leaching, electrospinning, supercritical fluid technology, particle aggregation, freeze-drying, and fiber extrusion and bonding.

Fiber Bonding

The fiber bonding method is used for the production of microfibrous scaffolds with high porous interconnectivity for rapid influx and efflux of nutrients within cells. The high surface area increases the efficiency.

Extrusion Method

In this method, a known quantity of microfibers is produced, arrayed in a Teflon mold that was previously customized. It is heated for a standard time, temperature, and compression, which helps procure fibers with a spatial random arrangement to prevent them from collapsing (Mikos & Temenoff, 2000). The desired scaffold of specific pore size and porosity can be obtained by altering the quantity and size of fibers and also by controlling the degree of compression of bundles. Starch-polycaprolactone fiber mesh scaffolds have been shown to improve cell adhesion and proliferation for bovine articular chondrocytes (Oliveira et al., 2007). When conjugated with fluid flow bioreactor culture, this resulted in mineralized tissue development similar to bone by increasing the differentiation of marrow stromal cells (Gomes et al., 2006). Starch-based scaffolds have been reported with promising results for application in vasculature engineering as they improve the growth of human endothelial cells (Santos et al., 2007).

Wet Spinning

In the wet-spinning technique, the powdered polymer is dissolved in a solvent which, when bathed with a non-solvent, causes immersion-precipitation to take place. This non-solvent-induced phase inversion technique allows the formation of a continuous polymer fiber. The homogeneous filament solution (solvent, polymer, additives), due to solvent–non-solvent exchange, causes polymer desolvation, leading to solidification (Puppi et al., 2010). Fibers produced using wet spinning have been seen to have various applications. They could be used as carriers for drug release applications, such as PLLA fibers (Gao et al., 2007) or chitosan (Denkbaş et al., 2000) fabricated with 5-fluorouracil or PCL fabricated with clodronate for treating bone disease and for anti-inflammatory reactions. Other materials for scaffolds could be silk fibroin (Yao et al., 2002).

Prototyping

Solid free-form fabrication or rapid prototyping helps in procuring a scaffold with controlled morphology (Yeong et al., 2004). Using computer-aided design, customized or desired shapes and structure of the scaffold can be produced, which means more reproducibility, flexibility, and versatility in creating precisely controlled scaffolds (Landers et al., 2002; Peltola et al., 2008). CT scans or MRI scans can be used to derive the models. 3D printers and plotters are the fastest prototyping systems among all others available, and their other strengths include ease of use and affordability (Pfister et al., 2004). The weakness of cell-seeding efficiency where the pore size is large is overcome by creating pore size gradients (Sobral et al., 2011) or using electrospun nanofibers in conjugation (Martins et al., 2009). Using 3D Bioplot, SPCL semi-tubular structures were produced for the treatment of spinal cord injuries by enhancing mechanical stability (Silva et al., 2010).

Electrospinning

Electrospinning enables the production of ultrafine fibers with diameters less than 1 μm and a high specific surface area. These fibers are randomly arranged like a mesh with properties such as high porosity and interconnectivity. The most critical part of this process is the selection of the solvent, which has two major roles: dissolving polymer molecules to form an electrified jet and then transferring it to a collector where, by vaporization of the solvent, polymer fibers remain. The random arrangement of the non-woven mesh resembles the network of collagen fibers present in the ECM of the human body, making this nanofiber mesh most suitable for the use of stem cells for cell adhesion and proliferation (Martins et al., 2007). ECM was electrospun with PCL. SEP/PCL nanofibers were enhanced using a coaxial electrospinning process that involves an auxiliary cylindrical electrode and a dual nozzle. When this blend was characterized, improved mechanical properties, hydrophilicity, and pore distribution were reported. Human dermal fibroblasts were better able to attach and proliferate onto this blend (Jim Jia et al., 2008; Kim et al., 2007, 2008). Electrospinning was also used to prepare ESM/PVA, ESM/PEO, and ESM/PCL nanofibers using catechin. Catechin helped in enhancing the insolubility of both blends via hydrogen bond interactions (Kang et al., 2010, 2012).

Particulate Leaching

Compression molding and particulate leaching are combined to fabricate 3D porous scaffolds. Polymer and leachable elements are loaded in a mold at high pressure and heated to a temperature higher than the polymer's glass transition temperature, which maximizes packaging. In the next cooling step, the porogen part in the polymer-porogen composite is selectively dissolved using a specific solvent. This process even helps control scaffold pore size and porosity (Correlo et al., 2009). Chitosan/polyester composites formed via this method have been reported to enhance cell attachment, proliferation, and even promote chondrogenic or osteogenic differentiation of BMC-9 cell lines (Costa-Pinto et al., 2008). They have also been shown to promote proliferation and differentiation in vivo for hBMSCs, resulting in bone formation (Costa-Pinto et al., 2012).

Supercritical Fluid

The supercritical fluid process is an eco-friendly, easy, and clean operation to procure solvent-free structures. The polymer solution is cast on an inert support, then dipped with cast film in a non-solvent of polymer (supercritical CO_2 as it dries the structure without collapsing). The phase inversion is thus created (Duarte et al., 2009). Various polymers have been made using this technique, for example PVA, PMMA, cellulose acetate, Nylon 6, polycarbonate/PEG, polycarbonate, and polysulfonate (Duarte et al., 2010). Even it can even be used as a drug carrier (Duarte et al., 2009).

Injectable Hydrogels

The development of injectable scaffolds has grabbed the attention of many scientists. It has shown promising results. Injectable polymers are solutions that can be easily injected into the body via syringe and have the characteristic property of transitioning from sol to gel state in response to temperature changes or any other external stimuli. Using this technique, a flowable polymer/gel/scaffold would fill the shape of the damaged tissue/organ and may also act as a carrier for growth factors, with the absence of residual solvents and no surgical operations required for implantation (Gutowska et al., 2001). It reduces discomfort for patients, as it is a minimally invasive technique. By mixing just prior to injection, the problems of cell attachment and biocompatibility are overcome. There are various ways of modifying existing biomaterials, such as altering solidification, biofunctionalization, porosity, mechanical properties, or even the biodegradation rate (Kretlow et al., 2007). A 3D network swollen by a solvent is called a gel. If the networks are connected via secondary forces, they are called physical gels; if covalent bonds connect them, they are called chemical gels (Guenet, 1993). Natural biomaterials are always preferred for injectable hydrogels due to their biodegradability, biocompatibility, and similarity to ECM. These could be chitosan, collagen, hyaluronic acid, fibrin, alginate, heparin, elastin, or chondroitin sulfate based hydrogels (Liu et al., 2017).

APPLICATIONS OF EGGSHELL MEMBRANE SCAFFOLDS IN TISSUE ENGINEERING

Traditionally, ESM has been used in Chinese medicine for healing injuries. It was also one of the important components of 'phoenix cloth' that was used as a dressing material for healing burns, corneal ulcers, and tympanic perforations. Some Sumo wrestlers in Japan still apply ESM to injuries as it is a natural medicine (Jun Jia et al., 2011; Sah & Pramanik, 2014). The modern-day applications of ESM are shown in Figure 7.3.

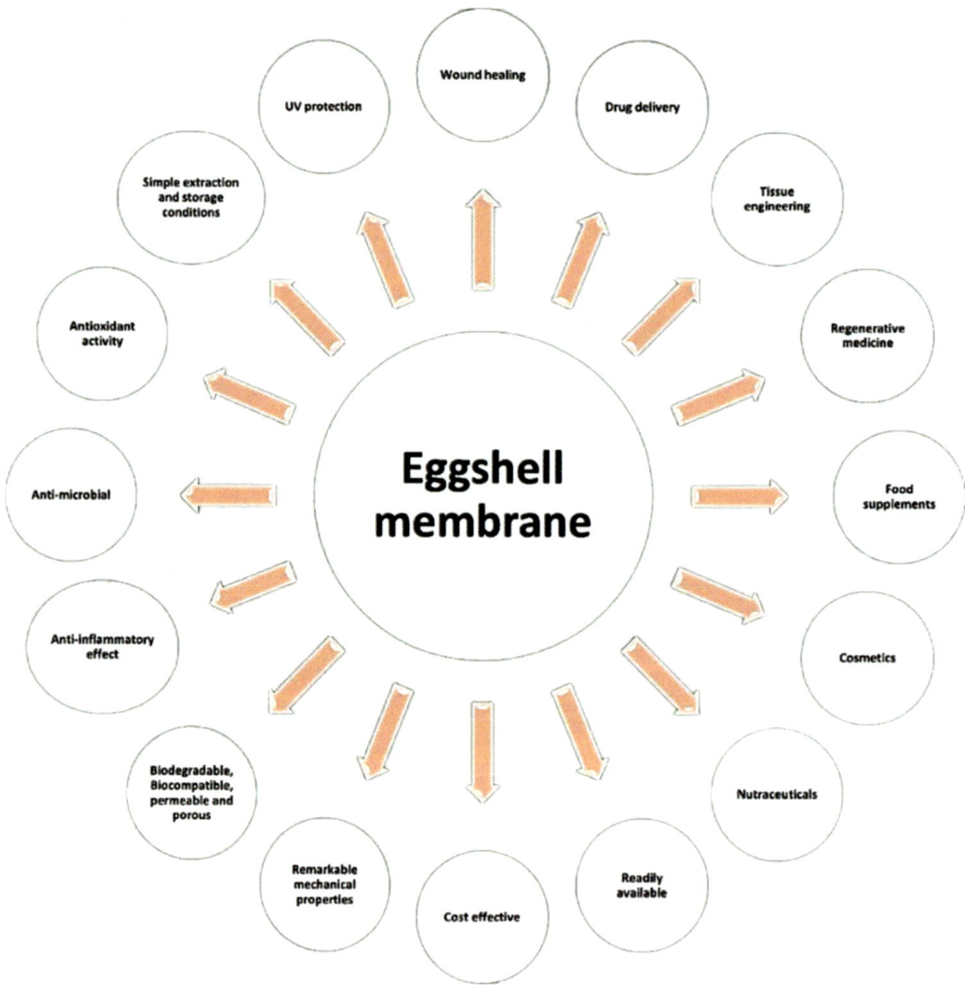

FIGURE 7.3 A brief summary of the various current and potential applications of eggshell membrane in tissue engineering (Mensah et al., 2023).

Skin Tissue Engineering

Both lipid and hydrated protein phases, along with permeability independent of pH, make it suitable for a skin graft model and even promise regeneration (Houk & Guy, 1988). It helps stimulate epithelial tissue regeneration, thereby promoting skin tissue regeneration. The application of Natural Eggshell Membrane (NEM®) as biological dressings is used for healing burnt skin due to the presence of high protein, hyaluronic acid, collagen, and glucosamine (Maeda & Sasaki, 1982). It was shown that ESM powder promotes MMP activity that helps in wound healing (Vuong et al., 2018). Another report found that ESM also helped in self-treating a traumatic upper lip (Zadik, 2007). Pepsin-solubilized collagen (PSC) was cross-linked with soluble ESM and was evaluated for adhesion of dermal fibroblasts (Ohto-Fujita et al., 2011). SEP was grafted on the surface of PCL nanofibers, which then enhanced adhesion and proliferation of human dermal fibroblasts (Jim Jia et al., 2008). Soluble ESM was used to improve the viscoelasticity of pepsin-solubilized collagen, which helped improve the membrane's flexibility. This was then evaluated using fibroblasts for attachment and proliferation (Ino et al., 2006). It was also found that ESM, via some free radical damaging pathway, helped in reducing facial wrinkles by inducing the expression of genes like Gclm, NqoI, Hmox1, and Gclc in keratinocytes (Jensen et al., 2016). Another research suggested that it even helps form a moisture layer that protects the skin and has anti-inflammatory and antimicrobial activity (Yoo et al., 2014). In humans, skin is made up of collagen type I, which aids in maintaining structure, and type III collagen, which provides flexibility. However, with age, the ratio of these two collagens changes. An interesting insight was observed when collagen composition was compared with ESM. Table 7.5 gives an idea of these compositions. These similarities show that ESM has a high affinity for dermal fibroblasts and could thus aid in providing skin with elasticity

TABLE 7.5
The Values Indicates the Proportion of Amino Acid (Number of Residues per about 1000) (Takigaura et al., 2009)

Amino Acid	Egg Shell Membrane	Collagen
Lysine	34	28
Histidine	39	5
Arginine	63	46
Aspartic acid	73	46
Threonine	51	17
Serine	50	37
Glutamic acid	105	70
Proline	73	129
Glycine	55	333
Alanine	25	112
Cystine	92	–
Valine	60	20
Methionine	38	6
Isoleucine	32	12
Leucine	48	23
Tyrosine	23	2
Phenylalanine	18	12
Hydroxy-proline	–	98
Hydroxy-lysine	–	6
Tryptophan	29	–

and flexibility, and thus possesses characteristics for skin regeneration. In another research, alkaline water-soluble eggshell membrane (ASESM) was used as an ECM replicate on a PMBN polymer. PMBN are artificial biointerfaces for cell membranes that facilitate stabilization of immobilized molecules and reduction in nonspecific protein adsorption. This was tested for human dermal fibroblasts and showed great results. Cells adhered to them but not to glycine-blocked PMBN. Cells even expressed genes related to ECM like collagen type III and decorin (Ohto-Fujita et al., 2011). SEP via the entrapment method, was immobilized on a PDL-LA membrane in HFIP. NIH3T3 cell culture showed improved cytocompatibility with this blend (Lu et al., 2009). SEP was prepared and conjugated with PSC via cross-links by a water-soluble carbodiimide using the performic acid method. This created a more stable and resilient structure with thicker collagen fibrils and enhanced flexibility. This artificial skin module was reported to increase cell adhesion and proliferation of fibroblast cells (Ino et al., 2006; Takahashi et al., 1996b).

Bone Tissue Engineering

Electrospinning was used to prepare SEP/PLGA nanofibers for periodontal tissue regeneration. The spongy layer aided in cell anchorage and propagation, whereas the dense layer blocked in-growth of connective tissue and maintained structural integrity (Jun Jia et al., 2012). The fabricated chitosan/silk fibroin/ESM hydrogel was reported to provide better chondrocyte anchorage, growth, and differentiation and promises to be used as a substitute for cartilage (Adali et al., 2019). Osteoblast in vitro culture can undergo mineralization given that the terminal ends of type X collagen are removed (Arias et al., 1997). It has also been used to treat stiffness and pain in knee osteoarthritis patients (Ruff et al., 2009). Electrospinning was used to prepare ESMP/PLGA nanofibers, which showed improved mechanical properties. The bead-free nanofibers were evenly distributed, and an interconnected porous construct was formed with an interwoven morphology. Biological assays with mouse fibroblast cells, L-929, showed improved adhesion, proliferation, and migration. This has shown great potential in GTR, specifically in periodontal therapy (Jun Jia et al., 2012). Another report supports the same claim that ESM can aid in periodontal regeneration due to its structure and composition (Jun Jia et al., 2011).

Cartilage Tissue Engineering

A study on meniscal cell attachment and proliferation was compared between raw ESM and autoclaved ESM. The study reported that the autoclaved ESM had its surface charge converted from cationic to neutral, with no mucilaginous coating on the surface. It even showed high resistance towards biodegradation. The autoclaved membrane supported higher meniscal cell adhesion, growth, cell proliferation index, and DNA content on the scaffold (Pillai et al., 2015). Processed ESM powder was combined with collagen. This helped in increasing the mechanical hardness of the scaffold, and the interconnected porous domain was uniformly distributed as dense structures. The study reported that this scaffold helped in increasing the adhesion and proliferative nature of fibroblast and skeletal cells. This affordable natural biomaterial could be of encouraging significance in the field of regenerative medicine (Rønning et al., 2020). SEP is very brittle; hence, PVA was blended with it to improve its mechanical properties. PVA concentration was positively correlated with the blend's mechanical properties, like tensile strength and elongation-at-break. With 10% SEP, the tensile strength was 28.9 MPa, and the elongation-at-break was 288%. PVA/SEP blends are transparent and homogeneous, which indicates a certain degree of miscibility, and with an increase in PVA content, PVA gets crystallized. With increased SEP in the blend, more spindle-shaped cells were observed with good attachment and even spreading out. Adding PVA up to 50% helps enhance the mechanical properties of SEP without compromising its biocompatibility, as observed with NIH3T3 cell culture (Yi et al., 2006). Even when the blend of SEP was made via co-electrospinning using PLA, increased mechanical properties were observed, and uniform fibres were formed. Its surface was found to be super hydrophilic. When tested with NIH3T3 cells, improved biocompatibility was observed with the blends compared to pure PLA (Xiong et al., 2010, 2012). The same results were seen using ESMP/chitosan blend, showing improved biocompatibility (Qi et al., 2009).

Nerve Tissue Engineering

ESM is reported to promote nerve regeneration and functional improvement in damaged nerves (Farjah et al., 2013). It was mixed with PLA to form a PLA/SEP composite in the ratio of 9:1. This showed improved bioavailability and NIH3T3 cell anchorage (Xiong et al., 2010). SEP was also immobilized on a PDLLA membrane, which improved biocompatibility and enhanced NIH3T3 fibroblast proliferation (Lu et al., 2009). As soluble ESM can easily be processed with other biomaterials like fibronectin or collagen, it could be used for nerve regeneration due to the presence of glycosaminoglycan and tubular collagen (Kitahara et al., 1998; Wang et al., 1998).

CONCLUSION

ESM, not only because it is easily available but also due to its composition, could be a revolutionary biomaterial in the field of tissue engineering and regenerative medicine. Tomorrow, it might be an important material for facial reconstruction or restoration of damaged body tissue. More research is still on the way toward translational applications for various clinical therapeutics. A green approach must be looked into to make industrial-scale production possible without compromising its quality. It can be used with other materials to form injectable gels for regenerative purposes. It can even be prepared in a way to create organoids. One study reported that ESM aids in downregulating the pro-inflammatory cytokine, TNF-α, which even widens its use as an anti-inflammatory product (Benson et al., 2012). ESM's ability to mimic ECM makes it a promising biomaterial. Further studies need to be done to turn this biowaste into a bioproduct. Today, it shows potential in the field of tissue engineering; tomorrow, it might be promising for cancer therapy or biomedical research. Further research will help in decoding its functions.

REFERENCES

Adali, T., Kalkan, R., & Karimizarandi, L. (2019). The chondrocyte cell proliferation of a chitosan/silk fibroin/egg shell membrane hydrogels. *International Journal of Biological Macromolecules*, *124*, 541–547. https://doi.org/10.1016/j.ijbiomac.2018.11.226

Arias, J. L., Nakamura, O., Fernández, M. S., Wu, J. J., Knigge, P., Eyre, D. R., & Caplan, A. I. (1997). Role of type X collagen on experimental mineralization of eggshell membranes. *Connective Tissue Research*, *36*(1), 21–33. https://doi.org/10.3109/03008209709160211

Bellis, S. L. (2011). Advantages of RGD peptides for directing cell association with biomaterials. *Biomaterials*, *32*(18), 4205–4210. https://doi.org/10.1016/j.biomaterials.2011.02.029

Benson, K. F., Ruff, K. J., & Jensen, G. S. (2012). Effects of natural eggshell membrane (NEM) on cytokine production in cultures of peripheral blood mononuclear cells: Increased suppression of tumor necrosis factor-α levels after in vitro digestion. *Journal of Medicinal Food*, *15*(4), 360–368. https://doi.org/10.1089/jmf.2011.0197

Chan, B. P., & Leong, K. W. (2008). Scaffolding in tissue engineering: General approaches and tissue-specific considerations. *European Spine Journal*, *17*(SUPPL. 4). https://doi.org/10.1007/s00586-008-0745-3

Correlo, V. M., Boesel, L. F., Pinho, E., Costa-Pinto, A. R., Alves Da Silva, M. L., Bhattacharya, M., Mano, J. F., Neves, N. M., & Reis, R. L. (2009). Melt-based compression-molded scaffolds from chitosan-polyester blends and composites: Morphology and mechanical properties. *Journal of Biomedical Materials Research - Part A*, *91*(2), 489–504. https://doi.org/10.1002/jbm.a.32221

Costa-Pinto, A. R., Correlo, V. M., Sol, P. C., Bhattacharya, M., Srouji, S., Livne, E., Reis, R. L., & Neves, N. M. (2012). Chitosan–poly(butylene succinate) scaffolds and human bone marrow stromal cells induce bone repair in a mouse calvaria model. *Journal of Tissue Engineering and Regenerative Medicine*, *12*(3), 181–204. https://doi.org/10.1002/term

Costa-Pinto, A. R., Salgado, A. J., Correlo, V. M., Sol, P., Bhattacharya, M., Charbord, P., Reis, R. L., & Neves, N. M. (2008). Adhesion, proliferation, and osteogenic differentiation of a mouse mesenchymal stem cell line (BMC9) seeded on novel melt-based chitosan/polyester 3D porous scaffolds. *Tissue Engineering – Part A.*, *14*(6), 1049–1057. https://doi.org/10.1089/ten.tea.2007.0153

Deb, P., Deoghare, A. B., Borah, A., Barua, E., & Das Lala, S. (2018). Scaffold development using biomaterials: A review. *Materials Today: Proceedings*, *5*(5), 12909–12919. https://doi.org/10.1016/j.matpr.2018.02.276

Denkbaş, E. B., Seyyal, M., & Pişkin, E. (2000). Implantable 5-fluorouracil loaded chitosan scaffolds prepared by wet spinning. *Journal of Membrane Science*, *172*(1–2), 33–38. https://doi.org/10.1016/S0376-7388(00)00314-8

Duarte, A. R. C., Caridade, S. G., Mano, J. F., & Reis, R. L. (2009). Processing of novel bioactive polymeric matrixes for tissue engineering using supercritical fluid technology. *Materials Science and Engineering C*, *29*(7), 2110–2115. https://doi.org/10.1016/j.msec.2009.04.012

Duarte, A. R. C., Mano, J. F., & Reis, R. L. (2009). Dexamethasone-loaded scaffolds prepared by supercritical-assisted phase inversion. *Acta Biomaterialia*, *5*(6), 2054–2062. https://doi.org/10.1016/j.actbio.2009.01.047

Duarte, A. R. C., Mano, J. F., & Reis, R. L. (2010). Novel 3D scaffolds of chitosan-PLLA blends for tissue engineering applications: Preparation and characterization. *Journal of Supercritical Fluids*, *54*(3), 282–289. https://doi.org/10.1016/j.supflu.2010.05.017

Farjah, G. H., Heshmatian, B., Karimipour, M., & Saberi, A. (2013). Using eggshell membrane as nerve guide channels in peripheral nerve regeneration. *Iranian Journal of Basic Medical Sciences*, *16*(8), 901–905. https://doi.org/10.22038/ijbms.2013.1347

Fisher, M. B., & Mauck, R. L. (2013). Tissue engineering and regenerative medicine: Recent innovations and the transition to translation. *Tissue Engineering - Part B: Reviews*, *19*(1), 1–13. https://doi.org/10.1089/ten.teb.2012.0723

Frantz, C., Stewart, K. M., & Weaver, V. M. (2010). The extracellular matrix at a glance. *Journal of Cell Science*, *123*(24), 4195–4200. https://doi.org/10.1242/jcs.023820

Gao, H., Gu, Y., & Ping, Q. (2007). The implantable 5-fluorouracil-loaded poly(l-lactic acid) fibers prepared by wet-spinning from suspension. *Journal of Controlled Release*, *118*(3), 325–332. https://doi.org/10.1016/j.jconrel.2006.12.028

Gautron, J., Hincke, M. T., Panheleux, M., Garcia-Ruiz, J. M., Boldicke, T., & Nys, Y. (2001). Ovotransferrin is a matrix protein of the hen eggshell membranes and basal calcified layer. *Connective Tissue Research*, *42*(4), 255–267. https://doi.org/10.3109/03008200109016840

Gomes, M. E., Holtorf, H. L., Reis, R. U. I. L., & Mikos, A. G. (2006). Influence of the porosity of starch-based fiber mesh scaffolds on the proliferation and osteogenic differentiation of bone marrow stromal cells cultured in a flow perfusion bioreactor. *Tissue Engineering*, *12*(4), 801–819.

Guenet, J. (1993). Thermoreversible gelation of polymers and biopolymers. *Polymer*, *34*(13), 3000.

Gutowska, A., Jeong, B., & Jasionowski, M. (2001). Injectable gels for tissue engineering. *Anatomical Record*, *263*(4), 342–349. https://doi.org/10.1002/ar.1115

He, Y., Gudmann, N. S., Willumsen, N., & Karsdal, M. A. (2019). Type X collagen. In *Biochemistry of Collagens, Laminins and Elastin: Structure, Function and Biomarkers*, pp. 91–97. https://doi.org/10.1016/B978-0-12-817068-7.00010-0

Houk, J., & Guy, R. H. (1988). Membrane models for skin penetration studies. *Chemical Reviews*, *88*(3), 455–472. https://doi.org/10.1021/cr00085a001

Ino, T., Hattori, M., Yoshida, T., Hattori, S., Yoshimura, K., & Takahashi, K. (2006). Improved physical and biochemical features of a collagen membrane by conjugating with soluble egg shell membrane protein. *Bioscience, Biotechnology and Biochemistry*, *70*(4), 865–873. https://doi.org/10.1271/bbb.70.865

Ishikawa, S.-I., Suyama, K., Arihara, K., & Itoh, M. (2002). Uptake and recovery of gold ions from electroplating wastes using eggshell membrane. *Bioresource Technology*, *81*(3), 201–206. https://doi.org/10.1016/S0960-8524(01)00134-1

Ishikawa, S.-I., Suyama, K., & Satoh, I. (1999). Biosorption of actinides from dilute waste actinide solution by eggshell membrane. *Applied Biochemistry and Biotechnology*, *78*(1–3), 521–534. https://doi.org/10.1385/abab:78:1-3:521

Jensen, G. S., Shah, B., Holtz, R., Patel, A., & Lo, D. C. (2016). Reduction of facial wrinkles by hydrolyzed water-soluble egg membrane associated with reduction of free radical stress and support of matrix production by dermal fibroblasts. *Clinical, Cosmetic and Investigational Dermatology*, *9*, 357–366. https://doi.org/10.2147/CCID.S111999

Jerosch, J. (2011). Effects of glucosamine and chondroitin sulfate on cartilage metabolism in OA: Outlook on other nutrient partners especially omega-3 fatty acids. *International Journal of Rheumatology, 2011.* https://doi.org/10.1155/2011/969012

Jia, J., Duan, Y. Y., Yu, J., & Lu, J. W. (2008). Preparation and immobilization of soluble eggshell membrane protein on the electrospun nanofibers to enhance cell adhesion and growth. *Journal of Biomedical Materials Research - Part A*, *86*(2), 364–373. https://doi.org/10.1002/jbm.a.31606

Jia, J., Guo, Z., Yu, J., & Duan, Y. (2011). A new candidate for guided tissue regeneration: Biomimetic eggshell membrane. *Iranian Journal of Medical Hypothesis and Ideas, 5*, 1–5.

Jia, J., Liu, G., Guo, Z. X., Yu, J., & Duan, Y. (2012). Preparation and characterization of soluble eggshell membrane protein/PLGA electrospun nanofibers for guided tissue regeneration membrane. *Journal of Nanomaterials*, *2012*. https://doi.org/10.1155/2012/282736

Kang, J., Chen, L., & Sukigara, S. (2012). Preparation and characterization of electrospun polycaprolactone nanofiber webs containing water-soluble eggshell membrane and catechin. *Journal of Fiber Bioengineering and Informatics*, *5*(2), 217–226. https://doi.org/10.3993/jfbi06201210

Kang, J., Kotaki, M., Okubayashi, S., & Sukigara, S. (2010). Fabrication of electrospun eggshell membrane nanofibers by treatment with catechin. *Journal of Applied Polymer Science*, *117*(5), 2042–2049. https://doi.org/10.1002/app.32035

Karantza, V. (2011). Keratins in health and cancer: More than mere epithelial cell markers. *Oncogene*, *30*(2), 127–138. https://doi.org/10.1038/onc.2010.456

Kim, G. H., Min, T., Park, S. A., & Kim, W. D. (2008). Coaxially electrospun micro/nanofibrous poly(ε-caprolactone)/eggshell- protein scaffold. *Bioinspiration and Biomimetics*, *3*(1). https://doi.org/10.1088/1748-3182/3/1/016006

Kim, G. H., Min, T., Park, S. A., Kim, W. D., & Koh, Y. H. (2007). Fabrication of a biocomposite reinforced with hydrophilic eggshell proteins. *Biomedical Materials*, *2*(4), 250–256. https://doi.org/10.1088/1748-6041/2/4/007

Kitahara, A. K., Suzuki, Y., Nishimura, Y., Suzuki, K., Kiyotani, T., Takimoto, Y., Nakamura, T., Shimizu, Y., & Endo, K. (1998). Evaluation of collagen nerve guide in facial nerve regeneration. *Journal of Artificial Organs*, *1*(1), 22–27. https://doi.org/10.1007/BF01340448

Konopka, J. B. (2012). N-acetylglucosamine functions in cell signaling. *Scientifica*, *2012*, 1–15. https://doi.org/10.6064/2012/489208

Kretlow, J. D., Klouda, L., & Mikos, A. G. (2007). Injectable matrices and scaffolds for drug delivery in tissue engineering. *Advanced Drug Delivery Reviews*, *59*(4–5), 263–273. https://doi.org/10.1016/j.addr.2007.03.013

Kular, J. K., Basu, S., & Sharma, R. I. (2014). The extracellular matrix: Structure, composition, age-related differences, tools for analysis and applications for tissue engineering. *Journal of Tissue Engineering*, *5*. https://doi.org/10.1177/2041731414557112

Landers, R., Pfister, A., Hübner, U., John, H., Schmelzeisen, R., & Mülhaupt, R. (2002). Fabrication of soft tissue engineering scaffolds by means of rapid prototyping techniques. *Journal of Materials Science*, *37*(15), 3107–3116. https://doi.org/10.1023/A:1016189724389

Laurent, U. B. G., Fraser, J. R. E., Engström-Laurent, A., Reed, R. K., Dahl, L. B., & Laurent, T. C. (1992). Catabolism of hyaluronan in the knee joint of the rabbit. *Matrix*, *12*(2), 130–136. https://doi.org/10.1016/S0934-8832(11)80054-5

Lee, R. J. (2007). Tissue engineering. In *Innovative Management of Atrial Fibrillation*, pp. 178–185. https://doi.org/10.1002/9780470994818.ch15

Leeming, D. J., & Karsdal, M. A. (2019). Type V collagen. In *Biochemistry of Collagens, Laminins and Elastin: Structure, Function and Biomarkers* (2nd ed., Vol. 1, Issue V). Elsevier Inc. https://doi.org/10.1016/B978-0-12-817068-7.00005-7

Li, N., Niu, L. na, Qi, Y. pin, Yiu, C. K. Y., Ryou, H., Arola, D. D., Chen, J. hua, Pashley, D. H., & Tay, F. R. (2011). Subtleties of biomineralisation revealed by manipulation of the eggshell membrane. *Biomaterials*, *32*(34), 8743–8752. https://doi.org/10.1016/j.biomaterials.2011.08.007

Liddington, R. C., & Ginsberg, M. H. (2002). Integrin activation takes shape. *Journal of Cell Biology*, *158*(5), 833–839. https://doi.org/10.1083/jcb.200206011

Liu, M., Zeng, X., Ma, C., Yi, H., Ali, Z., Mou, X., Li, S., Deng, Y., & He, N. (2017). Injectable hydrogels for cartilage and bone tissue engineering. *Bone Research*, *5*(January). https://doi.org/10.1038/boneres.2017.14

Lo, D. D., Zimmermann, A. S., Nauta, A., Longaker, M. T., & Lorenz, H. P. (2012). Scarless fetal skin wound healing update. *Birth Defects Research Part C - Embryo Today: Reviews*, *96*(3), 237–247. https://doi.org/10.1002/bdrc.21018

Lovvorn, H. N., Cheung, D. T., Nimni, M. E., Perelman, N., Estes, J. M., & Adzick, N. S. (1999). Relative distribution and crosslinking of collagen distinguish fetal from adult sheep wound repair. *Journal of Pediatric Surgery*, *34*(1), 218–223. https://doi.org/10.1016/S0022-3468(99)90261-0

Lu, J. W., Li, Q., Qi, Q. L., Guo, Z. X., & Yu, J. (2009). Surface engineering of poly(D,L-lactic acid) by entrapment of soluble eggshell membrane protein. *Journal of Biomedical Materials Research - Part A*, *91*(3), 701–707. https://doi.org/10.1002/jbm.a.32304

Lunge, S., Thakre, D., Kamble, S., Labhsetwar, N., & Rayalu, S. (2012). Alumina supported carbon composite material with exceptionally high defluoridation property from eggshell waste. *Journal of Hazardous Materials*, *237–238*, 161–169. https://doi.org/10.1016/j.jhazmat.2012.08.023

Macneil, J. H. (1998). *Waste egg shell components recovery*. https://patents.google.com/patent/WO1998041326A1/en

Maeda, K., & Sasaki, Y. (1982). An experience of hen-egg membrane as a biological dressing. *Burns*, *8*(5), 313–316. https://doi.org/10.1016/0305-4179(82)90029-8

Maeda, N., Ishii, M., Nishimura, K., & Kamimura, K. (2011). Functions of chondroitin sulfate and heparan sulfate in the developing brain. *Neurochemical Research*, *36*(7), 1228–1240. https://doi.org/10.1007/s11064-010-0324-y

Malaval, L., Wade-Guéye, N. M., Boudiffa, M., Fei, J., Zirngibl, R., Chen, F., Laroche, N., Roux, J. P., Burt-Pichat, B., Duboeuf, F., Boivin, G., Jurdic, P., Lafage-Proust, M. H., Amédée, J., Vico, L., Rossant, J., & Aubin, J. E. (2008). Bone sialoprotein plays a functional role in bone formation and osteoclastogenesis. *Journal of Experimental Medicine*, *205*(5), 1145–1153. https://doi.org/10.1084/jem.20071294

Martins, A., Araújo, J. V., Reis, R. L., & Neves, N. M. (2007). Electrospun nanostructured scaffolds for tissue engineering applications. *Nanomedicine*, *2*(6), 929–942. https://doi.org/10.2217/17435889.2.6.929

Martins, A., Chung, S., Pedro, A. J., Sousa, R. A., Marques, A. P., Reis, R. L., & Neves, N. M. (2009). Hierarchical starch-based fibrous scaffold for bone tissue engineering application. *Journal of Tissue Engineering and Regenerative Medicine*, *12*(3), 181–204. https://doi.org/10.1002/term

McKee, T. J., Perlman, G., Morris, M., & Komarova, S. V. (2019). Extracellular matrix composition of connective tissues: A systematic review and meta-analysis. *Scientific Reports*, *9*(1), 1–15. https://doi.org/10.1038/s41598-019-46896-0

Mecham, R. P., & Foster, J. A. (1978). A structural model for desmosine cross linked peptides. *Biochemical Journal*, *173*(2), 617–625. https://doi.org/10.1042/bj1730617

Mensah, R. A., Jo, S. B., Kim, H., Park, S. M., Patel, K. D., Cho, K. J., Cook, M. T., Kirton, S. B., Hutter, V., Sidney, L. E., Alves-Lima, D., Lin, H., Lee, J. H., Kim, H. W., & Chau, D. Y. (2022). The eggshell membrane: A potential biomaterial for corneal wound healing. *Journal of Biomaterials Applications*, *36*(5), 912–929. https://doi.org/10.1177/08853282211024040

Mensah, R. A., Salim, K., Peszko, K., Diop, S., Wong, T. H., & Chau, D. Y. (2023). The chicken eggshell membrane: A versatile, sustainable, biological material for translational biomedical applications. *Biomedical Materials*, *18*(4). https://doi.org/10.1088/1748-605X/acd316.

Mikos, A. G., & Temenoff, J. S. (2000). Formation of highly porous biodegradable scaffolds for tissue engineering. *Electronic Journal of Biotechnology*, *3*(2), 114–119. https://doi.org/10.2225/vol3-issue2-fulltext-5

Morello, R., & Esposito, P. W. (2012). Osteogenesis imperfecta. *JK Science*, *8*(3), 223–252. https://doi.org/10.5772/34775

Nakano, T., Ikawa, N. I., & Ozimek, L. (2003). Chemical composition of chicken eggshell and shell membranes. *Poultry Science*, *82*(3), 510–514. https://doi.org/10.1093/ps/82.3.510

Nitkin, R. M., & Rothschild, T. C. (1990). Agrin-induced reorganization of extracellular matrix components on cultured myotubes: Relationship to AChR aggregation. *Journal of Cell Biology*, *111*(3), 1161–1170. https://doi.org/10.1083/jcb.111.3.1161

Nusgens, B. V. (2010). Hyaluronic acid and extracellular matrix: A primitive molecule? *Annales de Dermatologie et de Venereologie*, *137*(SUPPL. 1), S3–S8. https://doi.org/10.1016/S0151-9638(10)70002-8

O'Brien, F. J. (2011). Biomaterials & scaffolds for tissue engineering. *Materials Today*, *14*(3), 88–95. https://doi.org/10.1016/S1369-7021(11)70058-X

O'Regan, A. W., Nau, G. J., Chupp, G. L., & Berman, J. S. (2000). Osteopontin (Eta-I) in cell-mediated immunity: Teaching an old dog new tricks. *Immunology Today*, *21*(10), 475–478. https://doi.org/10.1016/S0167-5699(00)01715-1

Ohto-Fujita, E., Konno, T., Shimizu, M., Ishihara, K., Sugitate, T., Miyake, J., Yoshimura, K., Taniwaki, K., Sakurai, T., Hasebe, Y., & Atomi, Y. (2011). Hydrolyzed eggshell membrane immobilized on phosphorylcholine polymer supplies extracellular matrix environment for human dermal fibroblasts. *Cell and Tissue Research*, *345*(1), 177–190. https://doi.org/10.1007/s00441-011-1172-z

Oliveira, J. T., Crawford, A., Mundy, J. M., Moreira, A. R., Gomes, M. E., Hatton, P. V., & Reis, R. L. (2007). A cartilage tissue engineering approach combining starch-polycaprolactone fibre mesh scaffolds with bovine articular chondrocytes. *Journal of Materials Science: Materials in Medicine*, *18*(2), 295–302. https://doi.org/10.1007/s10856-006-0692-7

Paulsson, M. (1992). The role of laminin in attachment, growth, and differentiation of cultured cells: A brief review. *Cytotechnology*, *9*(1–3), 99–106. https://doi.org/10.1007/BF02521736

Peltola, S. M., Melchels, F. P. W., Grijpma, D. W., & Kellomäki, M. (2008). A review of rapid prototyping techniques for tissue engineering purposes. *Annals of Medicine*, *40*(4), 268–280. https://doi.org/10.1080/07853890701881788

Pfister, A., Landers, R., Laib, A., Hübner, U., Schmelzeisen, R., & Mülhaupt, R. (2004). Biofunctional rapid prototyping for tissue-engineering applications: 3D bioplotting versus 3D printing. *Journal of Polymer Science, Part A: Polymer Chemistry*, *42*(3), 624–638. https://doi.org/10.1002/pola.10807

Pillai, M. M., Akshaya, T. R., Elakkiya, V., Gopinathan, J., Sahanand, K. S., Rai, B. K. D., Bhattacharyya, A., & Selvakumar, R. (2015). Egg shell membrane-a potential natural scaffold for human meniscal tissue engineering: An in vitro study. *RSC Advances*, *5*(93), 76019–76025. https://doi.org/10.1039/c5ra09959e

Puppi, D., Piras, A. M., Chiellini, F., Chiellini, E., Martins, A., Leonor, I. B., Neves, N., & Reis, R. (2010). Optimized electro- and wet-spinning techniques for the production of polymeric fibrous scaffolds loaded with bisphosphonate and hydroxyapatite. *Journal of Tissue Engineering and Regenerative Medicine*, *12*(3), 181–204. https://doi.org/10.1002/term.310

Qi, Q., Li, Q., Lu, J., Guo, Z., & Yu, J. (2009). Preparation and characterization of soluble eggshell membrane protein/chitosan blend films. *Chinese Journal of Polymer Science*, *27*(3), 387–392.

Rønning, S. B., Berg, R. S., Høst, V., Veiseth-Kent, E., Wilhelmsen, C. R., Haugen, E., Suso, H. P., Barham, P., Schmidt, R., & Pedersen, M. E. (2020). Processed eggshell membrane powder is a promising biomaterial for use in tissue engineering. *International Journal of Molecular Sciences*, *21*(21), 1–14. https://doi.org/10.3390/ijms21218130

Rose, M. L. H., & Hincke, M. T. (2009). Protein constituents of the eggshell: Eggshell-specific matrix proteins. *Cellular and Molecular Life Sciences*, *66*(16), 2707–2719. https://doi.org/10.1007/s00018-009-0046-y

Ruff, K. J., Winkler, A., Jackson, R. W., DeVore, D. P., & Ritz, B. W. (2009). Eggshell membrane in the treatment of pain and stiffness from osteoarthritis of the knee: A randomized, multicenter, double-blind, placebo-controlled clinical study. *Clinical Rheumatology*, *28*(8), 907–914. https://doi.org/10.1007/s10067-009-1173-4

Sachindra, N. M., Bhaskar, N., Sakhare, P. Z., & Mahendrakar, N. S. (2014). *A process for separation of shell membrane from wet egg shells*. http://www.allindianpatents.com/patents/258838-a-process-for-separation-of-shell-membrane-from-wet-egg-shells

Sah, M. K., & Pramanik, K. (2014). Soluble-eggshell-membrane-protein-modified porous silk fibroin scaffolds with enhanced cell adhesion and proliferation properties. *Journal of Applied Polymer Science*, *131*(8), 1–11. https://doi.org/10.1002/app.40138

Santos, M. I., Fuchs, S., Gomes, M. E., Unger, R. E., Reis, R. L., & Kirkpatrick, C. J. (2007). Response of micro- and macrovascular endothelial cells to starch-based fiber meshes for bone tissue engineering. *Biomaterials*, *28*(2), 240–248. https://doi.org/10.1016/j.biomaterials.2006.08.006

Sawai, T., Usui, N., Sando, K., Fukui, Y., Kamata, S., Okada, A., Taniguchi, N., Itano, N., & Kimata, K. (1997). Hyaluronic acid of wound fluid in adult and fetal rabbits. *Journal of Pediatric Surgery*, *32*(1), 41–43. https://doi.org/10.1016/S0022-3468(97)90089-0

Sayani, K., Dodd, C. M., Nedelec, B., Shen, Y. J., Ghahary, A., Tredget, E. E., & Scott, P. G. (2000). Delayed appearance of decorin in healing burn scars. *Histopathology*, *36*(3), 262–272. https://doi.org/10.1046/j.1365-2559.2000.00824.x

Silva, N. A., Salgado, A. J., Sousa, R. A., Oliveira, J. T., Pedro, A. J., Leite-Almeida, H., Cerqueira, R., Almeida, A., Mastronardi, F., Mano, J. F., Neves, N. M., Sousa, N., & Reis, R. L. (2010). Development and characterization of a novel hybrid tissue engineering-based scaffold for spinal cord injury repair. *Tissue Engineering - Part A*, *16*(1), 45–54. https://doi.org/10.1089/ten.tea.2008.0559

Singh, M., Foster, C. R., Dalal, S., & Singh, K. (2010). Osteopontin: Role in extracellular matrix deposition and myocardial remodeling post-MI. *Journal of Molecular and Cellular Cardiology*, *48*(3), 1–13. https://doi.org/10.1016/j.yjmcc.2009.06.015.Osteopontin

Sobral, J. M., Caridade, S. G., Sousa, R. A., Mano, J. F., & Reis, R. L. (2011). Three-dimensional plotted scaffolds with controlled pore size gradients: Effect of scaffold geometry on mechanical performance and cell seeding efficiency. *Acta Biomaterialia*, *7*(3), 1009–1018. https://doi.org/10.1016/j.actbio.2010.11.003

Soledad Fernandez, M., Moya, A., Lopez, L., & Arias, J. L. (2001). Secretion pattern, ultrastructural localization and function of extracellular matrix molecules involved in eggshell formation. *Matrix Biology*, *19*(8), 793–803. https://doi.org/10.1016/S0945-053X(00)00128-1

Soo, C., Hu, F. Y., Zhang, X., Wang, Y., Beanes, S. R., Lorenz, H. P., Hedrick, M. H., Mackool, R. J., Plaas, A., Kim, S. J., Longaker, M. T., Freymiller, E., & Ting, K. (2000). Differential expression of fibromodulin, a transforming growth factor-β modulator, in fetal skin development and scarless repair. *American Journal of Pathology*, *157*(2), 423–433. https://doi.org/10.1016/S0002-9440(10)64555-5

Takahashi, K., Shirai, K., & Hattori, M. (1996a). Conjugated film from reconstruction of collagen-soluble egg shell membrane protein matrix. *Nihon Chikusan Gakkaiho*, *67*(10), 862–868. https://doi.org/10.2508/chikusan.67.862

Takahashi, K., Shirai, K., Kitamura, M., & Hattori, M. (1996b). Soluble egg shell membrane protein as a regulating material for collagen matrix reconstruction. *Bioscience, Biotechnology and Biochemistry*, *60*(8), 1299–1302. https://doi.org/10.1271/bbb.60.1299

Takigaura, R., Sano, M., Sakamoto, M., & Murakami, S. (2009). *Fiber-treating liquid, modified cloth, and process for producing the same*. patent ID- TW200535151A

Tenti, P., & Vannucci, L. (2020). Lysyl oxidases: Linking structures and immunity in the tumor microenvironment. *Cancer Immunology, Immunotherapy*, *69*(2), 223–235. https://doi.org/10.1007/s00262-019-02404-x

Thoroski, J. (2003). *Eggshell processing methods and apparatus*. patent ID- US20040166213A1

Valenick, L. V., Hsia, H. C., & Schwarzbauer, J. E. (2005). Fibronectin fragmentation promotes α4β1 integrin-mediated contraction of a fibrin-fibronectin provisional matrix. *Experimental Cell Research*, *309*(1), 48–55. https://doi.org/10.1016/j.yexcr.2005.05.024

Varki, A. (2008). Sialic acids in human health and disease. *Trends in Molecular Medicine*, *14*(8), 351–360. https://doi.org/10.1016/j.molmed.2008.06.002

Vuong, T. T., Rønning, S. B., Ahmed, T. A. E., Brathagen, K., Høst, V., Hincke, M. T., Suso, H. P., & Pedersen, M. E. (2018). Processed eggshell membrane powder regulates cellular functions and increase MMP-activity important in early wound healing processes. *PLoS ONE*, *13*(8), 1–23. https://doi.org/10.1371/journal.pone.0201975

Wang, K. K., Nemeth, I. R., Seckel, B. R., Chakalis-Haley, D. P., Swann, D. A., Kuo, J. W., Bryan, D. J., & Cetrulo, C. L. (1998). Hyaluronic acid enhances peripheral nerve regeneration in vivo. *Microsurgery*, *18*(4), 270–275. https://doi.org/10.1002/(SICI)1098-2752(1998)18:4<270::AID-MICR11>3.0.CO;2-V

Xiong, X., Li, Q., Lu, J. W., Guo, Z. X., Sun, Z. H., & Yu, J. (2012). Fibrous scaffolds made by co-electrospinning soluble eggshell membrane protein with biodegradable synthetic polymers. *Journal of Biomaterials Science, Polymer Edition*, *23*(9), 1217–1230. https://doi.org/10.1163/092050611X576981

Xiong, X., Li, Q., Lu, J.-W., Guo, Z.-X., & Yu, J. (2010). Poly(lactic acid)/soluble eggshell membrane protein blend films: Preparation and characterization. *Journal of Applied Polymer Science*, *117*(4), 1955–1959. https://doi.org/10.1002/app.32181

Xu, Z., Neoh, K. G., & Kishen, A. (2010). A biomimetic strategy to form calcium phosphate crystals on type i collagen substrate. *Materials Science and Engineering C*, *30*(6), 822–826. https://doi.org/10.1016/j.msec.2010.03.014

Yang, Y. J., Jung, D., Yang, B., Hwang, B. H., & Cha, H. J. (2014). Aquatic proteins with repetitive motifs provide insights to bioengineering of novel biomaterials. *Biotechnology Journal*, *9*(12), 1493–1502. https://doi.org/10.1002/biot.201400070

Yao, J., Masuda, H., Zhao, C., & Asakura, T. (2002). Artificial spinning and characterization of silk fiber from. *Biochimie, 35*, 6–9.

Yeong, W. Y., Chua, C. K., Leong, K. F., & Chandrasekaran, M. (2004). Rapid prototyping in tissue engineering: Challenges and potential. *Trends in Biotechnology*, *22*(12), 643–652. https://doi.org/10.1016/j.tibtech.2004.10.004

Yi, F., Lu, J. W., Guo, Z. X., & Yu, J. (2006). Mechanical properties and biocompatibility of soluble eggshell membrane protein/poly(vinyl alcohol) blend films. *Journal of Biomaterials Science, Polymer Edition*, *17*(9), 1015–1024. https://doi.org/10.1163/156856206778365997

Yi, F., Yu, J., Guo, Z. X., Zhang, L. X., & Li, Q. (2003). Natural bioactive material: A preparation of soluble eggshell membrane protein. *Macromolecular Bioscience*, *3*(5), 234–237. https://doi.org/10.1002/mabi.200390030

Yoo, J., Park, K., Yoo, Y., Kim, J., Yang, H., & Shin, Y. (2014). Effects of egg shell membrane hydrolysates on anti-inflammatory, anti-wrinkle, anti-microbial activity and moisture-protection. *Korean Journal for Food Science of Animal Resources*, *34*(1), 26–32. https://doi.org/10.5851/kosfa.2014.34.1.26

Zadik, Y. (2007). Self-treatment of full-thickness traumatic lip laceration with chicken egg shell membrane. *Wilderness and Environmental Medicine*, *18*(3), 230–231. https://doi.org/10.1580/06-WEME-LE-082R1.1

Zeidler, G., Angeles, L., Base, D. D., Base, D. D., Patent, A. J., & Summit, B. R. (1991). *Method for shell removal from hard boiled eggs*. https://patents.google.com/patent/US5053238A/en

8 Quantitative Optical Phase Imaging Techniques

Interference-Based Basics and Working Principle

*Anusha Pillai[1], Anand K Asundi[2], and Ajeetkumar Patil[1]**

[1]Department of Atomic and Molecular Physics, Manipal Academy of Higher Education (MAHE), Manipal, Karnataka, India
[2]d'Optron Pte Ltd, Singapore
*Corresponding Author

INTRODUCTION TO OPTICAL PHASE IMAGING

The phase of a periodic signal is a real-valued scalar that describes the relative location of each point on the waveform within the span of each full period. Phase is expressed in terms of radians. Phase in wave optics is nothing but a two-dimensional complex component that is associated with a monochromatic coherent optical field [1]. When we compare phase and amplitude, the amplitude component is always better understood as the amplitude's square is intensity that is visible to the human eyes, hence being difficult to visualize and understand phase. Human eyes and most imaging sensors cannot detect the phase directly as the frequency of oscillation of light is around 10^{15} Hz which is much higher than the response time of human eyes, which is around 30 Hz, and most of the camera detectors can only reach up to 10^{8} Hz. Acquiring the phase information is important as it helps to retrieve the 3D quantitative information of the object, whereas amplitude can give qualitative 2D information through intensity variations. Thus, phase retrieval plays a very important role in many industrial and medical applications. The phase imaging methods, especially for biological cells and weakly absorbing semi-transparent samples, have a long history of development. Cell components such as the cytoplasm and most of the organelles weakly absorb light, and thus the contrast observed is very little. Contrast simply defines how clearly the sample is differentiated from the background. The contrast of an image is determined by the phenomenon of elastic scattering, which occurs when light waves interact with small particles or structures within a material, causing them to scatter in various directions. Unlike other types of scattering, elastic scattering does not alter the properties of the waves and is thus referred to as elastic. In optical microscopy, elastic scattering is commonly employed to visualize structures that are not visible to the naked eye, such as cells and organelles. Achieving optimal image contrast can be challenging since insufficient elastic scattering fails to reveal the shape and structures clearly, while excessive elastic scattering can result in the overall image being washed out due to multiple scattering. Both extremes in scattering conditions lead to poor image contrast. It's worth noting that samples with transparent and thin tissue slices exhibit weak elastic scattering, which means that the scattering is significantly weaker than the incident light. These structures primarily affect the phase component of the incident light field and are commonly referred to as phase objects [2]. In conventional bright-field microscopy, the phase information cannot be retrieved. This limitation is due to the microscope's ability to only detect variations in the amplitude of light. To address this issue, two main approaches are employed: labeled imaging

 DOI: 10.1201/9781003545712-8

techniques and label-free imaging methods. Labeled imaging techniques involve the use of stains and fluorescent tags. By applying these labels to the sample, phase objects can be converted into amplitude objects, allowing their visualization through the microscope. This method relies on modifying the sample with specific markers that can interact with light and produce detectable signals, enabling the observation of phase structures. On the other hand, label-free imaging methods utilize computational approaches to extract phase information without the need for exogenous labels. These techniques aim to directly capture and analyze the phase variations present in the sample. By leveraging advanced computational algorithms and image processing techniques, label-free imaging enables the visualization and characterization of phase objects without the requirement of staining or tagging. In summary, labeled imaging techniques involve converting phase objects into amplitude objects using stains or fluorescent tags, while label-free imaging methods employ computational approaches to directly extract and analyze the phase information from the sample. Both approaches provide valuable tools for observing and studying phase objects in microscopy.

Labeled Imaging

In the field of histopathology, stains are widely employed as a reliable method for visualizing and identifying cells and tissues, providing qualitative information. In the realm of cell biology, fluorescence microscopy is of great significance, utilizing diverse fluorescent tags to label cells and enabling detailed fluorescent imaging. Fluorescent imaging techniques allow for the incorporation of fluorescent molecules, quantum dots, fluorescence proteins, and other agents into the sample, playing a crucial role in visualizing specific structures or molecules within cells. These fluorescent labels enable researchers to obtain qualitative information about the location and distribution of targeted components. To further enhance the utility of fluorescence microscopy and obtain quantitative information from the sample, there are two primary approaches. Firstly, auto fluorescence can be utilized, which refers to the intrinsic fluorescence emitted by certain endogenous molecules within the sample. This inherent fluorescence can be exploited to quantify specific characteristics or components in the sample without the need for exogenous labels. Alternatively, researchers can opt to tag the sample with specific fluorescent molecules, quantum dots, fluorescence proteins, or similar agents. These tags are designed to selectively bind to target molecules or structures of interest, allowing for quantitative analysis and measurements. By precisely labeling and quantifying the fluorescence signal, researchers can gather quantitative data on the abundance, distribution, and dynamics of the labeled components within the sample. In summary, in histopathology, stains serve as a gold standard for qualitative observation and identification of cells and tissues. Fluorescence microscopy, on the other hand, utilizes various fluorescent tags to label cells, providing crucial tools for detailed fluorescent imaging in cell biology. To obtain quantitative information, researchers can leverage auto fluorescence or employ specific fluorescent tags, quantum dots, fluorescence proteins, and similar agents to label the sample and measure the fluorescence signal, enabling quantitative analysis of targeted components. Techniques like confocal microscopy and multiphoton microscopy are developed to overcome the major drawback of fluorescence imaging, which is the flare in the object, and these techniques have a high potential for optical sectioning; because of this capability one can observe the cell organelles in 3D. Also, there are many other microscopy techniques whose base is fluorescence; one such method is total internal reflection fluorescence microscopy (TIRFM). This technique has special property induced near the sample, near to the interface of two media having different refractive indices. The principle behind this technique is total internal reflection and formation of an evanescent wave. The field present on the surface is called the evanescent wave, which can in turn play an active role in fluorophore excitement. The advantage of this technique is that all the background fluorescence is reduced without the use of any pinhole. In addition, the cell is not illuminated inside, thus maintaining the healthy cell culture ambience; therefore, the survival of cell cultures while imaging is increased. The advantages of TIRFM are no background fluorescence, without focus, fluorescence is avoided and there is not much exposure of the light within the cell culture. Three-dimensional reconstruction of multi-angle TIRFM image stacks is a technique with high resolution dedicated completely to the biological phenomena that are happening up to a maximum of 800 nm below the surface of the cell. However, this technique is undemanding because the evanescent wave exponentially decreases, therefore allowing much lower exploration of the regions to be imaged. It limits the investigation of the biological phenomena near plasma the membrane. One limitation of traditional confocal microscopy and TIRFM is that it cannot go beyond the diffraction limit and thus researchers have developed and improved the optical imaging methods which has led to the emergence of super resolution fluorescence technique which includes structured illumination microscopy (SIM), stimulated emission depletion microscopy (STED), photo-activated localization microscopy (PALM), stochastic optical reconstruction microscopy (STORM), and points accumulation for imaging in nanoscale tomography (PAINT). Super-resolution methods like SIM enhance the resolution of imaging by illuminating the sample with a grid pattern, creating a moiré effect. This technique enables the transmission of fine structural details through the imaging optics, leading to a twofold improvement in spatial resolution compared to the traditional limit [3–5]. SIM represents an alternative super-resolution optical microscopy method that has found more widespread use in biomedical research over time. SIM involves the use of fringe patterns with varying orientations and phase shifts to illuminate specimens, recording the resulting moiré patterns in a sequential manner. Super-resolution images can then be reconstructed from these intensity patterns. SIM offers several advantages compared to other super-resolution techniques, including straightforward sample preparation, rapid imaging capabilities, and minimal invasiveness. Furthermore, advancements in electron devices such as detectors and wave-front modulators have led to significant developments and various extensions in the concept and implementation of SIM [3, 6–10]. STED microscopy gives us sub-diffraction resolution while keeping the important benefits of fluorescence such as high sensitivity, very good optical sectioning, and enhanced molecular specificity. In the case of STED, the highest resolution is obtained as it selectively quenches fluorophores used for labeling the samples. With the appropriate choice of type of excitation laser and the fluorophore, resolution can expand up to nanometer scale [11–18]. Also with STED, a beam of light has to excite a fluorescent molecule from ground state to excited state from where it gets de-excited by a photon to a lower state either by stimulated emission and spontaneous emission through florescence. To get the florophore into the ground state the stimulated emission overtakes the spontaneous emission, as the spontaneous emission is very much faster in the terms of nanoseconds. As the fluorescence window is very short, it requires many photons. That is to quench the complete florescence of a fluorescent molecule one will require light of much higher intensities, in other words saturation intensities. This might lead to bleaching of the sample and toxicity induced in the ample due to the much higher intensity light. This forms a major limitation in the STED leading to slow progress in the field of bioimaging. In addition, expensive and complex instrumentation is one such factor for its slow growth [19, 20]. Also, these above-mentioned techniques limit the illumination to the region of the sample lower than the diffraction limit, or in other words one can say it stochastically or temporally separate the fluorescence emission of single fluorophores so that their individual positions can be identified. Hence, the term single molecule localization microscopy has led to the developments of STORM, PALM, PAINT, etc. Another method in super resolution microscopy is PALM that has pushed the spatial resolution to much higher levels. PALM is a cutting-edge imaging technique that achieves high-resolution microscopy by precisely pinpointing fluorescent markers within a sample. Unlike traditional microscopy methods of its time, PALM could surpass the limitations of the diffraction limit. It achieves this by activating specific proteins within the sample through exposure to ultraviolet light, allowing their positions to be accurately determined. After prolonged ultraviolet exposure, these activated proteins lose their fluorescence, and the process is repeated until all the proteins in the specimen are precisely localized. Similar to other super-resolution techniques, PALM finds a wide range of applications and has made a significant impact on the field of biomedical research [19–22]. Super-resolution (SR) fluorescence microscopy comprises a category of optical imaging methods that can achieve finer spatial resolution than the diffraction limit allows. One commonly used SR technique, (STORM) relies on the principle of localizing single molecules. STORM typically delivers spatial resolution in the range of 20 to 30 nanometers, a significant improvement compared to conventional optical microscopy, offering ten-fold enhancement. While STORM provides superior spatial resolution using uncomplicated optical equipment and standard fluorescent dyes, it is susceptible to image artifacts and reduced image quality when sample preparation or imaging conditions are suboptimal. To obtain high-quality STORM images, it's essential to carefully optimize all three aspects: sample preparation, image acquisition, and image reconstruction [23–29]. In the case of STORM, which is again another variation of STED, it makes use of time-resolved selective switching of the fluorophores used for marking the sample. By selective switching, series of images are collected which are later combined to obtain images of nanometer (nm) scale resolution. In addition, by choosing the proper measurement optics one will be able to obtain the florescence from the below the sample surface which can be later reconstructed to produce 3D information [18]. Another method, called PAINT, is where super resolution can be achieved. Conventional PAINT relies on fluorescent molecules binding transiently and non-specifically to hydrophobic areas of a structure, gradually building a super-resolution image as each molecule is located. In all PAINT techniques, it's feasible to fine-tune the concentration of the interacting fluorescent molecules for optimization. DNA-PAINT, on the other hand, was developed to facilitate precise and specific binding of the fluorophore to the imaged structure. Fluorescence imaging has limitations, including issues like photobleaching and sample damage, typically stemming from highly focused illumination sources. Unlike the PALM and STORM methods, which are hampered by the fading of dye molecules over time, PAINT-based methods continuously renew the supply of fluorescent probes. This capability allows for extended imaging sessions, resulting in a higher concentration of localized points and the potential to achieve higher resolution images. Additionally, there are alternative techniques such as optical coherence tomography (OCT) that can also overcome limitations and achieve sub-micron level resolution [30–32]. One can achieve resolution beyond the diffraction limit without the use of sample labeling and the threats of photo bleaching could be avoided using these scattering techniques.

In addition, all forms of fluorescence microscopy that come under the category of extrinsic contrast techniques provide more molecular specificity and high spatial resolution. Use of exogenous labeling agents changes the cellular and molecular structures that includes cell permeabilization, chemical or immunostaining or genetic modification, therefore being very less likely to be used for imaging the live cells. Although, the fluorescence techniques require the use of tags, it won't be useful for the samples that are not fluorescent. Although most biological cells are transparent and do not change the amplitude of the light passing through them, they introduce phase delays due to different structural regions of non-uniform optical density that is the refractive index. Thus, retrieving phase qualitatively as well quantitatively became a new area of exploration for scientists in this field, in other words label-free imaging. It is also an ideal and non-invasive method to study live cell dynamics over any time scales. In addition, it is used to study the effect of drugs on cells, cell growth, cell proliferation, cell dry mass, single cell morphology, and motility of the biological cells. It becomes possible when we can measure minute structural and motion changes at the nanometer scale by quantifying variations in optical path length caused by cells, all without the need for physical contact or invasive methods.

LABEL-FREE IMAGING

Label-free imaging can be done in two ways, one is using interference and other is non-interferometric based. Under interference-based imaging, there is spatial light interferometry microscopy, gradient light interference microscopy, digital holography, etc. Non-interferometric-based techniques are nothing but computational imaging, where several algorithms are developed to retrieve the phase information. There are again different ways to retrieve the phase information, i.e., iterative methods, transport of intensity equation (TIE), etc. Within the category of interference-based techniques, we find spatial light interference microscopy (SLIM), a groundbreaking and exceptionally sensitive quantitative phase imaging (QPI) method, with the potential to revolutionize the study of biological structures and dynamics, offering capabilities previously unattainable. It is an advanced optical microscopy technique, which is a combination of phase contrast microscopy and the holography technique. Interference is a basic principle behind this technique. Therefore, there is the gradient light interferometry microscope (GLIM), which is the advanced version of the differential interference contrast microscope. In addition, there is digital holographic microscopy (DHM) that uses the principle of interference to obtain a hologram, a phase object from which the information is retrieved digitally. OCT is also an interferometric method; it is mostly used in medical imaging. We can obtain the 3D image of the internal structures of the tissues and the different layers within the tissue. While OCT is a powerful medical imaging technique with many applications it does suffer from drawbacks. One is the limited penetration depth. OCT can go up to tissues, it cannot image structures deep into body. The second point in OCT is that images are affected by various artifacts which are non-biological features that can interfere with the interpretation of the image. Examples of artifacts are motion artifacts due to patient movement and speckle noise caused by the interference pattern of the light waves. OCT is highly operator-dependent; it requires a skilled operator to perform a scan and interpret the results. The quality of the images depends on the position of the scan probe and the patient's cooperation. The cost is largely due to the specialized equipment required, such as the light source and the detection system. The drawbacks associated with OCT, such as noise, are common drawbacks for all the above-mentioned interference-based techniques because of their interferometric nature. Interference is the very sensitive phenomenon; even slight disturbances can lead to undesirable background that can be seen in the image. Further, to remove the noises, extra effort has to be made. The Interference-based methods are more prone to contamination in light path and sensitive to noise. The image here is calculated from the resulting interference pattern, even the slightest dust or any sort of contamination inside the cell chamber affects the light which could be observed in the final image as a broad-brush noise level. In particular, the Mach–Zehnder configuration used for DHM is especially weak as contamination can easily alter each path. The experimental setup involved with interferometric system is highly complex and requires a restricted environment to achieve stable measurements.

In order to overcome the restrictions, low coherent illumination techniques like spatial light Interference Microscopy (SLIM), etc. are used. Fusing common path configuration and broadband illumination methods reduces speckle noise issues and the problem caused due to mechanical vibrations. Nonetheless, these low coherent illumination techniques require a spatially coherent illumination to covenant precise and accurate phase measurements with the absence of halo artifacts. Inevitably, the maximum resolution achieved is limited by the coherent diffraction limit, even though the synthetic aperture achieves oblique illumination or structured illumination that broadens the imaging resolution to that of a traditional, incoherent counterpart. This group of techniques demands a very convoluted optical setup that is not usually available to most of the pathologists and biologists that avert their wide application in biomedicine. To overcome the limitations associated with interference-based, algorithms to retrieve phase can be engaged to achieve intensity measurements at different axially displaced planes without explicit manipulation of object and reference beams. These techniques are more computational algorithms, mathematical functions involved to retrieve the phase information, Therefore providing us with the label-free imaging setup with less cost and a compact system. This domain is again subclassified into wave front-based and phase retrieval. The wave front sensing-based involves a Shack–Hartmann sensor, pyramid warfront sensor, etc., and phase retrieval has TIE, DPC, etc. which are examined in detail in the upcoming sections.

CLASSIFICATION OF OPTICAL PHASE IMAGING TECHNIQUES

Broadly, the phase can be retrieved both in qualitative and quantitative ways. As shown in Figure 8.1 qualitative techniques are generally where phase can be visually observed that includes phase

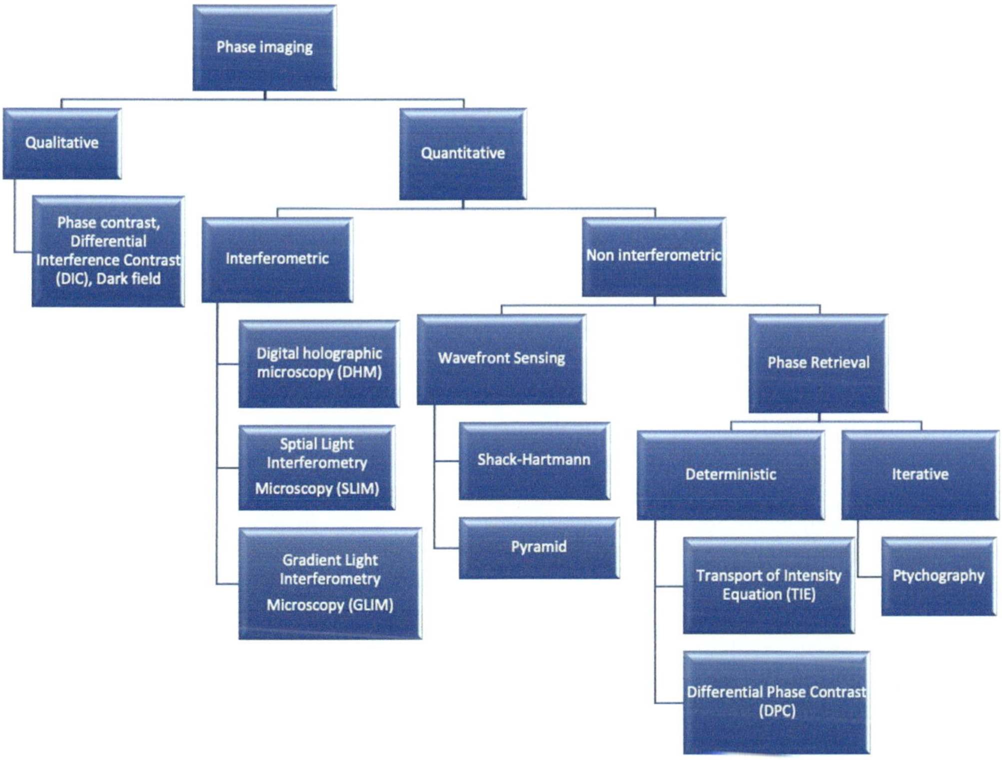

FIGURE 8.1 Classification of phase imaging techniques.

contrast and differential interference contrast (DIC). In quantitative, it can divide into two lineages, interferometric and non-interferometric techniques. In this chapter various interference-based optical phase imaging methods are discussed. Non-Interference-based QPI techniques are discussed in the next chapter.

Qualitative Optical Phase Imaging

In this type, one can only measure and obtain the amplitude information, the phase information cannot be retrieved. The phase changes can be seen visually in the form of contrast. This is more than two-dimensional. There are different qualitative imaging techniques, some of them are phase contrast microscopy, differential interference contrast microscopy, and dark field microscopy. The working principle of each method will be discussed in the further sections below.

Phase Contrast Microscope

In the case of transparent objects, the wave theory of light curiously enough has never been exclusively applied even though the only difference in either thickness or refractive index. This method paved way for the invention of the phase contrast (PC) microscope by Frederik (Frits) Zernike; he was awarded the Nobel Prize in 1953 [2]. The PC microscope and its related techniques has been a cornerstone of every cell biology laboratory.

The theory behind the working principle is that an illuminating wave passes through the sample and divides into two components. The first component (S) is the one that passes through the sample without any interaction (unretarded rays). In addition, the second component (D) gets scattered all over the sample and passes through (retarded rays). After passing the sample these two components interfere to produce a resultant wave (P). Mathematically this process in the phase contrast microscopy can be represented as Equation 8.1

$$P = S + D \tag{8.1}$$

The adapters in the PC microscope are the annular ring and phase plate. These convert the minute phase changes, which are caused due to the difference in the refractive index and the thickness of the sample into gray shades (amplitude changes). The beam passes through the annular ring and the condenser, thus illuminating the sample as shown in Figure 8.2. The principle behind the PC microscope is that phase and amplitude between the rays diffracted by the sample and unretarded rays are varied to obtain the contrast. Some of the light passes through the sample diffracted from them and some of the light passes from around the specimen unretarded. Due to the variation in refractive index and thickness of the sample the light passing through the sample, when it arrives at the image plane is shifted in phase from the unretarded light at about 90 degrees. When the annular ring is placed at front focal plane of the condenser lens and when the phase plate is placed at the back focal plane of the objective lens, it is observed that there is phase shift between diffracted and unretarded rays, leading to destructive interference. These phase differences can be seen in the form of amplitude difference through the eyepiece. This is called the positive or dark phase contrast [2]. The phase contrast microscope has led to a revolution in the field of biological imaging as the enhanced signal has led to observing the biological samples that are cells and tissues, especially the transparent samples without stains. The phase contrast microscope and its abilities have been combined with X-rays and have been used to study lung tissues [33]. In addition, phase contrast microscopes are combined with DHM to obtain quantitative information [34]. Thus, one can phase contrast with various methods to obtain better contrast and label-free imaging.

Differential Interference Contrast Microscope

The differential interference contrast (DIC) microscope operates on the principle of utilizing interference. It involves splitting a single light beam into two separate beams. One of these beams passes through the specimen, while the other serves as a reference beam that passes directly through a reference point. These two beams are then recombined, resulting in the formation of an interference pattern. Georges Nomarski made significant advancements in this technology in 1960, introducing modifications to the basic setup. One key enhancement involved the incorporation of a Wollaston prism, positioned in the high-magnification back focal plane of the microscope's high numerical aperture objectives. Additionally, he reduced the separation between the two beams to a size smaller than the resolution limit of the objective. As a result, when the two beams were combined, they not only generated an interference pattern but also produced an enhanced image of the specimen. In this method, the beam is initially a plane of polarized light as it gets split by the Wollaston prism and will have different polarization as shown in Figure 8.3.

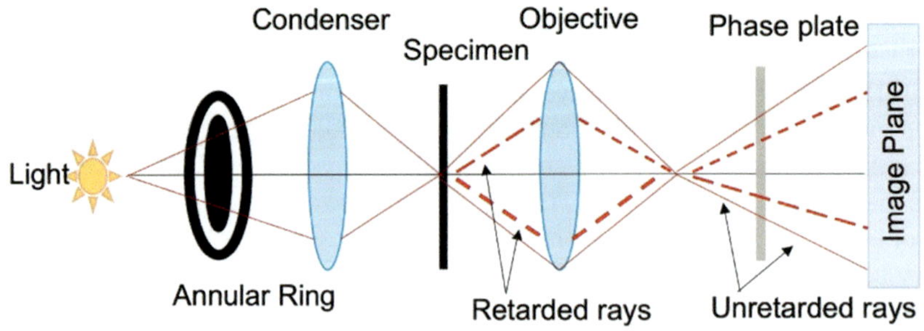

FIGURE 8.2 Schematic of a phase contrast microscope.

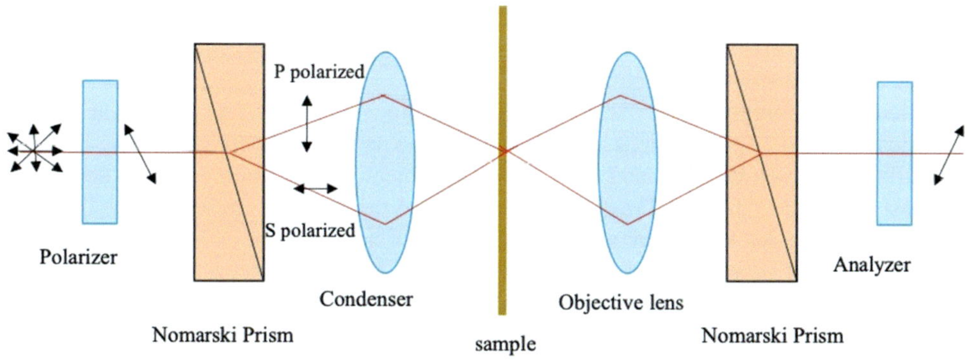

FIGURE 8.3 Schematic of a differential interference contrast microscope.

As the two beams traverse the specimen, they encounter differing refractive index variations, leading to unequal changes in their properties. Subsequently, the initial separation between the split beams is eliminated, and their polarization angle remains unaltered after passing through both the sample and the objective lens. In the intermediate image plane, these two beams combine, resulting in the creation of amplitude contrast. This contrast reveals the discrepancies in the optical path length caused by the sample. The system's high sensitivity to changes in the specimen's curvature and refractive index gradients is attributed to the fact that the beam separation is smaller than the diffraction limit, enabling it to discern even subtle alterations in the sample. The name differential is because of the differences in the optical path length of the two beams [2]. The intensity distribution in the DIC is given by Equation 8.2.

$$I^{DIC}\left(x',y'\right)=\frac{1}{2}\tilde{I}\left(x,y\right)1-COS\ [\frac{2\pi}{\lambda}\{\Gamma+d.\gamma(x',y')\bullet\cos(\theta(x',y')-\sigma)\ \}] \quad (8.2)$$

where $\tilde{I}$ (x, y) is intensity of the illumination beam before prism 1, γ (x', y') and θ (x', y') are gradient magnitude and azimuth, λ is wavelength. Differential interference contrast (DIC) microscopy offers several advantages, it eliminates phase halo and shading effects, producing pseudo three-dimensional (3D) images that appear genuinely 3D. However, it's important to note that the resulting image is not an exact replica of the sample. This can be appreciated by adjusting the Wollaston prism and observing how it affects the image's appearance. To gain a better understanding of a sample's structure, one can compare images obtained using phase contrast and DIC. DIC operates at the resolution limit of the objective lens, allowing for higher resolution compared to phase contrast. Optical sections acquired through DIC are notably clearer and provide better resolution along the z-axis than those obtained with phase contrast. In the field of bioimaging, DIC microscopy has been widely used for various studies. For instance, it has been applied to examine epithelial cells and in vivo wound healing. Additionally, DIC has been combined with infrared video microscopy for live cell imaging of brain slices and cellular organelles. While DIC is primarily a qualitative method, it can be enhanced by incorporating algorithms to obtain qualitative information. It's worth mentioning that DIC microscopy requires a microscope that can adapt to its optical requirements, and it cannot be used for samples that exhibit birefringence. These limitations should be considered when employing DIC.

Dark Field Microscope

Dark field microscopy is another technique invented to enhance the contrast. In this technique, the ocular lens and the objective lens used are same as that of ordinary light microscopes. Except for the condenser, the condenser lens used here in the dark field condenser has an opaque disk. This disk does not allow the central light which usually passes through and around the specimen to enter the objective lens directly. The condenser allows only those lights that are obliquely scattered from every angle, forming an inverted hollow cone of light (which has its apex centered in sample plane) to reach the sample as shown in Figure 8.4.

This microscope observes highly scattering or diffracting specimens. In the absence of sample, the numerical aperture of the condenser is greater than the numerical aperture of objective lens, in this case all the oblique rays cross and do not enter the objective lens leading to dark field of view. The contrast achieved in dark field microscopes (DFMs) is very high because only the light that is highly scattered or diffracted off the specimen enters the objective lens. DFM enables us to study marine samples, it is very useful in examine details like boundaries, edges etc. Contrast is not very accurate compared with PC and DIC. In addition, the images obtained are more vulnerable to distortion, inaccuracies, and degradation. A very thin sample will be more prone to show shadow artifacts in the image obtained. If water or oil is used in the slides, air bubbles are unavoidable leading to a more distorted image.

Although PC, DIC, and dark field produce high contrast advantages, in the case of PC and DIC both have an important drawback in that the intensity that is measured has an non-invertible, and have a nonlinear relationship with the phase of the sample [35]. The halo and shadow artifacts caused will be a hindrance while image processing and cell segmentation can solve it. In addition, being a qualitative approach, without the phase information being retrieved one cannot quantify

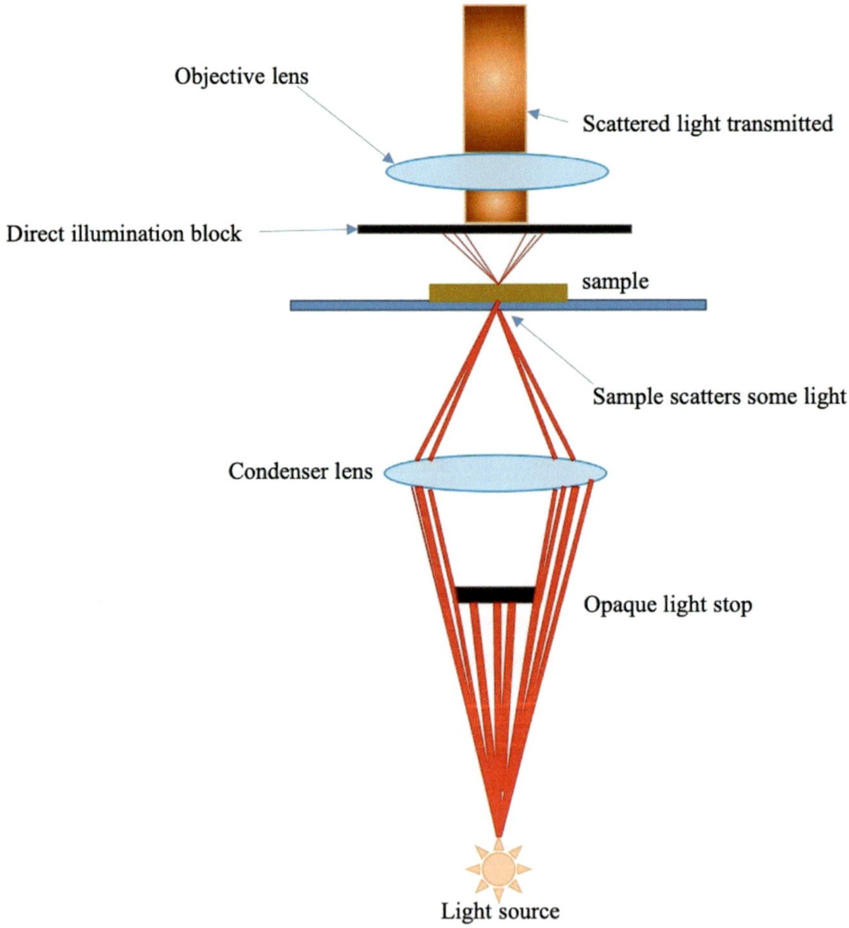

FIGURE 8.4 Schematic of a dark field microscope.

features like optical thickness, dry mass index, optical volume, etc. Due to these limitations and also the invention of digital sensors, their advancement has resulted in the emergence of phase-measuring leading QPI [36, 37], which combines the optics, theory of imaging, and the computational algorithms to retrieve the phase information [35].

QUANTITATIVE PHASE IMAGING

Conventional optical imaging techniques like phase contrast microscopy and differential interference contrast (DIC) microscopy, while non-invasive, do not offer the level of subcellular specificity required and are primarily qualitative in their nature. Even though the qualitative methods have a certain level of advancement, there is one important drawback, that is phase and intensity have a nonlinear relationship with each other which is noninvertible. Just as with the amplitude information, one cannot obtain quantitative features such as height, volume, refractive index, etc. Another drawback is that halo and shadow artifacts are formed and thus make it very complicated to process the images [38, 39]. This is where quantitative imaging of the biological cells and tissues are of importance. In quantitative phase imaging (QPI), not only the intensity and amplitude information but also the phase information can be retrieved. QPI is a technique, which converts the phase changes to pixel intensities using interference techniques [38, 40]. As mentioned earlier, QPI can be done in two ways; one is using interference and other is non-interferometric based. Under the interference-based method, there is spatial light interferometry microscopy, gradient light interference microscopy, digital holography, etc. Non-interferometric based techniques are nothing but computational imaging, where several algorithms are developed to retrieve the phase information. There are two ways this can be done; one is wavefront sensing and other using phase retrieval algorithms. Shack–Hartmann and pyramid are the most commonly used wavefront sensing methods. Phase retrieval algorithms again have two methods, one is iterative based which includes techniques like phytography and another method is deterministic which has techniques like TIR and DPC.

INTERFERENCE BASED

Qualitative techniques use methods like staining, fixing, and tagging while preparing the sample which in turn will damage the cell morphology and structure. Also, as mentioned earlier, the cells' quantitative features cannot be obtained with the qualitative approach [41–43]. When it comes to phase measurement, interferometry plays a crucial role. The basic principle of interference is nothing but the superimposing of two beams leading to a fringe pattern. This pattern is captured by the camera or digital devices through an intensity signal. Interferometry has several branches such as white light interferometry (SLIM, GLIM) and digital holographic microscopy (DHM). These will be discussed further in the sections below.

Spatial Light Interferometry Microscopy

Spatial light interferometry microscopy (SLIM), a new and exceptionally sensitive quantitative phase imaging (QPI technique, holds the potential to revolutionize research in biology and other fields by providing the means to investigate structure and dynamics with unprecedented precision. It is an advanced optical microscopy technique, which is a combination of phase contrast microscopy and the holography technique. Interference is a basic principle behind this technique. The SLIM setup contains the conventional phase contrast microscope and SLIM module attached to it. The condenser annulus is placed at the focal plane of the condenser lens, which collimates the light, therefore illuminating the specimen. The basic phase contrast microscope induces the delay in the phase between the light that is scattered and light that is unscattered about ¼ of the wavelength using a phase ring. The tube lens delivers the image to the image plane that is further processed by the SLIM module. The back focal plane of the objective lens is relayed onto the surface of spatial light modulator (SLM) by the Fourier lens L1 as shown in Figure 8.5.

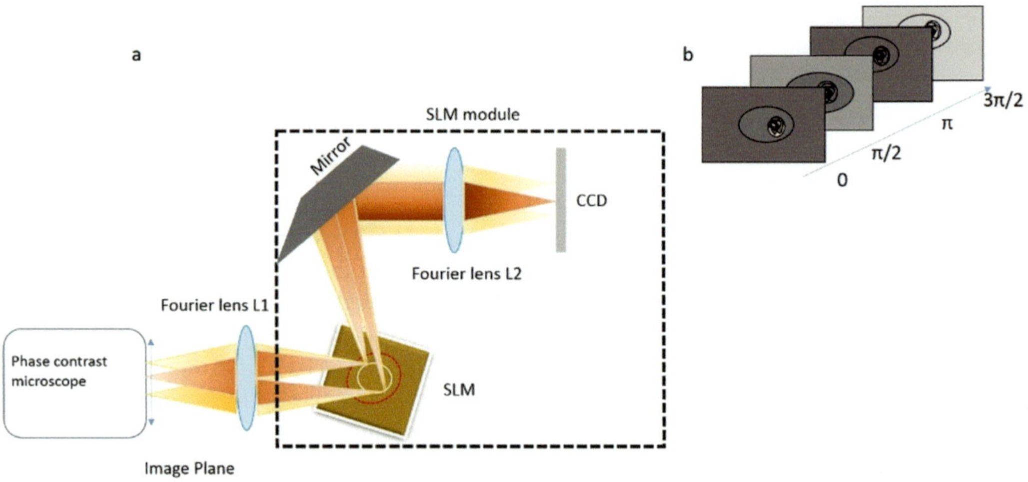

FIGURE 8.5 Schematic of a spatial light interferometry microscopy.

Different masks on the SLM are displayed so that the phase delay between the scattered and the unscattered beam is modulated accordingly. The final image is reconstructed at the charge-coupled device (CCD) plane by Fourier lens L2. The image plane is conjugated with the CCD plane. SLIM produces additional spatial modulation to the output given by the phase contrast microscope. Along with the $\Pi/2$ phase shift introduced between scattered and unscattered beams the SLIM module provides $\Pi/2$ phase shifts by increment and thus images are recorded. The signal interfering at the sample plane at the position r (x, y) is given by Equation 8.3.

$$I(r,\varnothing) = I_o + I_1(r) + 2\,|\,\Gamma_{01}(r)\,|\cos[(\mathrm{r}) + \varnothing] \tag{8.3}$$

where I_o and I_1 are intensity of reference and sample respectively, and $\Gamma_{01}(r)$ is the temporal cross-correlation function between the two fields evaluated at the origin. ϕ represents the phase shift between the scattered and unscattered fields. Here, four images are captured with the corresponding reference beam phase shifts $0, \Pi/2, \Pi, 3\Pi/2$ as shown in Figure 8.5(b). The phase delay [44] is calculated using by Equation 8.4

$$\Delta\phi(r) = \tan^{-1}\frac{I\left(r,-\frac{\pi}{2}\right) - I\left(r,\frac{\pi}{2}\right)}{I(r,0) - I(r,\pi)} \tag{8.4}$$

Speckle-free images are obtained using SLIM which enables spatially sensitive optical path measurement. In addition, as it uses common path interferometry it allows for temporally sensitive optical path measurement. As the image obtained from the SLIM technique is free of speckles, it paves the way for spatially sensitive measurements of the optical path length. Measurement of the optical path length with temporal sensitivity is achieved as SLIM uses the method of common path interferometry. As it uses broadband illumination, it offers many opportunities for spectroscopic studies. SLIM is widely used in the field of biology to study the cell dynamics, cell growth, proliferation, cell migration, etc. [45]. A lot of label-free cell imaging studies have been done with the help of SLIM [46] and also to study the intrinsic cellular structures of the transparent samples [47]. Researchers are very eager and confident about the efficiency of SLIM as it could bring a major turning point in optical microscopy as it could easily modify the phase contrast microscopes that are currently providing both qualitative and quantitative information of the nanoscale structures [47].

Gradient Light Interferometry Microscope

Gradient light interferometry microscopy (GLIM) is another advanced technique that is used to retrieve the depth information from both thin and thick samples that are unlabeled. This technique extracts the phase information from the samples by using the method of low coherence interferometry, which is further used to extract features such as cell volume, cell mass, surface area, etc. In GLIM the phase shifts between the two interfering waves are controlled so as to obtain multiple images and combine them, this technique also suppresses most of the incoherent background noise because of the multiple scattering [48]. GLIM is an excellent tool for in vitro fertilization as its label-free and has an advantage over the labeling agents that enhances the contrast and fluorophores that could harm the potency and efficiency of the embryo. As GLIM is exploited as an additional component to an established inverted microscope, we expect that the biological community will adopt it quickly. GLIM is a special case of imaging technique, which retrieves the phase information from the samples that are both thick and thin. GLIM is the advanced version of the differential interference contrast microscope. The conventional DIC microscope consists of the Wollaston principle which shears the light from the source, producing two orthogonally separated beams of light. The optical setup is similar to the SLIM technique but the differential interference contrast microscope replaces the phase contrast microscope as shown in Figure 8.6.

In GLIM, the analyzer that renders the two orthogonally split beams parallelly is absent, both the light beams enter into GLIM module. The light beams are Fourier transformed spatially by L1. Inside there is a spatial light modulator that is placed in a plane in such a way that its axis is aligned matching to the polarization of the one of the fields. SLM induces the phase shift of n /2 to one of the beams and the other is unmodified. The image on the CCD results from the Fourier transformation of the light beams by lens L2. The field at the CCD is a combination of two fields: $U_n(r) = U(r) + U(r+\delta r)e^{i\phi_n}$, where $\delta r = \delta x\hat{x}$ represents the spatial offset between the two fields, and U is the image field. The intensity for each phase shift, $I_n(r)$, i.e., $I_n(r) = |U_n(r)|^2$ expressed as In(r) = I(r) +I(r+δr) +2|γ(r, δ r) |cos [Δ ϕ(r) + ϕ n]. In these equations, $I(r)$ and ϕ(r) represent the intensity and phase of the image field, respectively. γ is the mutual intensity or the temporal cross-correlation function at zero delay between the two fields, where γ (**r**,δ **r**) = $\langle U^*(\mathbf{r})U(\mathbf{r}+\delta\,\mathbf{r})\rangle_t$. The parameter $\phi_n = n\pi/2$ is a modulated phase offset between the two fields, externally controlled by the spatial light modulator (SLM). The term $\Delta\,\phi$ is defined as $\Delta\,\phi = \phi\,(r+\delta\,r) - \phi\,(r) \approx \nabla\,(\phi)$ δ r, approximately equal to the gradient of the phase (∇ ϕ) multiplied by δr, representing the phase

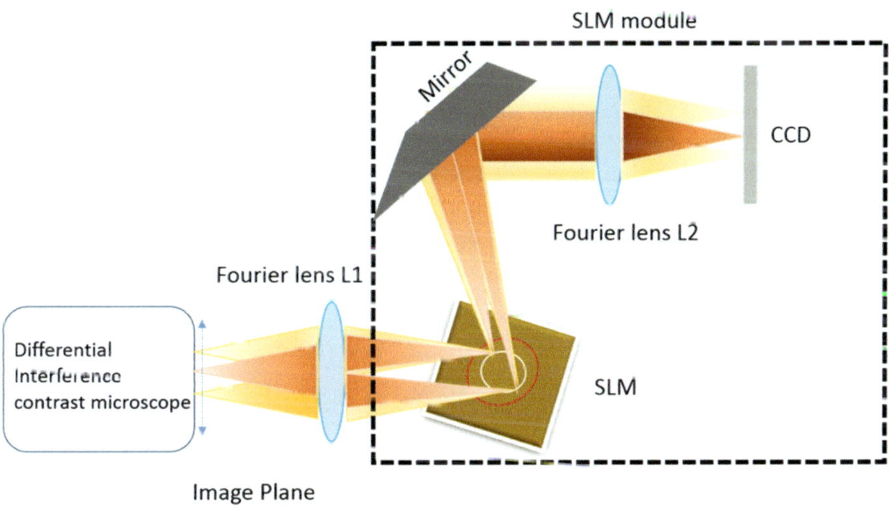

FIGURE 8.6 Schematic of a gradient light interferometry microscopy.

difference of interest. The spatial shift δr is introduced by the DIC prism and can be estimated experimentally. The quantity $\gamma(r, \delta r)$ is the mutual intensity, which is essentially the temporal cross-correlation function between the two fields, evaluated at zero delay, given by $\gamma\ (r, \delta\ r) = \langle U^{*}(r)U(r + \delta\ r)_{t}$. Combining the four intensity frames, we obtain the phase gradient as Equation 8.5

$$\nabla\ \phi(r) = \arg\ \{\ [I_1(r) - I_2(r)\ ],\ [I_3(r) - I_1(r)\ ]\ \}/\delta r \tag{8.5}$$

As GLIM uses common path white light methods that gives it the added benefit of giving efficiency like nanometer path length stability, speckle free images, and resolution that is not limited by diffraction. We could obtain the actual values of thin samples at the smallest condenser aperture and at the place of condenser aperture which is larger; this technique could be exploited as a method of tomography, providing an opportunity to get the time-lapse three-dimensional information of the samples that are thick. As it is a label-free technique, investigation of live cells, both thick and thin biological samples without any damage over both broad temporal and spatial scales are possible. GLIM has been employed in studying the variety of man-made and biological samples which cannot be studying in the transmission configuration for example semiconductor wafers, samples that are opaque and birefringent and bulk tissues [49]. Real-time tomography imaging of the thick samples also been studied using GLIM [48]. Differential phase information from the transparent samples is retrieved using GLIM [50]. This technique is not limited by photo bleaching and photo toxicity commonly associated with fluorescence microscopy. In addition, we are able to obtain three-dimensional information from unlabeled specimens and extraordinary optical sectioning of the samples. However, its application is limited as its lacks the specificity similar to other label-free images [49, 51].

Digital Holographic Microscopy

Digital holographic microscopy (DHM) is an efficient and potential technique in the field of biomedical as well as general imaging applications [52]. DHM has different types of techniques, which includes off-axis DHM, in-line DHM, quantitative phase microscopy by digital holography (DH-QPM), etc. These techniques are efficient instruments to track and profile micro-sized objects in 3D volumes. In the case of DHM it is possible to record the hologram that contains all the real time information, 3D structures of an object even without any mechanical focus adjustment, whereas in the case of conventional microscopy methods only 2D fixed images on a plane can be recorded and most of the information which is not in the focal plane is completely lost or missed. In addition, the data are available in the digital form for post processing [53, 54]. Holography is a 3D technique, which was invented by Dennis Gabor in 1947. The working principles involve two steps. One is recording the hologram and other is reconstruction. The light from the source is incident on a beam splitter and one of the beams passes through the object to be imaged, the other beam is reference beam. These two beams interfere and an interferometric pattern is produced which is recorded in the photographic plate as shown in Figure 8.7. The resulting interferometric pattern carries all the information of the object, i.e., mainly the phase and intensity information. In the case of digital holographic microscopy, the CCD replaces the photographic plate [19]. The CCD captures the interferogram obtained. Later the phase information and other 3D information are retrieved. In other words, digital holography is a two-step process where a 2D hologram is recorded and the images are reconstructed using specified algorithms. There are several approaches to retrieve the phase. In a paper by Marquet et al. [55] [159], the phase contrast in the images obtained from digital holographic microscopy is quantitatively related to its optical path difference and is given by Equation 8.6 [56]

$$(x, y) = (x, y).\overline{n_c}\ (x, y) - n_m \tag{8.6}$$

where $d\ (x, y)$ is the sample thickness, $n_c\ (x, y)$ is the mean z-integrated intracellular refractive index at the (x, y) position and n_m is the refractive index of the surrounding sample medium. The images are reconstructed from the single recorded hologram, the value of each pixel i of the reconstructed image is defined as Equation 8.7

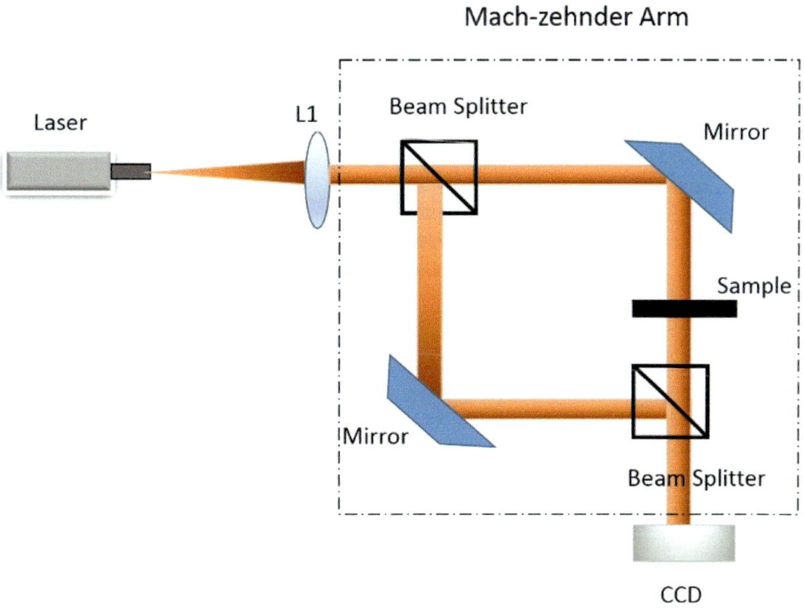

FIGURE 8.7 Schematic of a Mach–Zehnder interferometer-based holographic system.

$$\phi_i = \frac{2\Pi}{\lambda} \int_0^{d_i} n_c, i(z)\, dz + n_m (D - d_i)$$

$$= \frac{2\pi}{\lambda} \left[\left(\overline{n_{c,i}} - n_m \right) d_i + n_m D \right] \quad [7]$$

where λ is the wavelength, z is the axial co-ordinate, *di* is the thickness of the sample corresponding to the pixel *i*, $n_c, i(z)dz$ is representing the intracellular refractive index along the thickness *di*, *nm* is the refractive index of the medium. As they are using a flow cell here *D* represents the thickness of the flow chamber. Digital holography uses different configurations, such as off axis, quantitative phase microscopy using DHM, twin beam holography, in-line digital holography and Mach–Zehnder interferometer-based holographic system schematic Figure 8.7 shows a schematic of a Mach–Zehnder configuration.

Full field, label-free, non-contact, non-invasive, real time refocusing of an object is obtained in DHM without any mechanical focus adjustment and its efficiency has been proven in 3D imaging and tracking. In the recent years, DHM has established a strong position when it is considered potential for various applications. DHM is a combination of microscopy and holography, which further expands its application; scientists are not only using DHM in the optical field, but its applications are expanded into microbiology, medicine, marine, metrology, etc. The efficiency of DHM has been demonstrated with several studies of cells and structured surfaces [57], cell membrane fluctuations [58], imaging the phase profile of RBCs and HeLa cells, drug response as well as tumor growth and metastasis are also explored [59], cell dynamic detection have been explored by combining DHM and GS algorithms techniques [60]. It is believed that with further development of optical imaging along with computational methods, some of the features like performance, noise suppression, resolution enhancement must be improved.

In Table 8.1, we give an overview of the strengths and shortcomings of each technique, owing to their efficiency and applications to the field of biology.

TABLE 8.1
Comparison of Different Optical Phase Imaging Techniques

Imaging Technique	Advantages	Limitations	Resolution	References
Phase contrast	• High contrast • Works well for transparent samples • Label-free imaging	• Halo as shadow artifacts • Works well only at lower magnification • Halo affect increases with higher magnification. • Qualitative in nature • Lacks Specificy	• Axial ~2 µm • Lateral ~0.3 µm	[2, 36]
DIC	• Thick samples can be imaged with better depth of focus • Label-free imaging	• Pool contrast • Birefringent samples cannot be imaged • Qualitative in nature • Lacks specificity	• Axial ~2 µm • Lateral ~0.3 µm	[2, 3]
Dark field	• Diffraction limited resolution can be achieved • Label-free	• Qualitative in nature • Lacks specificity	• Axial ~2 µm • Lateral ~0.3 µm	[36]
SLIM	• Quantitative • Label-free	• Low contrast • Lacks specificity • Requires noise-free environment	• Axial ~2 µm • Lateral ~0.3µm	[36]
GLIM	• Speckle free • Label-free • Quantitntive in nature	• Lacks specificity • Complex and expensive system • Requires noise-free environment	• Diffraction limited resolution	[51]
DHM	• Label-free • Quantitative	• Speckle noise • Complex and expensive system • Requires noise free environment	• Axial ~2 µm • Lateral ~0.3 µm	[52]

SUMMARY

Optical phase imaging based on interferometry is a powerful technique for quantifying phase variations in transparent or semi-transparent samples. It provides precise measurements of properties like refractive index and thickness, and its high spatial resolution is invaluable for various scientific and industrial applications. However, it comes with certain drawbacks, including the need for a complex setup, vulnerability to vibrations, and intricate data processing. In contrast, non-interferometric methods offer simplicity and robustness, making them particularly advantageous for routine microscopy tasks and applications that require staining or labeling. Their inherent speed and resilience to environmental factors can often make them a better choice, especially in real-time observations or under less-than-ideal conditions. While interferometric methods are indispensable for many advanced studies, the versatility and accessibility of non-interferometric techniques underscore their importance in the microscopy field.

ACKNOWLEDGEMENTS

Ms. Anusha Pillai gratefully acknowledges the Dr. T.M.A. Pai scholarship provided by Manipal Academy of Higher Education. The authors are thankful to Manipal Academy of Higher Education for INTRAMURAL GRANT – MAHE/CDS/PHD/IMF/2019.

REFERENCES

1. Popescu G, Park YK, Choi W, Dasari RR, Feld MS, Badizadegan K. Imaging red blood cell dynamics by quantitative phase microscopy. *Blood Cells Mol Dis*. 2008;41(1):10–16.
2. Zernike F. Phase contrast, a new method for the microscopic observation of transparent objects. *Physica*. 1942;9(7):686–698.
3. Watanabe K, Palonpon AF, Smith NI, Chiu LD, Kasai A, Hashimoto H, et al. ARTICLE structured line illumination Raman microscopy. 2015. Available from: www.nature.com/naturecommunications
4. Gustafsson MGL, Shao L, Carlton PM, Wang CJR, Golubovskaya IN, Cande WZ, et al. Three-dimensional resolution doubling in wide-field fluorescence microscopy by structured illumination. *Biophys J*. 2008;94(12):4957–4970.
5. Gustafsson MGL. Surpassing the lateral resolution limit by a factor of two using structured illumination microscopy. *J Microsc*. 2000;198(2):82–87. Available from: https://onlinelibrary.wiley.com/doi/full/10.1046/j.1365-2818.2000.00710.x
6. SHEPPARD CJR. Super-resolution in confocal imaging. *Optik*. 1988;80(2):53–54.
7. Heintzmann R, Cremer CG. Laterally modulated excitation microscopy: improvement of resolution by using a diffraction grating. *Proc SPIE*. 1999;3568:185–196. Available from: https://www.spiedigitallibrary.org/conference-proceedings-of-spie/3568/0000/Laterally-modulated-excitation-microscopy--improvement-of-resolution-by-using/10.1117/12.336833.full
8. Gustafsson MGL. Surpassing the lateral resolution limit by a factor of two using structured illumination microscopy. *J Microsc*. 2000;198(2):82–87. Available from: https://onlinelibrary.wiley.com/doi/full/10.1046/j.1365-2818.2000.00710.x
9. Heintzmann R, Huser T. Super-resolution structured illumination microscopy. *Chem Rev*. 2017;117(23):13890–13908. Available from: https://pubmed.ncbi.nlm.nih.gov/29125755/
10. Saxena M, Eluru G, Gorthi SS. Structured illumination microscopy. *Adv Opt Photonics*. 2015;7(2):241–275. Available from: https://opg.optica.org/viewmedia.cfm?uri=aop-7-2-241&seq=0&html=true
11. Hell SW. Far-field optical nanoscopy. *Science*. 2007;316(5828):1153–1158. Available from: https://www.science.org/doi/10.1126/science.1137395
12. Wicker K. Super-resolution fluorescence microscopy using structured illumination. *Neuromethods*. 2014;86:133–165. Available from: https://link.springer.com/protocol/10.1007/978-1-62703-983-3_7
13. Sarunic M, Weinberg S, letters JIO, 2006 undefined. Full-field swept-source phase microscopy. opg.optica.org. Available from: https://opg.optica.org/abstract.cfm?uri=ol-31-10-1462
14. Castello M, Koho S, Tortarolo G, Vicidomini G, Diaspro A. Evaluating image resolution in stimulated emission depletion microscopy. *Optica*. 2018;5(1):32–35. Available from: https://opg.optica.org/viewmedia.cfm?uri=optica-5-1-32&seq=0&html=true
15. Galiani S, Harke B, Vicidomini G, Lignani G, Benfenati F, Diaspro A, et al. Strategies to maximize the performance of a STED microscope. *Opt Express*. 2012;20(7):7362–7374. Available from: https://opg.optica.org/viewmedia.cfm?uri=oe-20-7-7362&seq=0&html=true
16. Hell SW, Wichmann J. Breaking the diffraction resolution limit by stimulated emission: stimulated-emission-depletion fluorescence microscopy. *Opt Lett*. 1994;19(11):780–782.
17. Schneider J, Zahn J, Maglione M, Sigrist SJ, Marquard J, Chojnacki J, Kräusslich HG, Sahl SJ, Engelhardt J, Hell SW. Ultrafast, temporally stochastic STED nanoscopy of millisecond dynamics. *Nat Methods*. 2015;12(9):827–830.
18. Vicidomini G, Bianchini P, Diaspro A. STED super-resolved microscopy. *Nat Methods*. 2018;15(3):173–182. Available from: https://www.nature.com/articles/nmeth.4593
19. Klein T, Proppert S, Sauer M. Eight years of single-molecule localization microscopy. *Histochem Cell Biol*. 2014;141(6):561–575. Available from: https://link.springer.com/article/10.1007/s00418-014-1184-3
20. Hein B, Willig KI, Hell SW. Stimulated emission depletion (STED) nanoscopy of a fluorescent protein-labeled organelle inside a living cell. *Proc Natl Acad Sci U S A*. 2008;105(38):14271–14276.
21. El Schich Z, Nilsson E, Gerdtsson AS, Wingren C, Wingren AG. Interfacing antibody-based microarrays and digital holography enables label-free detection for loss of cell volume. *Future Sci OA*. 2015;1(3). Available from: https://www.future-science.com/doi/10.4155/fso.14.2
22. Photoactivated Localization Microscopy (PALM): an optical technique for achieving 10-nm resolution. 2010. Available from: www.cshprotocols.org
23. Bates M, Jones SA, Zhuang X. Preparation of photoswitchable labeled antibodies for STORM imaging. *Cold Spring Harb Protoc*. 2013;2013(6):540–541. Available from: https://europepmc.org/article/med/23734027

24. Bates M, Jones SA, Zhuang X. Stochastic Optical Reconstruction Microscopy (STORM): a method for superresolution fluorescence imaging. 2013. Available from: http://cshprotocols.cshlp.org/
25. Bates M, Huang B, Dempsey GT, Zhuang X. Multicolor super-resolution imaging with photo-switchable fluorescent probes. *Science*. 2007;317(5845):1749. Available from: /pmc/articles/PMC2633025/
26. Betzig E, Patterson GH, Sougrat R, Lindwasser OW, Olenych S, Bonifacino JS, et al. Imaging intracellular fluorescent proteins at nanometer resolution. *Science*. 2006;313(5793):1642–1645. Available from: https://www.science.org/doi/10.1126/science.1127344
27. Dani A, Huang B, Bergan J, Dulac C, Zhuang X. Neurotechnique superresolution imaging of chemical synapses in the brain. *Neuron*. 2010;68:843–856.
28. Dempsey GT, Bates M, Kowtoniuk WE, Liu DR, Tsien RY, Zhuang X. Photoswitching mechanism of cyanine dyes. *J Am Chem Soc*. 2009;131(51):18192–18193. Available from: https://pubs.acs.org/doi/full/10.1021/ja904588g
29. Dempsey GT, Vaughan JC, Chen KH, Bates M, Zhuang X. Evaluation of fluorophores for optimal performance in localization-based super-resolution imaging. *Nat Methods*. 2011;8(12):1027. Available from: /pmc/articles/PMC3272503/
30. Dubois A, Vabre L, Boccara AC, Beaurepaire E. High-resolution full-field optical coherence tomography with a Linnik microscope. *Appl Opt*. 2002;41(4):805–812. Available from: https://opg.optica.org/viewmedia.cfm?uri=ao-41-4-805&seq=0&html=true
31. Shirazi MF, Kim P, Jeon M, Kim J. Full-field optical coherence tomography using Galvo filter-based wavelength swept laser. 2016. Available from: www.mdpi.com/journal/sensors
32. Huang B, Wang W, Bates M, Zhuang X. Three-dimensional super-resolution imaging by stochastic optical reconstruction microscopy. *Science*. 2008;319(5864):810. Available from: /pmc/articles/PMC2633023/
33. Bae Kim G, Yoon YJ, Joo Shin T, Shik Youn H, Gho YS, Lee SJ. X-ray imaging of various biological samples using a phase-contrast hard X-ray microscope. Available from: https://analyticalsciencejournals.onlinelibrary.wiley.com/doi/10.1002/jemt.20601
34. Palacios F, Font O, Palacios G, Ricardo J, Escobedo M, Gomes LF, et al. Phase and polarization contrast methods by use of digital holographic microscopy: applications to different types of biological samples. In: *Holography - basic principles and contemporary applications*. 2013. Available from: https://www.intechopen.com/chapters/44919
35. Zuo C, Li J, Sun J, Fan Y, Zhang J, Lu L, et al. Transport of intensity equation: a tutorial. *Opt Lasers Eng*. 2020;135:106187.
36. Chen X, Kandel ME, Popescu G. Spatial light interference microscopy: principle and applications to biomedicine. *Adv Opt Photonics*. 2021;13(2):353. Available from: /pmc/articles/PMC9048520/
37. Haldar JP, Wang Z, Popescu G, Liang ZP. Deconvolved spatial light interference microscopy for live cell imaging. *IEEE Trans Biomed Eng*. 2011;58(9):2489. Available from: /pmc/articles/PMC3286342/
38. Park YK, Depeursinge C, Popescu G. Quantitative phase imaging in biomedicine. *Nat Photon*. 2018;12(10):578–589. Available from: https://www.nature.com/articles/s41566-018-0253-x
39. Abbott BP, Abbott R, Abbott T, Abernathy MR, Acernese F, Ackley K, Adams C, Adams T, Addesso P, Adhikari RX, Adya VB. Observation of gravitational waves from a binary black hole merger. *Phys Rev Lett*. 2016;116(6):061102..
40. Popescu G. *Quantitative phase imaging of cells and tissues*. New York: McGraw-Hill Education; 2011.
41. Technology OLP in, 1980 undefined. Electronic speckle pattern interferometry. iopscience.iop.org. Available from: https://iopscience.iop.org/article/10.1088/0305-4624/11/1/303/meta
42. Wang WC, Hwang CH, Lin SY. Vibration measurement by the time-averaged electronic speckle pattern interferometry methods. *Appl. Opt*. 1996, 35, 4502–4509.
43. Kuznetsova Y, Brueck S, Neumann A, Brueck SRJ. Imaging interferometric microscopy Resolution Enhancement for Optical Nanolithography View project Imaging interferometric microscopy. 2002. Available from: https://www.researchgate.net/publication/3967798
44. Min E, Kandel ME, Ko CJ, Popescu G, Jung W, Best-Popescu C. Label-free, multi-scale imaging of ex-vivo mouse brain using spatial light interference microscopy. *Sci Rep*. 2016;6(1):1–12. Available from: https://www.nature.com/articles/srep39667
45. Popescu G. *Quantitative phase imaging of cells and tissues*. New York: McGraw-Hill Education; 2011.
46. Park YK, Depeursinge C, Popescu G. Quantitative phase imaging in biomedicine. *Nat Photon*. 2018;12(10):578–589. Available from: https://www.nature.com/articles/s41566-018-0253-x
47. Wang Z, Millet L, Mir M, Ding H, Unarunotai S, Rogers J, et al. Spatial light interference microscopy (SLIM). *Opt. Express* 2011, 19, 1016–1026.

48. Popescu G. Gradient light interference microscopy (GLIM) for studying thick 3D cellular systems. In: *2017 OPJ-OSA joint symposia on nanophotonics and digital photonics (2017)*, paper 31pOD1. 2017:31pOD1. Available from: https://opg.optica.org/abstract.cfm?uri=OPJ-2017-31pOD1
49. Kandel ME, Hu C, Naseri Kouzehgarani G, Min E, Sullivan KM, Kong H, et al. Epi-illumination gradient light interference microscopy for imaging opaque structures. *Nat Commun*. 2019;10(1):1–9. Available from: https://www.nature.com/articles/s41467-019-12634-3
50. Wang Y, Kandel ME, Fanous MJ, Hu C, Chen H, Lu X, et al. Harmonically decoupled gradient light interference microscopy (HD-GLIM). *Opt Lett*. 2020;45(6):1487. Available from: /pmc/articles/PMC7716386/
51. Nguyen TH, Kandel ME, Rubessa M, Wheeler MB, Popescu G. Gradient light interference microscopy for 3D imaging of unlabeled specimens. Available from: www.nature.com/naturecommunications
52. Min J, Yao B, Trendafilova V, Ketelhut S, Kastl L, Greve B, et al. Quantitative phase imaging of cells in a flow cytometry arrangement utilizing Michelson interferometer-based off-axis digital holographic microscopy. *J Biophotonics*. 2019;12(9):e201900085. Available from: https://onlinelibrary.wiley.com/doi/full/10.1002/jbio.201900085
53. Antal O, Hackler L, Shen J, Mán I, Hideghéty K, Kitajka K, et al. Combination of unsaturated fatty acids and ionizing radiation on human glioma cells: Cellular, biochemical and gene expression analysis. *Lipids Health Dis*. 2014;13(1):1–15. Available from: https://lipidworld.biomedcentral.com/articles/10.1186/1476-511X-13-142
54. Cirenajwis H, Smiljanic S, Honeth G, Hegardt C, Marton LJ, Oredssona SM. Reduction of the putative CD44+CD24- breast cancer stem cell population by targeting the polyamine metabolic pathway with PG11047. *Anticancer Drug*. 2010;21(10):897–906. Available from: https://journals.lww.com/anti-cancerdrugs/Fulltext/2010/11000/Reduction_of_the_putative_CD44_CD24__breast_cancer.3.aspx
55. Marquet P, Rappaz B, Magistretti PJ, Cuche E, Emery Y, Colomb T, et al. Digital holographic microscopy: a noninvasive contrast imaging technique allowing quantitative visualization of living cells with subwavelength axial accuracy. *Opt Lett*. 2005;30(5):468–470.
56. Rappaz B, Breton B, Shaffer E, Turcatti G. Digital holographic microscopy: a quantitative label-free microscopy technique for phenotypic screening. *Comb Chem High Throughput Screen*. 2014 Jan;17(1), 80–88. doi: 10.2174/13862073113166660062. PMID: 24152227; PMCID: PMC3894694.
57. Kemper B, Von Bally G. Digital holographic microscopy for live cell applications and technical inspection. *Appl Opt*. 2008;47(4):A52–A61. Available from: https://opg.optica.org/viewmedia.cfm?uri=ao-47-4-A52&seq=0&html=true
58. Anand A, Chhaniwal V, Mahajan S, Trivedi V, Singh A, Leitgeb R, et al. Self-referencing digital holographic microscope for dynamic imaging of living cells. 2014. Available from: www.msubaroda.ac.in
59. Hellesvik M, Øye H, Aksnes H. Exploiting the potential of commercial digital holographic microscopy by combining it with 3D matrix cell culture assays. *Sci Rep*. 2020;10(1):1–14. Available from: https://www.nature.com/articles/s41598-020-71538-1
60. Jiang Y, Li H, Pang Y, Ling J, Wang H, Yang Y, et al. Cell image reconstruction using digital holography with an improved GS algorithm. *Front Physiol*. 2022;13:1040777.

9 Quantitative Optical Phase Imaging Techniques

Non-Interference-Based Basics and Working Principle

Anusha Pillai[1], *Anand K Asundi*[2], *and Ajeetkumar Patil*[1*]

[1]Department of Atomic and Molecular Physics, Manipal Academy of Higher Education (MAHE), Manipal, Karnataka, India
[2]d'Optron Pte Ltd, Singapore
*Corresponding Author

INTRODUCTION TO NON-INTERFERENCE-BASED OPTICAL PHASE IMAGING

Non-interferometric-based optical phase imaging methods that utilize computational techniques have ushered in a new era in microscopy. These methods address several challenges associated with traditional interferometry by introducing innovative ways to capture and analyze the phase information of transparent or semi-transparent samples. This advancement has broadened the scope of optical phase imaging, making it more accessible and versatile for researchers in various fields. The key feature of computational imaging methods is their ability to acquire phase information without the need for complex and sensitive interferometric setups. Instead, they rely on computational algorithms to process multiple images acquired under different illumination conditions. By systematically varying the illumination or sample position, a sequence of images is collected, each with different phase information encoded in the captured data. This dataset is then subjected to sophisticated computational algorithms, such as phase retrieval algorithms, to reconstruct the sample's phase profile. One of the remarkable advantages of computational methods is their ability to work with a variety of microscopy modalities, including bright field, dark field, differential interference contrast (DIC), and holographic microscopy. This versatility allows researchers to choose the imaging modality that best suits their specific application, making these methods adaptable to a wide range of biological, material, and industrial samples. Computational optical phase imaging offers several benefits, such as simplicity; the experimental setup is often simpler and less sensitive to environmental disturbances like vibrations, making it more user-friendly and suitable for routine microscopy tasks; speed, these methods can capture phase information in real-time, which is crucial for dynamic observations, such as live cell imaging; quantitative imaging, like interferometric methods, computational techniques provide quantitative phase information, allowing the measurement of refractive index variations, thickness, and other sample properties; enhanced accessibility, researchers without specialized interferometric equipment can access and benefit from the advantages of phase imaging using computational methods. In summary, non-interferometric-based optical phase imaging methods that leverage computational approaches have expanded the horizons of microscopy. Their simplicity, versatility, and ability to provide quantitative phase information offer significant advantages, making them important tools in modern microscopy, whether for biological investigations, materials analysis, or industrial quality control. These methods bridge the gap between traditional interferometry and standard microscopy, making advanced phase imaging techniques more accessible to a broader scientific and industrial

DOI: 10.1201/9781003545712-9

audience. As discussed in the previous chapter on the different interference-based quantitative phase imaging (QPI), we can see the experimental setups required for the interference-based QPI are complex and very sensitive to external noise and disturbances. Hence overall cost, maintenance, and expenses are more. This problem can be overcome using the non-interferometric approach. There are two types mainly: one is wavefront sensing and the second is phase retrieval. Different methods under each type of non-interferometric technique are discussed in the sections below.

WAVE FRONT SENSING

Wave front sensing is an approach that investigates retrieving the aberrations in a wave front that are usually formed by the delay in the phase due to the sample. This technique has its own advantages; some of the advantages are higher speed, sensitivity, and temporal resolution. This technique does not require complex instrumentation like the other interferometry methods. In the case of interferometry, because of the reference arm, they are much affected by the vibrations and other disturbances in the optical path. Here in this case, a reference arm is not required; therefore, one need not worry about the variations and other disturbances caused in the optical path. Single image acquisition is a technique used by wave front sensing which results in very high temporal resolution. The spatial resolution is not very high as the light coming from each of the phase measurement spots is spread over many pixels on a digital camera sensor. In the case of digital holographic microscope (DHM), something similar happens where the resolution is lost in the process of cropping the side band in the Fourier space during the imaging and reconstruction. This loss in spatial resolution can be compensated by using high NA objectives for imaging. Wave front sensing can be used for imaging high-speed cell dynamics by obtaining accurate phase information. Even in the case of other QPI methods like DHM, where the spatial resolution is improved with fitting procedures, the presence of background fluctuations and both spatial and temporal noise that is caused due to the recovery of phase by numerical integration will cause certain effects on cell segmentation. One can perform measurements of dry mass of cells even when the density is high with utmost accuracy. Some of the types of wave front sensing techniques are discussed below.

SHACK–HARTMANN WAVE FRONT SENSOR

Shack–Hartmann is a type of wave front sensing method; it is mainly used for measuring the shape of an aberrated wave front. Johannes Franz Hartmann invented it. The original set up had a Hartmann mask that is nothing but an array of pinholes, rather than lenses. Later in the 1960s, Roland Shack and Ben Platt introduced lens arrays [1]. The sensitivity of the sensor increased substantially with the use of lens arrays, allowing the wave front sensing at much lower intensity levels that is very crucial in the applications like astronomy. A Shack–Hartmann wave front sensor is used in the fields of medicine and industry. The experimental setup consists of a micro lens array and an image sensor that is mounted in the focal plans of array of micro lenses. The sensors that have a Hartmann mask in them are called Hartmann wave front sensors. The basic measurement principle is depicted in Figure 9.1. The sensor has array of lenses with the same focal lengths. Each lens is focused on to the charge-coupled device (CCD) array. In Figure 9.1 we see that across each lens there is a tilt of the wave front, here it is visualized for a one-dimensional case. The tilt here is measured from the position of the focal spot onto the image plane i.e., Δx. Such displacements that are similar are observed in the y directions as well i.e., Δy. With the help of such set of discrete tilts, the wave front phase aberrations are approximated. As only the tilts are measured, this sensor cannot detect the discontinuous steps in the wave front. With the help of information obtained from $\frac{\Delta x}{\Delta y}$ which is called as slopes one can measure the approximate phase of the wave front [2].

This explains the basic working principle of a Shack–Hartmann wave front sensor, there several approaches [3] and advanced developments observed for improvement of the system [4–11]. One

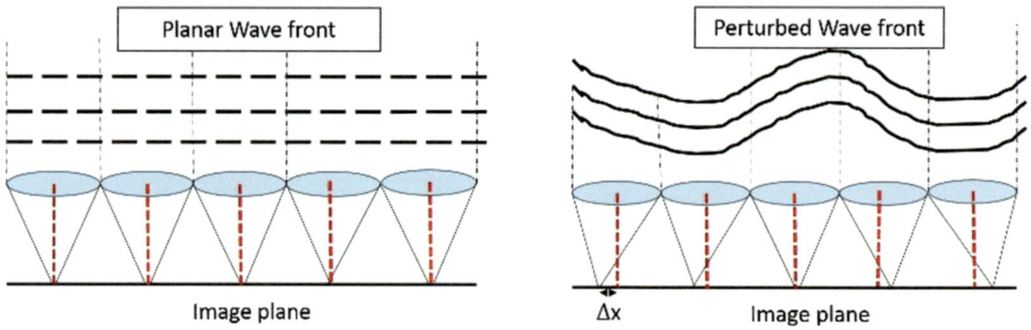

FIGURE 9.1 Schematic showing the working principle of a Shack–Hartmann wave front sensor.

of the advantages of this method over interferometry is that it can performed with an incoherent source. As we mentioned before, the absence of a reference arm gives it an advantage of not having a vibration isolation. Highly aberrated wave fronts can be measured using this method, whereas the traditional interferometry fails in this aspect. The application of this sensor is widely seen in metrology because of its simple and versatile measuring methods of the complex wave fronts. Adding to this, its applications are observed even in other areas of optics. One major application of this sensor is seen in measuring the ocular aberrations in ophthalmology. There are plenty of devices available for assessing the eye. Most of the devices available are associated with refractive surgery laser that allows the aberration corrections and thus enhances the eyesight. Other optics such as contact lenses with a correction with respect to the respective eye problem has slowly been explored but not commercialized widely. The Shack–Hartmann sensor is also involved with the adaptive optics to view the relatively unexplored parts of the retina. With the help of this device, scientists can view and understand the morphology of the retina. This device also helps in understanding the differences between healthy and unhealthy retinas. The Shack–Hartmann sensor has had a direct impact on the vision of millions of people worldwide [12].

Applications of this method can also be observed from the field of astronomy to diagnose the quality of the beam [13, 14], optical testing [15, 16], atmospheric laser communication [17, 18], and microscopic aberration correction [19, 20]. Although it has its benefits, the limited physical size of the micro-lens results in the underutilization of the full pixel resolution of the image sensor, causing a reduction in spatial resolution [21].

Pyramid Wave Front Sensor

The pyramid technique is a type of wave front sensor where it measures the optical aberrations in the wave front. The name pyramid is given as it uses a pyramidal prism that has a large apex angle.

The PyWFS is a pyramid wave front sensor (WFS) performing optical Fourier filtering thanks to a glass pyramid located in the focal plane, as shown in Figure 9.2. This pyramid splits the electromagnetic (EM) field into four beams, each producing four different filtered images of the entrance pupil. The slope of the wave front is derived from the pixel intensity that is shown below as equations.

$$S(x) = \frac{(I_1 + I_4) - (I_2 + I_3)}{(I_1 + I_2 + I_3 + I_4)}$$

$$S(y) = \frac{(I_1 + I_2) - (I_3 + I_4)}{(I_1 + I_2 + I_3 + I_4)}$$

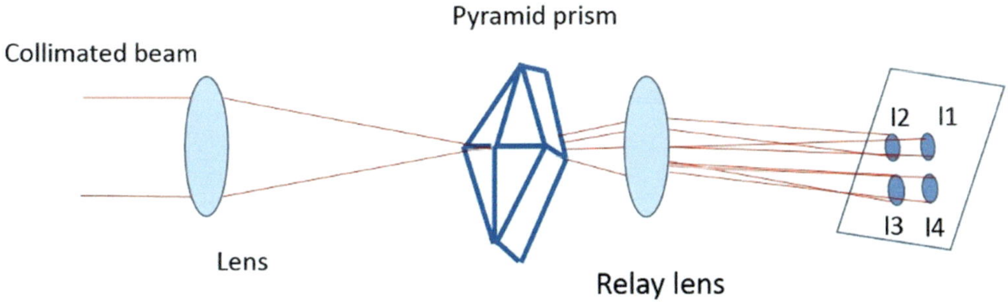

FIGURE 9.2 Schematic showing the working principle of a pyramid wave front sensor.

Four identical pupil images are formed when the aberrations are absent with the help of a prism with four faces, whose tip is centred at the peak of the point spread function [22]. If the optical aberrations are present, then there will be variations in the intensity distributions among the pupils. One can record the intensity distribution in the pupil image for obtaining the local wave front gradients. From these obtained gradients, one can evaluate and measure the wave front aberrations.

This processing requires a pixel-wise difference; also, the normalization requires accurate pupil mapping on the detector with no differential distortion or rotation among the four pupil images. Common distortions due to the upstream adaptive optics (AO) system are cancelled out by the difference and have no impact on the slope measurement [23].

Fauvarque et al. [24] proposed the pyramid wave front sensor as a convolution system. Furthermore, there are lot of other approaches and methods implemented to enhance the working of the PyWFS [25–29]. The PyWFS has a lot of advantage over the Shack–Hartman wave front sensors as it has higher sensitivity in the closed-loop applications [30]. Adjustable sensitivity and pupil sampling also has an upper hand in the case of the pyramid sensor. The PyWFS is applied widely in the field of ophthalmology [31]. It is used for high-resolution imaging of other structures inside the retina [29, 32], in the field of adaptive optics [33]. The main disadvantage of the pyramid sensor is that it does not have any moving component inside it, because of this the dynamic imaging will be difficult [21]. Modulation demands for improvisation; some changes such as using an extended source can help us obtain improved modulation.

DIFFERENTIAL PHASE CONTRAST (PHASE RETRIEVAL DETERMINISTIC)

Differential phase-contrast (DPC) is an imaging technique that focuses on the quantitative approach with enhanced spatial resolution and high optical sectioning and the speckle noise is considerably reduced. The traditional differential phase contrast method gives us an image whose intensity is proportional to the rate of change of phase across the phase modulated by the intensity of light transmitted. One important property of a differential phase contrast microscope is its response to a phase gradient. Consider an object with

$$t(x) = e^{j\varnothing x}$$

where φ is the phase gradient. We thus have $T(m) = \delta(m - \frac{\varnothing}{2\pi}$ which contains one spatial frequency component. Therefore, the image of the phase gradient is given by.

$I(x)\ C(\frac{\varnothing}{2\pi}; \frac{\varnothing}{2\pi}$ it indicates it's an independent position and completely specified by C (m; m)

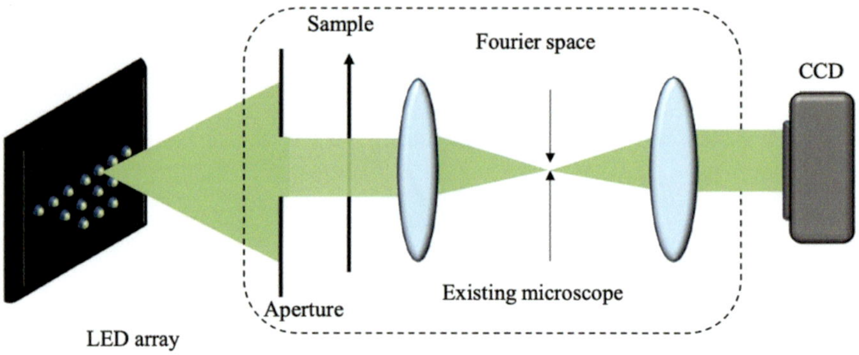

FIGURE 9.3 Schematic showing the working principle of differential phase contrast.

This is the working principle of traditional DPC microscope that was executed in the year 1984 by Sheppard et al. [34]. There are different developments observed further where the phase contrast is obtained by the sample with different illumination scheme of light shown in the Figure 9.3 [35].

The recently developed quantitative differential phase contrast (DPC) method retrieves the intricate transmittance characteristics of a sample by utilizing coded-illumination measurements and a phase retrieval optimization. DPC enables spatial resolution that is twice as fine as the coherent diffraction limit and can be readily applied using an LED array-based coded-illumination source on a standard microscope [35]. Conventional differential phase contrast (DPC) measurements involve taking four intensity images, each acquired using a semi-circular illumination pattern at varying rotation angles [36]. The time-multiplexed nature of the measurements assumes that the sample remains during acquisition. However, live biological samples may exhibit non-stationary behaviour, defined as moving more than one pixel during the acquisition time. When only a single measurement is needed (single-shot), the sensor's exposure time can be adjusted to ensure relatively stationary behaviour. However, for multi-shot methods, the acquisition time is limited by the sensor's readout time. Although each individual measurement may have an appropriate exposure time to maintain the stationary assumption, motion occurring between measurements in multi-shot DPC acquisition can introduce errors in the reconstructed complex-field. To address this issue and estimate motion in DPC images, various approaches have been introduced, referred to as motion-resolved DPC (mrDPC). This approach involves acquiring an additional colour-multiplexed measurement simultaneously with a constant coded-illumination pattern. A navigator measurement is used, employing one colour channel of the source LEDs to display a constant illumination pattern (a half-circle), ensuring consistent spatial frequency contrast. A colour camera then separates the navigator and DPC measurements without making assumptions about the sample's dispersion or colour. Non-rigid motion can be estimated from the navigator measurements and corrected in the DPC measurements before phase retrieval. This method allows for the recovery of quantitative phase images for each time point in the data capture, resulting in temporal resolution equivalent to single-shot methods. When it comes to weakly interacting phase objects, the best technique to obtain the image contrast is DPC. Several ex vivo imaging studies have been done employing DPC on elegans [37], live breast cancer cells [38], retinal cells [39], HeLa cells [40], umbilical cord cells [41], and sperm motility analysis [42] and for various QPI imaging of biological cells [43] and biological nanoparticles [44]. Associating DPC with STEM also allows us to directly characterize the local electromagnetic fields of the single atoms at different levels [36]. However, despite the advantages, the camera here has a large pixel size and thus intensity cannot be sampled or digitized, if we use the image sensor with a smaller pixel size or a camera with high magnification one can solve this under-sampling issue but at the expense of small field of view (FOV) [40].

TRANSPORT OF INTENSITY EQUATION (TIE) (PHASE RETRIEVAL DETERMINISTIC)

Quantitative phase imaging using transport of intensity equation (TIE) [45–47] needs lower-intensity measurements and is very much compatible with commercially accessible microscopic parts along with Köhler illumination. Figure 9.4 shows the experimental setup of TIE.

Figure 9.5 shows three images, i.e., one focused image and the other two are slightly defocused images of a polystyrene bead of size 8 μm with the defocus distance of 3.5 μm, captured in the bright field mode using TIE system.

The greatest advantage of QPI using computational methods is their non-interferometric nature. Along with this advantage of TIE, when the object is illuminated with partially coherent light emanating from optical condenser where the angular part of the illumination put up to the transverse resolution, in turn provides appreciable spatial resolution over coherent diffraction limit. The important and most crucial feature of any QPI imaging it gives us the information of the phase. The phase of an object, cell, and for any matter carries the information about its shape, size etc. In the case of biological cells if we are able, to retrieve the phase successfully we will be able to obtain information about cell volume, cell thickness, cell dry mass, etc. In this case, we are using the transport of intensity equation to retrieve the phase.

$$\left(\frac{\partial}{\partial z}\right)\ln E(x,z) = -\Delta W(x,z) - \nabla W(x,z) \bullet \nabla \ln E(x,z) \tag{6}$$

Where the operators $\Lambda = \nabla^2$ and ∇ are operating only on the lateral coordinates x, i.e., $\nabla = (\partial/\partial x, \partial/\partial y)^T$, Teague found this outcome. Capturing one focused image and two slightly

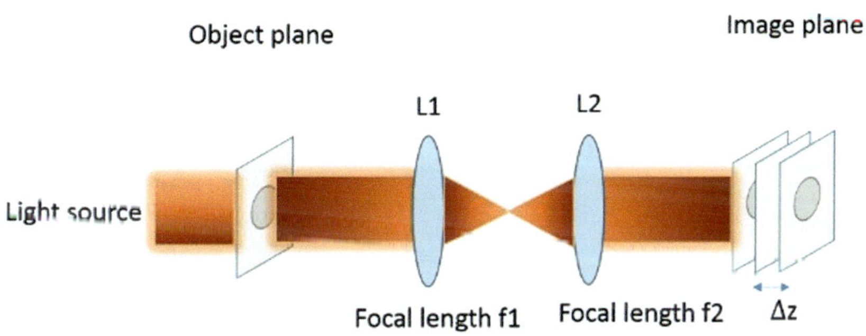

FIGURE 9.4 The experimental setup of the transport of intensity equation.

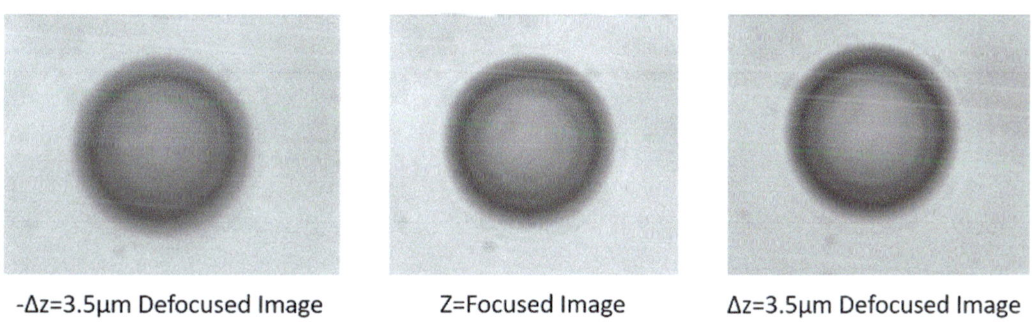

FIGURE 9.5 Focused and defocused images of a polystyrene bead.

out-of-focus images with an out-of-focus distance optimized to specific sample as shown in Figure 9.6 To produce a well-behaved Weak object transfer function (WOTF) function the defocus distance Δz should be chosen wisely (the Fresnel number is large compared with unity) [48]. The transport of intensity equation will be the solution for the phase imaging method. The reconstruction algorithm of TIE is like of that of conventional TIE. Fast Fourier transform (FFT) and discrete cosine transform are the most used reconstruction algorithms. Here we used FFT to reconstruct the phase information from the intensity images. Figure 9.6(a) shows the intensity derivative of the three images along z direction, 9.6(b) shows the reconstructed phase image, 9.6(c) shows the height profile of the bead, 9.6(d) shows the 3D image of the bead.

We have recorded the image of the polystyrene bead in the phase contrast mode; we observed that the boundaries of the sample are more defined in the phase contrast mode. Figure 9.7(a) shows the intensity derivative of the three-phase contrast images along the z-direction, 9.7(b) shows the reconstructed phase image, 9.7(c) shows the height profile of the bead, 9.7(d) shows the 3D image of the bead.

TIE rises the rigid beam coherent required for interferometry expanding its applications to X-ray diffraction, electron beam microscopy, and neutron radiography where the partial coherent illumination is extremely important and unavoidable. Apart from these applications, TIE provides superior QPI quality, rescuing the image deterioration caused due to speckle noise One major drawback of TIE on entering into dynamic phase measurements is that it has to capture images at least two different axial planes by either moving the detector or the sample which is definitely going to disturb the whole process of image acquisition, also, it increases the quantifying time and troubles in real-time measurements. Therefore, using a spatial light modulator or an electrically tunable lens [49] can be placed at the rear aperture plane to capture the images at different axial planes with different depths without moving any optical parts. Such a QPI system has promised real-time analysis and phase measurements with nanometre sensitivity and millisecond temporal resolution [50]. These techniques are used in analyzing the morphology changes due to drug induction, the engulfment process carried out by macrophages and imaging cellular dynamics of breast cancer cells. Even though TIE prompts and provides an excellent result, translating TIE into clinical and biological laboratories is still a huge challenge due to its complex and bulky configuration and highly expensive hardware (ETL, SLM) [50].

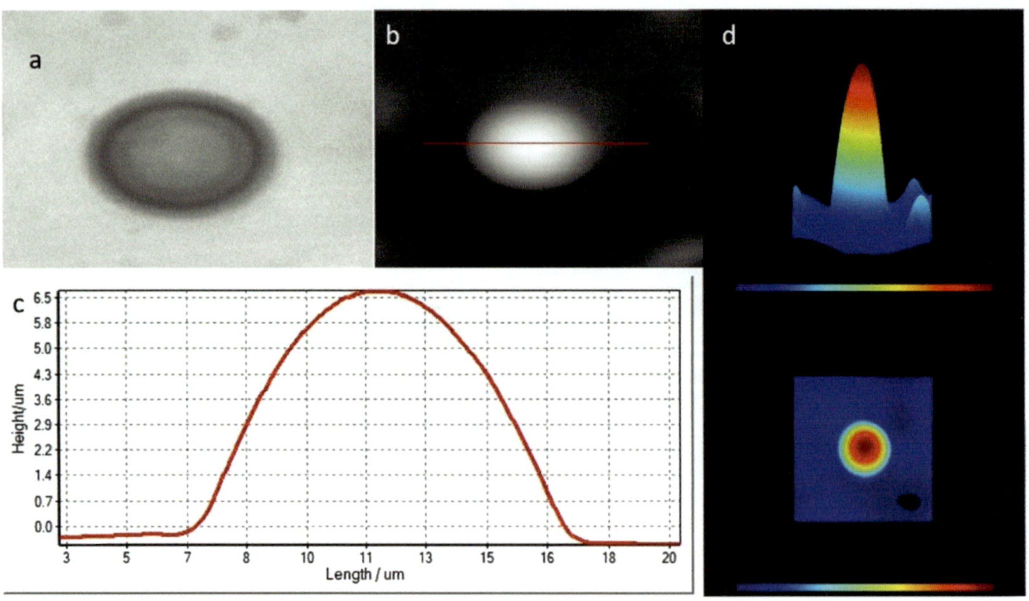

FIGURE 9.6 (a) Intensity derivative image. (b) Phase image. (c) Height profile. (d) 3D profile.

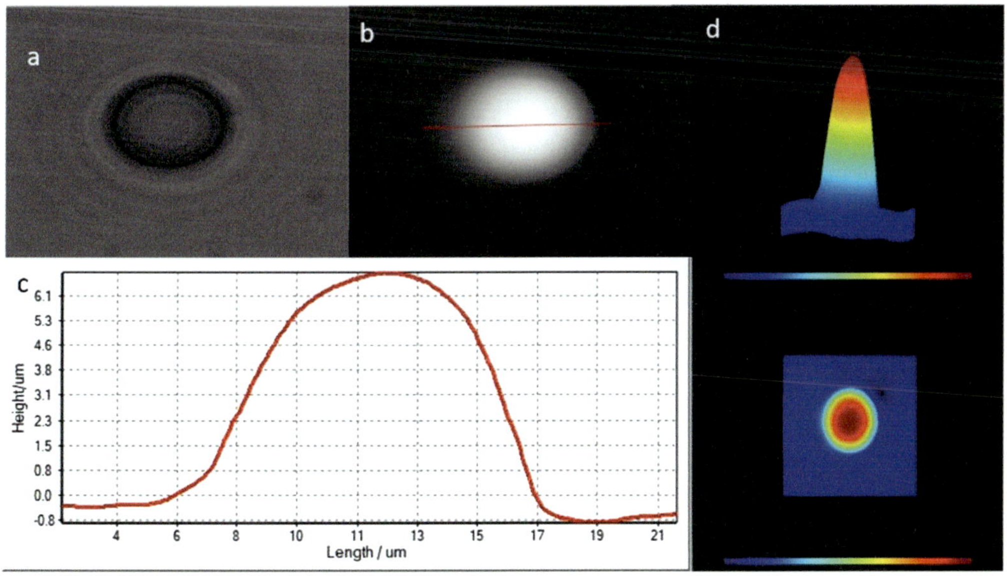

FIGURE 9.7 (a) Intensity derivative image. (b) Phase image. (c) Height profile. (d) 3D profile.

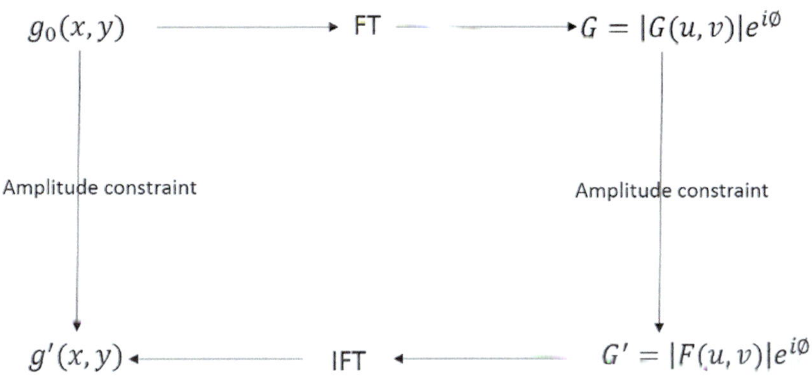

FIGURE 9.8 The working process of Gerchberg–Saxton algorithm.

PHASE RETRIEVAL (ITERATIVE)

Iterative phase retrieval originated in the field of electron microscopy. The Gerchberg–Saxton algorithm [51, 52] it is a first iterative type of algorithm, in this intensity image of an object and the far-field diffraction pattern of the object is inconsistent with the phase map of the object. Figure 9.8 shows the working process of the Gerchberg–Saxton algorithm. First, an initial guess g0(x, y) is input, and the amplitude G(u, v) is obtained by the Fourier transform. Next, the amplitude G(u, v) is replaced by a known function F(u, v) to form a new function. Then, the estimated square function g′ (x, y) is obtained by the inverse Fourier transform, and the input value g(x, y) of a new iteration is formed with known constraints [53].

This algorithm provided a correct solution and thus it has application in areas such as X-ray diffraction imaging [54–57], adaptive optics [58, 59], and quantitative phase microscopy [60–63].

As the scattering of the object is very weak in the case of X-ray imaging, the rays which are scattered will be dominant than the signal of our interests. To address this issue, a beam stopper is introduced to shield the detector from the detrimental effects of diffracted rays, but this comes at the cost of losing low spatial frequencies [64]. This issue is resolved by adding an electron microscope image of the sample so that the low spatial frequency data that were missing was recreated. The Gerchberg–Saxton (GS) algorithm has its share of limitations [59–61]. One of these limitations is the potential for convergence issues during early iterations or the risk of getting trapped in local optima. To tackle these challenges, various modifications have been proposed to enhance the convergence of the GS algorithm. In 1973, Misell introduced a crucial modification to the original GS algorithm. He expanded its scope beyond examining intensity distributions at the image and Fourier planes. Misell's approach allowed the study of diffraction patterns at different propagation distances and introduced extra intensity constraints to improve the accuracy and convergence of the iterative algorithm. Misell's algorithm is considered more practical and user-friendly than the original GS algorithm, offering a fresh perspective on improving iterative phase retrieval methods. Beyond Misell's approach, several new algorithms have emerged that work with multiple propagation distances [63, 64], multiple illumination wavelengths [65], phase modulation [56, 66], and speckle illumination modulation [65, 67]. These approaches aim to overcome the limitations of the GS algorithm and expand its capabilities. Another challenge related to the GS algorithm pertains to solving inverse problems and ensuring solution uniqueness. In cases involving small objects, obtaining near-field images may not be feasible, rendering GS algorithm-based reconstruction impossible. For 2D signals [68] and their Fourier magnitude, there is not a unique correlation, which means there can be multiple solutions to the problem. Bates provided a solution to this issue, showing that with knowledge of the object's spatial characteristics, one can determine the object using its auto-correlation function. In 1982, Fienup offered a theoretical explanation for solving phase issues through convex optimization principles. He modified the GS algorithm based on cybernetics concepts, resulting in the hybrid input–output (HIO) algorithm. This led to improved analysis and enhancements in object reconstruction. Various techniques, including hybrid projection reflection, difference map [66], relaxed averaged alternating reflections, and charge flipping [67, 69] have been introduced, but implementing these algorithms requires significant skill and experience. Marchesini introduced further modifications to the algorithm, proposing a solution that converges in all cases, known as the shrink-wrap algorithm [69]. This method can be used to reconstruct the atomic distribution of nanocrystals. Table 9.1

Another approach in iterative phase retrieval methods is ptychography, originating in electron microscopy and invented by Rodenburg and Faulkner in 2004. It combines elements of the GS and HIO algorithms and introduces a new method known as ptychographic iterative engine (PIE) [69–71]. In this experimental approach, the object is illuminated by light limited by diffraction patterns, and this process continues until the entire region of interest has been scanned. The result is a set of sub-aperture diffraction patterns that overlap spatially, serving as intensity constraints. The complex amplitude distribution from the scanned region can be reconstructed using GS and HIO algorithms. The ptychographic iterative method is more stable and reliable than GS and HIO, as it benefits from a wealth of spatially overlapping data. Furthermore, it significantly improves the non-convexity issue. This method has become essential in optical phase imaging and microscopy techniques [72]. It addresses challenges related to direct phase reconstruction and overcomes the limitations of interferometric methods, representing an evolution in imaging. Ptychographic imaging has found application in methods such as X-ray diffraction imaging, electron microscopy, and adaptive optics [73]. However, two major drawbacks of this approach are that it requires a large number of iterations for a reliable solution and a substantial amount of intensity data, which hinders its application in high-speed and real-time imaging [74].

TABLE 9.1
Comparison of Non-Interferometric-Based Optical Phase Imaging Techniques

Imaging Technique	Advantages	Limitations	Resolution	References
Shack–Hartmann	Label-free Quantitative Non-inerferometric	Complex and expensive system	Lateral resolution is proportional to 1/NA axial resolution is proportional to $1/NA^2$	[20]
Pyramid	Label-free Quantitative Non-interferometric	Complex and expensive system Cannot work in closed loop applications	Axial resolution 4.5 μm	[32]
DPC	Label-free Quantitative Non-interferometric	Complex and expensive hardware	Abbe diffraction-limited lateral $\lambda\ (2 \times NA) = 385$ nm Abbe diffraction limited axial resolution. $\lambda/(1-\sqrt{1-\text{NA}_2}) = 2083$ nm	[36]
TIE	Label-free Quantitative Non-interferometric	Limited resolution	Diffraction limited resolution	[48]
Iterative GS algorithm	Label-free Quantitative Non-interferometric	Large number of iterations Large amount of data	Diffraction limited resolution	[67]
Ptychography	Label-free Quantitative Non-interferometric	Large number of itera ions	Diffraction limited resolution	[69–72, 75]

Table 9.1 presents an overview of the strengths and limitations of each technique, considering their efficiency and applications in the field of biology. An ideal optical phase imaging instrument is expected to offer quantitative measurements, excellent selectivity and specificity, high axial and spatial resolution, and sensitivity while also providing strong contrast. Furthermore, it should support in vivo, label-free, real-time imaging of live cells. Ideally, an optical phase imaging device should be compact, portable, and cost-effective. We can observe that each technique excels in one aspect but may have drawbacks in other areas. None of the instruments fully meet all the criteria expected of an ideal device. To address this issue, significant efforts are being made to develop multi-modal imaging techniques that combine the strengths of different methods to achieve a more comprehensive imaging solution. Numerous studies have explored multimodal approaches in the field, such as the work by Kim YS and others [76], which involves combining quantitative phase imaging with fluorescence techniques. This integration enhances both spatial and temporal resolution and provides molecular specificity. As a result, this method yields quantitative parameters that significantly contribute to a deeper understanding of pathological cells and tissues. In another study by Rajput SK and colleagues [77], a combination of non-interferometric methods, like the TIE, with fluorescence has been implemented. This approach simplifies the experimental setup and facilitates phase information retrieval through algorithms The technique involves stacking defocused fluorescence images with TIE to obtain phase information. They also highlight the potential for real-time imaging enabled by electrically tunable lenses. To manage and extract valuable insights from the data obtained through QPI, scientists have explored the integration of QPI with artificial intelligence (AI). This synergy between QPI and AI has opened exciting possibilities for advancing biomedical applications [78].

SUMMARY

Here we have delved into different types of optical phase imaging methods, starting with the qualitative approach and then the quantitative method and their classifications based on interference in the previous chapter, followed by non-interference-based methods in the current chapter. There are strengths and shortcomings to each technique owing to their applications in biology. Interference-based imaging techniques, such as holography and interferometry, offer the advantage of high spatial resolution and the ability to provide quantitative phase information, making them particularly useful for detailed examinations of transparent samples. However, these methods come with the drawback of requiring complex setups, which can be sensitive to environmental factors, like vibrations and temperature fluctuations. Additionally, they may have limited compatibility with certain sample types or necessitate challenging sample preparations. In contrast, non-interference-based imaging techniques, like bright field microscopy, offer simplicity and versatility, making them suitable for a wide range of samples and real-time imaging of dynamic processes. Yet, they generally provide lower spatial resolution and may lack the capability to offer quantitative phase data. The choice between interference-based and non-interference-based techniques depends on the specific imaging requirements and the nature of the samples under investigation. Combining different modalities can help overcome the situation of not meeting all the criteria an ideal optical phase instrument requires to fulfil its applications in the field of biomedicine. When it comes to qualitative, one cannot obtain phase information; in the case of quantitative, the phase is retrieved. In the case of imaging transparent samples, one can combine QPI with phase contrast, which will enhance the contrast as well as provide quantitative information like optical height, volume, cell dry mass, irregularities, etc., thus enabling in-depth study of biological cells. Similarly, if one requires improved resolution, one can combine DHM with TIE; also, TIE can be combined with super-resolution techniques. The development in the imaging provided by TIE is much higher in comparison to all the above-mentioned techniques. In the biomedical field, label-free imaging using TIE plays a crucial role. TIE also can be combined with fluorescent techniques, thus enhancing its efficiency to a much higher level. Such multi-modal approaches are being developed to make optical phase imaging much more efficient.

With all these discussions and developments of optical phase imaging techniques and multi-modal approaches in various fields and areas, one can believe it is a very good replacement for most of the limitations in the optical phase imaging methods in the fields of biomedical research. However, the application of multi-modal imaging has not been fully explored. Several issues must be addressed so that correlative QPI approaches can be effectively developed and established in biological labs and clinics.

Further, AI must assist the upcoming field of research. Multimodal approaches, when combined with AI, will have much efficiency in diagnosing diseases based on the information of the morphology and molecules of a single cell; this combination, in turn, paves a new way for medical study and diagnosis.

ACKNOWLEDGEMENTS

Ms. Anusha Pillai gratefully acknowledges the Dr. T.M.A. Pai scholarship provided by Manipal Academy of Higher Education. The authors are thankful to Manipal Academy of Higher Education for INTRAMURAL GRANT – MAHE/CDS/PHD/IMF/2019.

REFERENCES

1. R. S.-S. M. of O. S. of America, undefined 1971, and undefined 1971, "Production and use of a lenticular Hartmann screen," *cir.nii.ac.jp*, Accessed: Jul. 08, 2023. https://cir.nii.ac.jp/crid/1573387450670986240.

2. S. Mauch and J. Reger, "Real-time spot detection and ordering for a Shack-Hartmann wavefront sensor with a low-cost FPGA," *IEEE Trans Instrum Meas*, vol. 63, no. 10, pp. 2379–2386, Oct. 2014, doi: 10.1109/TIM.2014.2310616.
3. A. Dubra and V. Akondi, "Shack-Hartmann wavefront sensor optical dynamic range," *Opt Express,* vol. 29, no. 6, pp. 8417–8429, Mar. 2021, doi: 10.1364/OE.419311.
4. M. Konnik and J. De Doná, "Waffle mode mitigation in adaptive optics systems: a constrained Receding Horizon Control approach," 201AD, doi: 10.1109/ACC.2013.6580355.
5. S. H. Baik, S. K. Park, C. J. Kim, and B. Cha, "A center detection algorithm for Shack–Hartmann wavefront sensor," *Opt Laser Technol*, vol. 39, no. 2, pp. 262–267, Mar. 2007, doi: 10.1016/J.OPTLASTEC.2005.08.005.
6. X. Yue, Y. Yang, F. Xiao, H. Dai, C. Geng, and Y. Zhang, "Optimization of virtual Shack-Hartmann wavefront sensing," *Sensors*, vol. 21, no. 14, Jul. 2021, doi: 10.3390/S21144698.
7. Z. Yang *et al.*, "Generalized Hartmann-Shack array of dielectric metalens sub-arrays for polarimetric beam profiling," *Nat Commun*, vol. 9, no. 1, pp. 1–7, Nov. 2018, doi: 10.1038/s41467-018-07056-6.
8. T. Endo, Y. Miwa, J. Suzuki, T. Ando, T. Takanezawa, and Y. Ezaki, "Improving techniques for Shack-Hartmann wavefront sensing: dynamic-range and frame rate," in *Optical and space surveillance technologies conference*, 2018, Accessed: Jul. 14, 2023. www.amostech.com.
9. A. K. Pandey, T. Larrieu, G. Dovillaire, S. Kazamias, and O. Guilbaud, "Shack-Hartmann wavefront sensing of ultrashort optical vortices," *Sensors*, vol. 22, no. 1, Jan. 2022, doi: 10.3390/S22010132.
10. N. Dubey, R. Kumar, and J. Rosen, "COACH-based Shack-Hartmann wavefront sensor," in *OSA imaging and applied optics congress 2021 (3D, COSI, DH, ISA, pcAOP) (2021), paper DW5C.1*, p. DW5C.1, Jul. 2021, doi: 10.1364/DH.2021.DW5C.1.
11. B. Vohnsen, A. C. Martins, S. Qaysi, and N. Sharmin, "Hartmann-Shack wavefront sensing without a lenslet array using a digital micromirror device *Appl. Opt.* 57, pp. E199–E204, 2018.
12. B. C. Platt and R. Shack, "History and principles of Shack-Hartmann wavefront sensing," *J Refractive Surgery*, vol. 17, no. 5, 2001, doi: 10.3928/1081-597X-20010901-13.
13. B. Schaefer and K. R. Mann, "Determination of beam parameters and coherence properties of laser radiation using an extended Hartmann-Shack wavefront sensor," *Appl Opt*, vol. 4926, pp. 150–161, 2002, doi: 10.1117/12.481685.
14. B. Schäfer, M. Lübbecke, and K. Mann, "Hartmann-Shack wave front measurements for real time determination of laser beam propagation parameters," *Rev Sci Instrum*, vol. 77, no. 5, 2006, doi: 10.1063/1.2198795/349004.
15. J. Pfund, N. Lindlein, J. Schwider, R. Burow, T. Blumel, and K. E. Elssner, "Absolute sphericity measurement:? a comparative study of the use of interferometry and a Shack–Hartmann sensor," *Opt Lett*, vol. 23, no. 10, pp. 742–744, 1998, doi: 10.1364/OL.23.000742.
16. J. E. Greivenkamp, D. G. Smith, R. O. Gappinger, and G. A. Williby, "Optical testing using Shack-Hartmann wavefront sensors," *Proc SPIE*, vol. 4416, pp. 260–263, 2001, doi: 10.1117/12.427063.
17. J. Gonglewski, B. Spielbusch, B. Pierson, and D. Dayton, "Atmospheric structure function measurements with a Shack–Hartmann wave-front sensor," *Opt Lett*, vol. 17, no. 24, pp. 1737–1739, 1992, doi: 10.1364/OL.17.001737.
18. J. C. Ricklin and F. M. Davidson, "Atmospheric turbulence effects on a partially coherent Gaussian beam: implications for free-space laser communication," *JOSA A*, vol. 19, no. 9, pp. 1794–1802, 2002, doi: 10.1364/JOSAA.19.001794.
19. M. J. Booth, "Adaptive optics in microscopy," doi: 10.1098/rsta.2007.0013.
20. J. W. Cha, J. Ballesta, and P. T. C. So, "Shack-Hartmann wavefront-sensor-based adaptive optics system for microscopy," *J Biomed Opt*, vol. 15, p. 46022, 2010, doi: 10.1117/1.3475954.
21. C. Zuo *et al.*, "Transport of intensity equation: a tutorial," *Opt Lasers Eng*, vol. 135, p. 106187, Dec. 2020, doi: 10.1016/J.OPTLASENG.2020.106187.
22. R. Ragazzoni, "Pupil plane wavefront sensing with an oscillating prism," *J Mod Optic*, vol. 43, no. 2, pp. 289–293, 2009, doi: 10.1080/09500349608232742.
23. M. Van Kooten and O. Lardière, "Double-pyramid wavefront sensors: tolerance relaxation and cheaper alternatives using achromatic double-roof prisms comet Hale-Bopp view project NFIRAOS view project," doi: 10.26698/AO4ELT5.0042.
24. O. Fauvarque, B. Neichel, T. Fusco, J. S.- Optica, and undefined 2016, "General formalism for Fourier-based wave front sensing," *opg.optica.org*, Accessed: Nov. 28, 2024. https://opg.optica.org/abstract.cfm?uri=optica-3-12-1440

25. M. Carbillet, C. Vérinaud, B. Femenía, A. Riccardi, and L. Fini, "Modelling astronomical adaptive optics – I. The software package caos," *Mon Not R Astron Soc*, vol. 356, pp. 1263–1275, 2005, doi: 10.1111/j.1365-2966.2004.08524.x.
26. N. Anugu *et al.*, "Design and development of a high-speed Visible Pyramid Wavefront Sensor for the MMT AO system."Proceedings Volume 11448, *Adaptive Optics Systems* VII; 114485J (2020) https://doi .org/10.1117/12.2576353
27. V. Akondi, S. Castillo, and B. Vohnsen, "Multi-faceted digital pyramid wavefront sensor," *Opt Commun*, vol. 323, pp. 77–86, Jul. 2014, doi: 10.1016/J.OPTCOM.2014.03.004.
28. V. Hutterer, E. T. Quinto, R. Professor, J. Kepler, and T. Wissenschaften, "Model-based wavefront reconstruction approaches for pyramid wavefront sensors in Adaptive Optics Doktorin der Technischen Wissenschaften in the Doctoral Program," 2018, Accessed: Jul. 14, 2023. www.jku.at
29. E. Brunner, J. Shatokhina, M. F. Shirazi, W. Drexler, R. Leitgeb, A. Pollreisz, C. K. Hitzenberger, R. Ramlau, and M. Pircher, "Retinal adaptive optics imaging with a pyramid wavefront sensor," *Biomed. Opt. Express* 12, 5969–5990 (2021)
30. S. Esposito and A. Riccardi, "Astronomy & astrophysics pyramid wavefront sensor behavior in partial correction adaptive optic systems," *A&A*, vol. 369, pp. 9–12, 2001, doi: 10.1051/0004-6361:20010219.
31. R. Ragazzoni, E. Diolaiti, and E. Vernet, "A pyramid wavefront sensor with no dynamic modulation," Accessed: Jul. 14, 2023. www.elsevier.com/locate/optcom
32. E. Brunner *et al.*, "Retinal adaptive optics imaging with a pyramid wavefront sensor," *Biomed Opt Express*, vol. 12, no. 10, pp. 5969–5990, Oct. 2021, doi: 10.1364/BOE.438915.
33. V. Chambouleyron *et al.*, "Pyramid wavefront sensor optical gains compensation using a convolutional model," *Astron Astrophys*, vol. 644, p. A6, Dec. 2020, doi: 10.1051/0004-6361/202037836.
34. C. Sheppard, D. K. Hamilton, and C. J. R. Sheppard, "Differential phase contrast in scanning optical microscopy," *J Microsc*, vol. 133, pp. 27–39, 1984, doi: 10.1111/j.1365-2818.1984.tb00460.x.
35. G. Zheng, C. Kolner, and C. Yang, "Microscopy refocusing and dark-field imaging by using a simple LED array," *Opt Lett*, vol. 15, no. 20, pp. 3987–3989, 2011.
36. M. Chen *et al.*, "3D differential phase contrast microscopy," *Biomed Opt Express*, vol. 7, no. 10, pp. 3940–3950, Oct. 2016, doi: 10.1364/BOE.7.003940.
37. Z. F. Phillips, M. Chen, and L. Waller, "Single-shot quantitative phase microscopy with color-multiplexed differential phase contrast (cDPC)," *PLoS One*, vol. 12, no. 2, Feb. 2017, doi: 10.1371/JOURNAL.PONE.0171228.
38. Y.-H. Chuang, J. A. Yeh, Y.-Z. Lin, Y. Luo, S. Vyas, and Y.-Y. Huang, "Multi-wavelength quantitative differential phase contrast imaging by radially asymmetric illumination," *Opt Lett*, vol. 44, no. 18, pp. 4542–4545, Sep. 2019, doi: 10.1364/OL.44.004542.
39. C. Moser, D. Carpentras, M. Künzi, and T. Laforest, "Effect of backscattering in phase contrast imaging of the retina," *Opt Express*, vol. 26, no. 6, pp. 6785–6795, Mar. 2018, doi: 10.1364/OE.26.006785.
40. C. Zuo, J. Sun, J. Zhang, Q. Chen, and Y. Fan, "Wide-field anti-aliased quantitative differential phase contrast microscopy," *Opt Express*, vol. 26, no. 19, pp. 25129–25146, Sep. 2018, doi: 10.1364/OE.26.025129.
41. P. Casteleiro Costa, P. Ledwig, A. Bergquist, J. Kurtzberg, and F. E. Robles, "Noninvasive white blood cell quantification in umbilical cord blood collection bags with quantitative oblique back-illumination microscopy," *Transfusion*, vol. 60, no. 3, pp. 588–597, Mar. 2020, doi: 10.1111/TRF.15704.
42. W. Lee *et al.*, "Color-coded LED microscopy for quantitative phase imaging: Implementation and application to sperm motility analysis," *Methods*, vol. 136, pp. 66–74, Mar. 2018, doi: 10.1016/J.YMETH.2017.11.010.
43. H.-H. Chen, Y.-Z. Lin, and Y. Luo, "Isotropic differential phase contrast microscopy for quantitative phase bio-imaging," 2018, doi: 10.1002/jbio.201700364.
44. C. Yurdakul *et al.*, "High-throughput, high-resolution interferometric light microscopy of biological nanoparticles," *ACS Nano*, vol. 14, 2020, doi: 10.1021/acsnano.9b08512.
45. N. Streibl, "Phase imaging by the transport equation of intensity," *Opt Commun*, vol. 49, no. 1, pp. 6–10, Feb. 1984, doi: 10.1016/0030-4018(84)90079-8.
46. S. S. Kou, L. Waller, G. Barbastathis, and C. J. R. Sheppard, "Transport-of-intensity approach to differential interference contrast (TI-DIC) microscopy for quantitative phase imaging," *Optics Letters*, vol. 35, no. 3, pp. 447–449, 2010, doi: 10.1364/OL.35.000447..
47. C. Zuo *et al.*, "Transport-of-intensity phase imaging using Savitzky-Golay differentiation filter - theory and applications," *Opt Express*, vol. 21, no. 5, pp. 5346–5362, Mar. 2013, doi: 10.1364/OE.21.005346.

48. C. Zuo *et al.*, "High-resolution transport-of-intensity quantitative phase microscopy with annular illumination," *nature.com*, doi: 10.1038/s41598-017-06837-1.
49. C. Zuo, Q. Chen, W. Qu, A. A.-O. express, and undefined 2013, "High-speed transport-of-intensity phase microscopy with an electrically tunable lens," *opg.optica.org*, Accessed: Oct. 15, 2024. https://opg.optica.org/abstract.cfm?uri=oe-21-20-24060
50. L. Lu, J. Sun, J. Zhang, Y. Fan, Q. Chen, and C. Zuo, "Quantitative phase imaging camera with a weak diffuser," *Front Phys*, vol. 7, no. May, p. 452287, May 2019, doi: 10.3389/FPHY.2019.00077/BIBTEX.
51. G. R. W., "A practical algorithm for the determination of plane from image and diffraction pictures," *Optik*, vol. 35, no. 2, pp. 237–246, 1972, Accessed: Jul. 14, 2023. https://cir.nii.ac.jp/crid/1572261550522209664
52. G. R. W., "Phase determination from image and diffraction plane pictures in the electron microscope," *Optik*, vol. 34, pp. 275–284, 1971, Accessed: Jul. 14, 2023. https://cir.nii.ac.jp/crid/1571417124634973824
53. T. Zhao and Y. Chi, "Modified Gerchberg–Saxton (G-S) algorithm and its application," *Entropy*, vol. 22, no. 12, p. 1354, Nov. 2020, doi: 10.3390/E22121354.
54. J. Miao, P. Charalambous, J. Kirz, and D. Sayre, "Extending the methodology of X-ray crystallography to allow imaging of micrometre-sized non-crystalline specimens," *Nature*, vol. 400, no. 6742, pp. 342–344, Jul. 1999, doi: 10.1038/22498.
55. J. M. Zuo, I. Vartanyants, M. Gao, R. Zhang, and L. A. Nagahara, "Atomic resolution imaging of a carbon nanotube from diffraction intensities," *Science*, vol. 300, no. 5624, pp. 1419–1421, May 2003, doi: 10.1126/SCIENCE.1083887/SUPPL_FILE/ZUO.SOM.PDF.
56. S. Eisebitt *et al.*, "Lensless imaging of magnetic nanostructures by X-ray spectro-holography," *Nature*, vol. 432, no. 7019, pp. 885–888, Dec. 2004, doi: 10.1038/nature03139.
57. R. A. Gonsalves and R. Chidlaw, "Wavefront sensing by phase retrieval," *Proc SPIE*, vol. 0207, pp. 32–39, Dec. 1979, doi: 10.1117/12.958223.
58. O. Guyon, "Limits of Adaptive Optics for high contra,st imaging,"· *Astrophys J*, vol. 629, no. 1, pp. 592–614, 2005, doi: 10.1086/431209/FULLTEXT/.
59. Y. Zhang, G. Pedrini, and W. Osten, "Wave front reconstruction from a sequence of interferograms recorded at different planes," *Opt Lett*, vol. 30, no. 8, pp. 833–835, Apr. 2005, doi: 10.1364/OL.30.000833.
60. O. Mudanyali *et al.*, "Compact, light-weight and cost-effective microscope based on lensless incoherent holography for telemedicine applications," *Lab Chip*, vol. 10, no. 11, pp. 1417–1428, May 2010, doi: 10.1039/C000453G.
61. A. Anand, G. Pedrini, W. Osten, and P. Almoro, "Wavefront sensing with random amplitude mask and phase retrieval," *Opt. Lett.* vol. 32, pp. 1584–1586, 2007.
62. P. Almoro, G. Pedrini, and W. Osten, "Complete wavefront reconstruction using sequential intensity measurements of a volume speckle field," *Appl Opt*, vol. 45, no. 34, pp. 8596–8605, Dec. 2006, doi: 10.1364/AO.45.008596.
63. C. C. Wackerman and A. E. Yagle, "Use of Fourier domain real-plane zeros to overcome a phase retrieval stagnation," *JOSA A*, vol. 8, no. 12, pp. 1898–1904, Dec. 1991, doi: 10.1364/JOSAA.8.001898.
64. G. Lu, Z. Zhang, F. T. S. Yu, and A. Tanone, "Pendulum iterative algorithm for phase retrieval from modulus data," *Proc SPIE*, vol. 33, no. 2, pp. 548–555, Feb. 1994, doi: 10.1117/12.153152.
65. R. H. T. Bates, "Uniqueness of solutions to two-dimensional fourier phase problems for localized and positive images," *Comput Vis Graph Image Process*, vol. 25, no. 2, pp. 205–217, Feb. 1984, doi: 10.1016/0734-189X(84)90103-8.
66. D. R. Luke, "Relaxed averaged alternating reflections for diffraction imaging," *Inverse Probl.*, vol. 21, no. 1, pp. 37–50, 2005, doi: 10.1088/0266-5611/21/1/004..
67. S. Marchesini, "A unified evaluation of iterative projection algorithms for phase retrieval," · *Review of Scientific Instruments*, vol. 78, no. 1, 2007, doi: 10.1063/1.2403783/349838.
68. Y. Zhang, G. Pedrini, and W. Osten, "Wave-front reconstruction from a sequence of interferograms recorded at different planes," *Opt Lett*, vol. 30, no. 8, pp. 833–835, 2005, doi: 10.1364/OL.30.000833.
69. R Hegerl, W. Hoppe "Dynamische Theorie der Kristallstrukturanalyse durch Elektronenbeugung im inhomogenen Primärstrahlwellenfeld - - 1970 - reports of the Bunsen Society for Physical Chemistry - Wiley Online Library." Accessed: Jul. 14, 2023. https://onlinelibrary.wiley.com/doi/abs/10.1002/bbpc.19700741112
70. S. Marchesini *et al.*, "X-ray image reconstruction from a diffraction pattern alone," *Phys Rev B*, vol. 68, no. 14, 2003, doi: 10.1103/PhysRevB.68.140101 .
71. H. M. L. Faulkner and J. M. Rodenburg, "Movable aperture lensless transmission microscopy: a novel phase retrieval algorithm," doi: 10.1103/PhysRevLett.93.023903.

72. P. Thibault, M. Dierolf, A. Menzel, O. Bunk, C. David, and F. Pfeiffer, "High-resolution scanning X-ray diffraction microscopy," *Science*, vol. 321, no. 5887, pp. 379–382, Jul. 2008, doi: 10.1126/SCIENCE.1158573/SUPPL_FILE/THIBAULT.SOM.PDF.
73. A. M. Maiden and J. M. Rodenburg, "An improved ptychographical phase retrieval algorithm for diffractive imaging," *Ultramicroscopy*, vol. 109, no. 10, pp. 1256–1262, Sep. 2009, doi: 10.1016/J.ULTRAMIC.2009.05.012.
74. P. Thibault, M. Dierolf, O. Bunk, A. Menzel, and F. Pfeiffer, "Probe retrieval in ptychographic coherent diffractive imaging," *Ultramicroscopy*, vol. 109, no. 4, pp. 338–343, Mar. 2009, doi: 10.1016/J.ULTRAMIC.2008.12.011.
75. H. M. L. Faulkner and J. M. Rodenburg, "Error tolerance of an iterative phase retrieval algorithm for moveable illumination microscopy," *Ultramicroscopy*, vol. 103, no. 2, pp. 153–164, May 2005, doi: 10.1016/J.ULTRAMIC.2004.11.006.
76. Y. S. Kim *et al.*, "Focus: medical technology: combining three-dimensional quantitative phase imaging and fluorescence microscopy for the study of ell pathophysiology," *Yale J Biol Med*, vol. 91, no. 3, p. 267, Sep. 2018, Accessed: Jul. 14, 2023. /pmc/articles/PMC6153632/
77. S. K. Rajput *et al.*, "Three-dimensional fluorescence imaging using the transport of intensity equation," *J Biomed Opt*, vol. 25, no. 3, p. 032004, Nov. 2019, doi: 10.1117/1.JBO.25.3.032004.
78. Y. J. Jo *et al.*, "Quantitative phase imaging and artificial intelligence: a review," *IEEE J Quantum Electron*, vol. 25, no. 1, Jan. 2018, doi: 10.1109/JSTQE.2018.2859234.

10 Machine-Learning Approach in Nonlinear Optical Microscopy

Manvitha Amin[1], Manikanth Karnati[1], Guan-Yu Zhuo[2], Yury V Kistinev[3], and Nirmal Mazumder[1]*

[1]Department of Biophysics, Manipal School of Life Sciences, Manipal Academy of Higher Education, Manipal, Karnataka, India
[2]Institute of New Drug Development, China Medical University, Taichung, Taiwan
[3]Laboratory of Laser Molecular Imaging and Machine Learning, Tomsk State University, Tomsk, Russia
*Corresponding Author

INTRODUCTION

The imaging of live cells in the surrounding tissue is a major goal and problem in the biological sciences. Nonlinear optical microscopy (NLO) empowers label-free observation of tissue and cell architecture, in addition to chemical and molecular labels for generating contrast and allowing visualization of cells in three dimensions over time (Adur et al., 2014). NLO by multiphoton excitation has become a prominent and effective tool for the three-dimensional (3D) modeling of biological specimens, integrating multiple excitation and emission approaches. Nonlinear optical technologies have become popular tools for biomolecular detection and medical diagnostics in cells and tissues at the micrometer to nanometer level (Li et al., 2018). The advancement of these optical approaches promotes and improves fundamental research in biology, medicine, pharmaceuticals, and other domains. In contrast to the optical approach, radiation is often used for imaging modalities such as computed tomography, plain radiography, X-ray, magnetic resonance imaging, and other nuclear imaging technologies. These procedures, however, are expensive and generate radiation. Optical imaging technologies can give sensitive, real-time, noninvasive, accurate, and cost-effective molecular information on human tissues for disease detection and diagnosis (Parodi et al., 2020). Nonlinear optical techniques provide several benefits over other optical imaging technologies, including depth penetration, noninvasiveness, ultrahigh resolution, and high sensitivity (Liang et al., 2017). Fluorescence microscopy is another typical approach for increasing the amount of information obtained from an essential biological sample. Highly specialized light-emitting probes, including semiconductor dye components, fluorescent proteins, and nanoparticles chemically attached to precise biological molecules such as proteins, DNA, and phospholipids, are delivered to living or fixed specimens in this procedure. It is feasible to limit the signal released from longer wavelengths by igniting the fluorescent species with a light or a laser, sensing it in a point scanning manner or widefield, and renovating pictures. Fluorescence-based approaches, on the other hand, have limitations due to the necessary short excitation wavelengths along with constant lasers, which induce scattering and absorption, photobleaching of fluorophores reducing by penetrating deep in thick materials and collection of signals (Oheim et al., 2006). With the introduction of NLO, these constraints might be circumvented by utilizing the interaction of the multiphoton process caused by pulsed lasers with infrared wavelengths. Due to the intrinsically low light scattering and absorption of infrared excitation wavelengths, NLO microscopy enables tissue-deep penetration (greater than 500 μm). Additionally, NLO techniques based on excitation eliminate the need for staining, allowing for the extraction of rich morphological

DOI: 10.1201/9781003545712-10

architectural or molecular information from a sample with nonlinear characteristics along with different chemical configurations. When particularly compared to equivalent linear imaging methods, NLO microscopy has a unique advantage in that it allows for label-free deep optical sectioning. Nonlinear processes, in contrast to linear processes, may generate substantial high-pulsed excited light with relatively low energy on a live sample (Adur et al., 2014). Near-infrared excitation light is commonly employed in nonlinear excitation microscopy and has the added benefit of lowering phototoxicity. Infrared light can be used to reduce dispersion in living cells and tissues. The penetration depth of nonlinear microscopes is increased as a result of all of these phenomena. Various biological NLO imaging methods have been advanced, such as two-photon excited fluorescence (TPEF), second harmonic generation (SHG), third harmonic generation (THG), coherent anti-Stokes Raman scattering (CARS), and stimulated Raman scattering (SRS). All these technologies have been employed in laser-scanning microscopes and are reliant on nonlinear interactions between light and excitable molecules (Parodi et al., 2020). Substantially larger tissue penetration up to the millimeter range due to near-infrared excitation and, as an outcome, reduced phototoxicity and less dispersion imposed on tissues and cells owing to the intrinsic confocal excitation of fluorophores in the focal plane and the absence of a detector pinhole results in enhanced optical sensor effectiveness; a wider excitation range of fluorochromes, allowing multicolor imaging using the same emitted light; the ability to combine fluorescence microscopy with other detection methods, such as higher harmonic generation microscopy (HHGM) and fluorescence lifetime iImaging (FLIM) (Palczewska et al., 2020).

The intensely localized activation of the two-photon absorption mechanism, as well as the decreased influence of light scattering in the material, results in two-photon microscopy. It allows the visual representation of both exogenous substances such as semiconductors, dye molecules, and fluorescent proteins including YFP, RFP, and GFP, as well as endogenous fluorophores. It is frequently used in fluorescence lifetime imaging studies, yielding results that are suited for metabolic and hybrid research. Excitation light with only a wavelength twice that of one-photon excitation is used in two-photon excitation, whereas excitation light with a wavelength three times that of one-photon excitation is used in three-photon imaging (Benninger et al., 2013). Excitation light with a longer wavelength loses fewer photons owing to dispersion in the biological sample. Furthermore, because three-photon excitation is reliant on the laser intensity of third power, potent excitation is restricted to a very tiny focused size, lowering out-of-focus fluorescence from superficial depths and enhancing the signal-to-background leverage ratio. The detection of noncentrosymmetric structures, such as myosin filaments, collagen fibers, and microtubules, as well as their spatial distribution in unaltered biological specimens, can be done by SHG (Campagnola et al., 2011). Water–protein and water–lipid interfaces, including internal and external membranes, as well as extracellular fluidic components, elicit THG microscopy, a label-free scatter technique. THG enables the identification of cellular and molecular processes in 3D of cultured tissues and animals, complementing multiparameter fluorescent lifetime microscopy and second harmonic generation in biological sciences. Coherent Raman scattering (CRS) microscopy has also been shown to be capable of exceeding the pace limitations for vibrational imaging. It is classified into two types: CARS and SRS (Weng et al., 2017). These processes frequently coexist in a sample because of the identical excitation circumstances with a Raman shift, which is the difference in frequency corresponding to the molecule frequency of vibration. SRS as well as CARS methods frequently coexist in a sample. One of the benefits is these signals' ability to provide insight into the molecular architecture and chemical configuration, which is recognized in the Raman spectrum's peak (Tipping et al., 2016).

TECHNIQUES OF NONLINEAR OPTICAL MICROSCOPY

Two-Photon Flourescence

Two-photon fluorescence is a second-order nonlinear optical process based on Maria Göppert Mayer's 1931 doctoral thesis idea of two-photon absorption, which was proven for the first time with the introduction of lasers 30 years later. Two-photon excitation is a fluorescence mechanism in which a fluorophore absorbs two photons at the same time. In this case, an electronic shift from the ground state to a higher electronic state is caused by the coupled action of two or more photons interacting with particles at the same time. When the sum quantity of these photons' energies equals the energy difference between the fundamental and excited states, this process is possible (Benninger et al., 2013). As a result, the most intense excitation is contained within a compact volume, and no out-of-range focus light is created, obviating the necessity of spatial filters to block unnecessary background light. Two-photon microscopy allows for enhanced imaging deep into thick samples. This amplified efficiency in such samples is enabled by three mechanisms; two-photon microscopy's excitation light scatters less than confocal microscopy's equivalent excitation light, in two-photon microscopy, excitation and fluorescence emission scattering is less deleterious, as no excitation light is absorbed outside the focus, extra light reaches the center (Crowe et al., 2014). Resolution of the image the numerical aperture (NA_0) of the optical system, and the wavelength of light λ employed in two-photon fluorescence microscopy determine the lateral Δ ρ and axial Δ z spatial resolution.

$$\Delta\rho = \frac{1.22\lambda}{2\sqrt{2}NA_0} \quad \Delta z = \frac{1.5n\lambda}{NA_0^2}$$

When evaluating the high spatial resolution provided by each modality, two important issues must be considered: two-photon excitation has a wavelength that is nearly twice as high as one-photon excitation. As a consequence, two-photon microscopy's lateral and axial resolution will be lower than that of an ideal confocal microscope with much the same microscope objective, and the confocal microscope equations above are only valid for an "ideal" confocal microscope with a completely closed pinhole. The lateral and axial resolutions will be marginally lower than in the optimal confocal setup because this is not typically the case in actual confocal microscopy (Soltanian-Zadeh et al., 2019). In TPEF, imaging thin and thick samples is also crucial. Another aspect of TPEF microscopy is fundamental depth. At increasing imaging depths, scattering leads excitation photons to miss the focus and fluorescence emission photons to miss the confocal pinhole. Outside-of-focus fluorescence, on the other hand, may scatter through the confocal pinhole, raising the background and lowering the signal-to-background ratio. In two-photon microscopy, scattering prevents excitation photons from reaching the focus at greater imaging depths, but these scattered photons do not activate fluorescence. In vivo tissue imaging at subcellular as well as cellular levels is possible using multiphoton laser microscopy (MPM). Multiphoton microscopy provides structural and functional label-free tissue imaging by using autofluorescence of endogenic fluorophores and the SHG tissue matrix components. As a result, MPM is employed to calculate the tissue's "SHG-to-AF Aging Index of Dermis" value. It is a strong method for visualizing fibrillar collagen in a variety of biological materials. It delivers high-dimensional visual data with no discernible disease-specific features (Kistenev et al., 2019). Machine learning is a good method that does not need the assessment of hidden relationships by an expert. Nave Bayes and boosting, support vector machines (SVM), random forest, and artificial neural networks are the most prominent supervised learning methods. The SVM approach involves creating a feature hyperplane that indicates the biggest distinction between different classes of data. If the optimal boundary among the classes has a very complicated figure, the input data is transformed using nonlinear integral transformation. Extracting feature techniques are essential in data analysis and preprocessing to generate and categorize region of interest (ROI) images of mathematical models. Textural, graph, gradient, and morphological aspects are examples of these approaches. The clarity and direction of different computer vision edge detector techniques can be used to determine the sharpness and orientation of collagen fibers. Modern techniques rely on first-order (Sobel Prewitt operators), second-order (Gaussian), and mathematical morphology. Convolution using a Sobel operator is a well-known technique in computer vision for emphasizing visual edges. A method for lymphedema diagnosis employs MPM and machine learning to estimate collagen disarray in vivo in human tissue. At depths of 0–80 μm, collagen fibers are seldom visible. In this issue, the SHG signal to AF signal ratio, often known as the collagen-to-elastin ratio, was found to be much greater. Detectors of the image were employed to produce a measurable estimate of collagen disarray in lymphedema tissue. For SHG tissue image extraction, the histogram of oriented gradients (HOG) approach was applied. With a shift in the distribution's median to bigger values, the image total edge length distribution was observed to grow broader (Campagnola et al., 2011). The bagging technique was carried out and the voting system was used, which was centered on a group categorization of numerous SHG photographs. Researchers were able to obtain 96% accuracy as a result of this. Future studies should focus on determining the sensitivity threshold of this method throughout the onset and progression of lymphedema.

Second Harmonic Generation

Second harmonc generation (SHG) is a coherent nonlinear process in which two incident photons interacting with a medium at their fundamental frequency are directly transformed into a single photon with precisely the same amount of energy at twice the frequency, without photon absorption or reemission. Since two photons of frequency ω are turned into one photon of frequency 2ω, SHG is known as the frequency-doubling phenomenon (Cisek et al., 2021). SHG can only be produced by noncentrosymmetric materials at both the molecular and macromolecular scales; hence, microcrystalline structures are necessary. The NLO concept states that an incident electric field $E\omega \rightarrow$ with a frequency causes a second order polarisation $P\!\rightarrow 2\omega$, i at 2ω in the ith direction defined as

$$\overrightarrow{P_{2\omega,i}} = \varepsilon_0 \left[\chi^{(1)}\vec{E} + \chi^{(2)}\vec{E}\vec{E} + \chi^{(3)}\vec{E}\vec{E}\vec{E} + .. \right.$$

Vacuum permittivity = ε_0
The nth-order nonlinear susceptibility tensor = χ (n)

The first portion of the equation represents light reflection as well as absorption, and the second is the second harmonic generation. Also, because term $\chi^{(2)}$ is dependent on the polarization of the excitation source, the SHG emission is polarization-sensitive.

The non-null coefficients dij = 1/2 $\varepsilon_0 \chi_{ij}(n)$ in a medium with hexagonal symmetry.

The nonlinear optical characteristics of the material are represented by these coefficients, which add to zero for inversion symmetry. Kleinman demonstrated SHG in crystalline quartz in 1962, and it has since been widely employed to produce shorter wavelengths in frequency-doubled pulsed lasers. The defining aspect of SHG is its sensitivity to media symmetry; second harmonic is only allowed in media with broken spatial inversion symmetry and in crystals within the dipole approximation. Crystals that are ferroelectric and piezoelectric are noncentrosymmetric resources without spatial inversion symmetry. The noncentrosymmetric structures of ferroelectric crystals, for example, are the result of spontaneous polarization generated from the position of the dipole electrical moments. Because there is no spatial inversion symmetry at the border of centrosymmetric media, the interfaces along with the surface of the media are also second harmonic active. Its activity may also be observed in media that lack temporal inversion symmetry, are subjected to a magnetic field, or have magnetic ordering (Friedl et al., 2007). The second harmonic is energy-saving since it does not necessitate an excited state, is highly directed, and also maintains the laser light's coherence. When a medium construction is ordered at the wavelength of the laser scale and lacks a center of inversion, it can give an efficient SHG signal. For biophysical as well as biological imaging applications, SHG has developed as a powerful NLO technique. In the research article, the researchers fostered a deep learning (DL) approach using a profound convolutional neural network (CNN) to gauge the flexible properties of collagenous tissues straightforwardly from noninvasive microscopy pictures. It primarily worked to employ DL algorithms to directly estimate nonlinear anisotropic elastic characteristics from tissue microscope pictures. This study demonstrates that deep CNN can also function effectively with a short dataset by combining verified and unverified learning methods and applying data growth approaches. The performance of the CNN may be improved further when additional images and mechanical testing data are obtained. The CNN employed in this study was created expressly for this purpose (Haft-Javaherian et al., 2019). The CNN's first to fifth blocks function as automated feature extractors, encoding the input picture into a feature vector of 64 digits. The first convolution layer's filters represent various local fiber network patterns, while the subsequent convolution layers' filters represent various combinations of these patterns at various locations and length scales. The CNN's sixth block connects the feature vector to elastic characteristics for classification and regression. Since the full CNN describes the nonlinear link between tissue structure and elastic properties, this technique is structural: mechanics inference from structural characteristics collected by CNN. This work proved the viability of applying DL approaches to analyze GLBP elastic characteristics using microscopy pictures in a rapid and noninvasive manner (Soltanian-Zadeh et al., 2019). The use of PCA to attempt to meet equi-biaxial stress–rain curves and quantify overall stiffness, followed by the use of a configurable deep CNN design to extract relevant patterns of tissues and deliver greater rigidity, and finally to anticipate PCA parameters of nonlinear anisotropic stress–strain curves and tackle difficulty of tiny datasets in the field of biomechanics. Researchers used a combination of unverified DL, verified learning, and data expansion. The growth of NLO processes in bioimaging has been considered. Polar dye molecules can highlight both endogenously SHG-active components in biological tissues and plasma membranes under an SHG microscope. This might allow for real-time monitoring of how toxins cause harm to cell membranes. Nonlinear optical flow cytometry is being developed, and combining SHG-based tests with flow cytometry might have an intriguing use (Hristu et al., 2018). Endogenous SHG bioimaging has been extensively developed and will be used in a variety of diagnostics. The advent of exogenous SHG triggering agents for the membrane, apart from amphiphilic polar pigments, will assist in the expansion of SHG imaging. In addition to fluorescent labeling in TPF imaging, the use of SHG-active nanoparticles for SHG labeling is intriguing for SHG imaging.

Third Harmonic Generation

In third harmonc generation (THG) microscopy, when the fundamental wavelength, normally over 1200 m or higher, irradiates a material having inherent nonuniformity due to refractive index mismatch, it results in a third-order NLO process. When the excitation wavelength's frequency is tripled, transforming three photons' energy output into a single emitted signal with one-third of the excitation wavelength and three times the intensity, THG occurs. Because of the Gouy phase difference of the fundamental beam, nonlinear signals generated at regions of symmetry concerning the focus overlap disruptively in THG imaging, resulting in null signal from a homogeneous material (Masihzadeh et al., 2015). Optical heterogeneities in the laser-illuminated material, such as those that occur at the interfacial region of two optically different elements, are revealed by the THG signal. Unlike SHG microscopy, THG microscopy does not require exact symmetries in the material to generate a signal. Structural interfaces, such as those developed by THG, involve third-order nonlinear susceptibility $\chi3$, both of which are inherent physical features of solids and liquids. As a result, THG is more adaptable than SHG since it does not need the object to be captured to have a certain asymmetry. THG bridges the gap in resolution between mesoscale imaging and high-resolution ultrasonography by recording three-dimensional tissue organization and function and includes nanoscopy, which employs scanning electron microscopy to study 3D tissue organization and function by sensing tissue and cell interfaces in a label-free and noninvasive method. Because of its accessibility in technological implementation in current multiphoton image processing systems and its diverse use with SHG and fluorescence imaging, THG has a lot of promise to become a common tool in infrared multiphoton microscopy (Weigelin et al., 2016). In the future, collecting signals from forward THG will improve in vitro applications in transparent samples, allowing researchers to understand cell–matrix interactions in tissue remodeling, microvesicle biology, and label-free methods in complex in vitro systems. When combined with chemically or biologically encoded markers, THG is very useful for intravital microscopy in small animal models. The optical third-order nonlinear susceptibilities of lipid concentrations inside cellular structures differ from those in an aqueous system like the cytoplasm of the cell. The signal from THG in the cell membrane produces a "biomarker" for lipid imaging. THG is also abundant in very absorbent materials. The hfRPE cells' light-absorbing pigment granules provide contrast for RPE cell microscopy. The THG signals from the cell membrane originate in lipids within the plasma membrane. THG may be formed by the TPAF signaling mostly from retinyl esters inside the ribosomes in albino animals. Researchers demonstrated that lipid-specific THG mapping of the entire retina might be achieved by utilizing a custom-built fiber laser system in an article (Masihzadeh et al., 2015). The wavelength used in laser design was chosen to strike the right balance among ocular spectral transmission, which falls dramatically about 1100 nm, reduced damage threshold due to decreased absorption of melanocytes in the retinal pigment epithelium at near-infrared wavelengths, and suitability with commercial laser technology and optical coherence tomography (OCT), as these are used securely for ocular refractive surgery. It may be as simple as installing a suitable chromatic filter in the emitting channel to add the THG modality to already existing multiphoton microscopes (Gavgiotaki et al., 2020). THG is typically collected and produced in a forward detector arrangement. In an epi-detection arrangement, however, weaker THG can be detected. Similarly, recording scattered and reflective signals from various locations inside the eye might be used to do in vivo THG scanning of the retina. This approach may be used to ex vivo image unfixed, undamaged animal tissues. Using a laser line at 1300 nm, THG has a wide range of applications in cellular and molecular investigation, focusing on immunological cell functionality, tissue structure, hemostasis and cancer investigations, and the increasing use of THG in clinical therapy.

Coherent Anti-Stokes Raman Scattering Microscopy

Coherent anti-Stokes Raman scattering microscopy (CARS) is based on intrinsic molecular vibrational optical contrast and can generate high-resolution, real-time, selective, and label-free pictures of biological materials. In CARS, two rays of light are commonly employed to concurrently excite the sample; one is referred to as the pump beam, and the other is referred to as the Stokes beam. Four simultaneous coherent Raman events will occur when the frequency difference between the two light beams matches the target sample's chemical bond vibrational frequency (Weng et al., 2017). In CARS, a pump field at frequency $\omega\ p$, a Stokes field at frequency $\omega\ s$, and a probe field at frequency $\omega\ pr$ (normally the same as the pump) are concurrently interacting with the sample. The pump and Stokes beams are temporally and spatially overlapped before being focused on a sample with a high NA microscope objective. They collaborate to investigate specific chemical vibrations and generate signals via the wave-mixing mechanism (Tipping et al., 2016). A 2D picture is obtained by raster scanning the laser focus throughout the sample at video-rate speed. By shifting the focus point on the z-axis, the tiny focal volume also enables 3D optical sectioning at varying depths inside the sample. The electron in the sample is shifted from its equilibrium location by high-energy ultrafast lasers, resulting in induced polarization that is no longer linearly dependent on the size of the applied field (Li et al., 2020). To incorporate the nonlinear electron movements, this polarization can be expressed as a power series in the field. CARS is a third-order NLO imaging method. CARS microscopy is an appealing method for atherosclerosis research. In an atherosclerosis study, it was shown that CARS-based multimodal NLO microscopy can visualize the many pathological components of atherosclerotic lesions and even so, may be used as an in situ histology tool without the labeling requirements of traditional techniques (Weng et al., 2017). This technique provided the morphology and a quantitative tool for evaluating lesions and the lipid and collagen percentage contents, whose accumulation would influence the mechanical characteristics of atherosclerotic arteries.

Stimulated Raman Scattering

Stimulated Raman scattering (SRS) microscopy is a high-resolution imaging technology that leverages the inherent contrast of vibrational fragments to produce chemical mapping of living tissues and cells. SRS imaging has shown tremendous potential in many biological and medicinal applications because of its subcellular spatial resolution and label-free scanning capability, in tissue diagnosis, and drug imaging (Li et al., 2021). SRS was initially discovered in 1962, but its significance as a Raman imaging platform has only lately been realized. Two arriving fields with pump frequencies $\omega\ p$ and Stokes frequencies $\omega\ s$ stimulate the sample in SRS. To give a linear spectrum frequency analysis on the time axis, both laser pulses are positively chirped using prisms, glasses, gratings, or other materials in this configuration. Excitation-detection in parallel is another approach for rapidly capturing dynamic events in cells or living animals. SRS imaging, often referred to as multiplex SRS photography, has significantly accelerated the capture of Raman spectra, bringing forth the requirement for considerable instrument upgrades. Epi detection is often needed; in vitro investigations using SRS imaging are inherently signal-limited. Backscattered light from the sample is acquired by epi imaging, resulting in a significantly lower signal than transmission mode (Manifold et al., 2019). Many in vitro investigations use denoising methods to boost picture SNR. CNNs have shown a lot of potential as a denoising method. These models are CNN-based or denoise pictures with inherent compression faults or noise; algorithms have been used. Gaussian noise was created, and it regularly outperformed blind denoising tests. Nonetheless, the completely connected nature of the most widely utilized CNNs for denoising necessitates a significant amount of time and work. Training takes a long time, and huge training materials are required for effectiveness (Suzuki et al., 2019). Furthermore, these profound photos with relatively narrowband noise are used to train learning denoising algorithms. The developed U-Net-based method is regarded as a specialized tool for enhancing SRS mapping. Deep learning can increase SRS sensing capabilities in biological organisms. Future research aimed at directly building a CNN on the low-power noise profile SRS might improve the generalized deep-learning denoising method.

CONCLUSION

NLO enables the study of cell and tissue architecture without the use of labels. Because of the intrinsic low light scattering and absorption of infrared excitation wavelengths, NLO microscopy enables deep tissue penetration. On a live sample, NLO methods yield high-pulsed stimulated light with relatively little energy (Parodi et al., 2020). Near-infrared excitation light is commonly used in nonlinear excitation microscopy. SHG, TPEF, CARS, SRS, and THG microscopy are the key NLO methods for efficient observation of label-free and living biotic material. Each of these approaches makes use of various features of the light–matter interaction and may be used to scan biological specimen characteristics under unlabeled settings (Benninger et al., 2013). TPEF is considered one of the primary NLO methods capable of exciting endogenous compounds and cofactors as well as exogenous fluorophores at high depths. When paired with a rapid sensor, including a single photon counter that is time-correlated, two photon excitation fluorescence-fluorescece lifetime imaging (TPEF-FLIM) may be used to quantify the lifespan of autofluorescent molecules (Crowe et al., 2014). The SHG technique, on the other hand, is extremely beneficial for detecting label-free signals in noncentrosymmetric media such as myosin, microtubules, and collagen fibrils. On the other hand, THG enables the detection of surfaces in media with different refraction indices, such as lipid-water associated in cellular compartments. Vibrational CRS techniques activate the vibrational resonant frequencies of the primary chemical bonds with a distinct and recognized fingerprint to analyze the chemical composition of the materials. Finally, SRS and CARS allow for the representation of the biochemical composition of biological material both at the tissue and single-cell levels by detecting the vibrational band of the sample at each point and thereby quantifying the amount of lipids, proteins, and DNA. Because of the range of biological material that may be seen, from single cellular constituents to heterogeneous tissues without the requirement for staining in a non-destructive and unaltered manner, multi-spectral nonlinear imaging and multi-modal methods will be of momentous importance for tissue engineering applications (Manifold et al., 2019). Since a result would be in the center, the need for a speedy and label-free analytical tool is a long way from being met. Indeed, preoperative cellular evaluations and surgical observations would greatly minimize the danger of tissue injury owing to imprecise removal and ambiguity in pathological area diagnosis.

Acknowledgments: We thank the Global Innovation and Technology Alliance (GITA), Department of Science and Technology (DST), India [Project Number – GITA/DST/TWN/P-95/2021], and the Indian Council of Medical Research (ICMR), (Project Number-ITR/Ad-hoc/43/2020-21, ID No. 2020-3286) Government of India for financial support. We also thank the Manipal School of Life Sciences, Manipal Academy of Higher Education, Manipal, Karnataka, India for the infrastructure and support provided during this study.

Conflicts of interest: The authors declare that they have no conflicts of interest with the contents of this chapter.

REFERENCES

Adur, J., Carvalho, H. F., Cesar, C. L., & Casco, V. H. (2014). Nonlinear optical microscopy signal processing strategies in cancer. *Cancer Informatics*, *13*, CIN-S12419.

Benninger, R. K., & Piston, D. W. (2013). Two-photon excitation microscopy for the study of living cells and tissues. *Current Protocols in Cell Biology*, *59*(1), 4–11.

Bueno, J. M., Ávila, F. J., & Artal, P. (2016). *Second harmonic generation microscopy: A tool for quantitative analysis of tissues*. InTech. doi: 10.5772/63493

Campagnola, P. J., & Dong, C. Y. (2011). Second harmonic generation microscopy: Principles and applications to disease diagnosis. *Laser & Photonics Reviews*, *5*(1), 13–26.

Cisek, R., Joseph, A., Harvey, M., & Tokarz, D. (2021). Polarization-sensitive second harmonic generation microscopy for investigations of diseased collagenous tissues. *Frontiers in Physics, 9*, 480.

Crowe, S. E., & Ellis-Davies, G. C. (2014). Longitudinal in vivo two-photon fluorescence imaging. *Journal of Comparative Neurology*, *522*(8), 1708–1727. doi: 10.1002/cne.23502. PMID: 24214350; PMCID: PMC4136386.

Friedl, P., Wolf, K., Harms, G., & von Andrian, U. H. (2007). Biological second and third harmonic generation microscopy. *Current Protocols in Cell Biology*, *34*(1), 4–15.

Gavgiotaki, E., Filippidis, G., Tsafas, V., Bovasianos, S., Kenanakis, G., Georgoulias, V., ... & Athanassakis, I. (2020). Third harmonic generation microscopy distinguishes malignant cell grade in human breast tissue biopsies. *Scientific Reports*, *10*(1), 1–13.

Haft-Javaherian, M., Fang, L., Muse, V., Schaffer, C. B., Nishimura, N., & Sabuncu, M. R. (2019). Deep convolutional neural networks for segmenting 3D in vivo multiphoton images of vasculature in Alzheimer disease mouse models. *PloS One*, *14*(3), e0213539.

Hristu, R., Eftimie, L. G., Stanciu, S. G., Tranca, D. E., Paun, B., Sajin, M., & Stanciu, G. A. (2018). Quantitative second harmonic generation microscopy for the structural characterization of capsular collagen in thyroid neoplasms. *Biomedical Optics Express*, *9*(8), 3923–3936.

Kistenev, Y. V., Nikolaev, V. V., Kurochkina, O. S., Borisov, A. V., Vrazhnov, D. A., & Sandykova, E. A. (2019). Application of multiphoton imaging and machine learning to lymphedema tissue analysis. *Biomedical Optics Express*, *10*(7), 3353–3368.

Li, R., Wang, X., Zhou, Y., Zong, H., Chen, M., & Sun, M. (2018). Advances in nonlinear optical microscopy for biophotonics. *Journal of Nanophotonics*, *12*(3), 033007.

Li, S., Li, Y., Yi, R., Liu, L., & Qu, J. (2020). Coherent anti-stokes Raman scattering microscopy and its applications. *Frontiers in Physics*, *8*, 515.

Li, Y., Shen, B., Li, S., Zhao, Y., Qu, J., & Liu, L. (2021). Review of stimulated Raman scattering microscopy techniques and applications in the biosciences. *Advanced Biology*, *5*(1), 2000184.

Liang, L., Liu, M., & Sun, W. (2017). A deep learning approach to estimate chemically-treated collagenous tissue nonlinear anisotropic stress-strain responses from microscopy images. *Acta Biomaterialia*, *63*, 227–235.

Manifold, B., Thomas, E., Francis, A. T., Hill, A. H., & Fu, D. (2019). Denoising of stimulated Raman scattering microscopy images via deep learning. *Biomedical Optics Express*, *10*(8), 3860–3874.

Masihzadeh, O., Lei, T. C., Domingue, S. R., Kahook, M. Y., Bartels, R. A., & Ammar, D. A. (2015). Third harmonic generation microscopy of a mouse retina. *Molecular Vision*, *21*, 538.

Oheim, M., Michael, D. J., Geisbauer, M., Madsen, D., & Chow, R. H. (2006). Principles of two-photon excitation fluorescence microscopy and other nonlinear imaging approaches. *Advanced Drug Delivery Reviews*, *58*(7), 788–808.

Palczewska, G., Boguslawski, J., Stremplewski, P., Kornaszewski, L., Zhang, J., Dong, Z., ... & Palczewski, K. (2020). Noninvasive two-photon optical biopsy of retinal fluorophores. *Proceedings of the National Academy of Sciences*, *117*(36), 22532–22543

Parodi, V., Jacchetti, E., Osellame, R., Cerullo, G., Polli, D., & Raimondi, M. T. (2020). Nonlinear optical microscopy: From fundamentals to applications in live bioimaging. *Frontiers in Bioengineering and Biotechnology*, *8*, 1174.

Soltanian-Zadeh, S., Sahingur, K., Blau, S., Gong, Y., & Farsiu, S. (2019). Fast and robust active neuron segmentation in two-photon calcium imaging using spatiotemporal deep learning. *Proceedings of the National Academy of Sciences*, *116*(17), 8554–8563.

Suzuki, Y., Kobayashi, K., Wakisaka, Y., Deng, D., Tanaka, S., Huang, C. J., ... & Ozeki, Y. (2019). Label-free chemical imaging flow cytometry by high-speed multicolor stimulated Raman scattering. *Proceedings of the National Academy of Sciences*, *116*(32), 15842–15848.

Tipping, W. J., Lee, M., Serrels, A., Brunton, V. G., & Hulme, A. N. (2016). Stimulated Raman scattering microscopy: An emerging tool for drug discovery. *Chemical Society Reviews*, *45*(8), 2075–2089.

Weigelin, B., Bakker, G. J., & Friedl, P. (2016). Third harmonic generation microscopy of cells and tissue organization. *Journal of Cell Science*, *129*(2), 245–255.

Weng, S. (2017). Integrating coherent anti-stokes Raman scattering imaging and deep learning analytics for high precision, real time, label free cancer diagnosis (Doctoral dissertation, Rice University, Houston, TX).

Weng, S., Xu, X., Li, J., & Wong, S. T. (2017). Combining deep learning and coherent anti-Stokes Raman scattering imaging for automated differential diagnosis of lung cancer. *Journal of Biomedical Optics*, *22*(10), 106017

11 Animal Models of Obsessive-Compulsive Disorder

*Feyba Jijimon, Sonam Fathima Mehak, and Gireesh Gangadharan**

Department of Ageing Research, Manipal School of Life Sciences, Manipal Academy of Higher Education, Manipal, Karnataka, India

*Corresponding Author

INTRODUCTION

Obsessive-compulsive disorder (OCD) is a psychological condition with multifaceted etiology, characterized by obsessions and compulsions. It is defined by the presence of substantial distress which interferes with an individual's ability to function (Lang et al., 2009). OCD encompasses a broad range of symptoms, including symmetry, contamination, hoarding, impulsivity, religious, and sexual obsessions, followed by a set of irresistible ritualistic compulsive behaviors performed to relieve the anxiety associated with the obsessions (Leckman et al., 1997). In any individual, the primary symptoms associated with OCD can alter over time. Disorders like autism, Parkinson's disease, Tourette's syndrome, obsessive-compulsive personality disorder, body dysmorphic disorder, and trichotillomania have symptoms similar to OCD (Stein, 2002). The presence of these overlapping sets of symptoms hints directly at an interpretation that OCD may be influenced by specific neural circuits.

Research into the precise causes of OCD has consistently identified abnormalities in the cortico-striato-thalamo-cortical (CSTC) circuit as the hallmark of the disease (Saxena and Rauch, 2000). Hyperactivity in various regions within the CSTC circuit has been thought to underlie the pathophysiology of OCD. The CSTC loop comprises projections that originate from a specific region of the frontal cortex, terminate in the striatum, and then travel by "direct" or "indirect" pathways through the basal ganglia to the thalamus, before returning to that same cortical region to close the loop. The basal ganglia play a vital role in regulating the CSTC network through the interplay of two populations of D1 and D2 dopamine receptor-expressing GABAergic medium spiny neurons – the D1 MSN and D2 MSN – which constitute the direct and indirect pathways of the basal ganglia. The striatum is the major input region of the basal ganglia and is composed of the caudate, putamen, and nucleus accumbens (NAc). The primary output regions of the basal ganglia consist of the globus pallidus interna (GPi) and the substantia nigra pars reticulata (SNr). These structures receive input from D1 medium spiny neurons (MSNs) within the basal ganglia and subsequently transmit inhibitory signals to the thalamus. This forms the direct pathway, increased activity of which is hypothesized to underlie the pathology of OCD. GPi/SNr receives the D2 MSNs through the globus pallidus externa (GPe) and the subthalamic nucleus (STN), which form the indirect pathway (Ahmari and Dougherty, 2015; Goodman, Storch, and Sheth, 2021). Structural and functional imaging research on the dysfunctions of the CSTC loop has focused on the orbitofrontal cortex (OFC), anterior cingulate cortex (ACC), ventromedial prefrontal cortex (vmPFC), dorsolateral prefrontal cortex (dlPFC), striatum, thalamus, and temporolimbic regions (Gan et al., 2017; Goodman, Storch, and Sheth, 2021).

 DOI: 10.1201/9781003545712-11

While this simplifies the neural circuitry underlying the obsessions and compulsions, there has been a conspicuous absence of direct experimental support for exploring the treatment strategies to overcome the proposed resistance to selective serotonin reuptake inhibitors (SSRIs) in OCD. OCD frequently co-occurs with other psychiatric disorders, and therefore it is important to take into consideration the clinical features prior to determining the treatment choices. SSRIs are the primary treatment choice for OCD, and cognitive behavioral therapy (CBT), especially exposure response therapy (ERP), has been shown to be effective in the management of OCD. Recently, deep brain stimulation (DBS) has become an alternative to lesional and capsulotomy studies in OCD (Alonso, Cuadras et al., 2015). However, a notable proportion of OCD patients fail to completely respond to these treatments (Gershkovich, Wheaton, and Simpson, 2017). The main focus of the available treatments is to reduce the activity of overactive brain regions associated with OCD.

There are several rodent models developed to understand the pathophysiology of OCD. Animal models of a disease or disorder aim at simulating specific medical conditions in non-human animals like rodents, in order to mimic certain aspects of the disease. Valid animal models should exhibit human-like symptoms, measurable and observable behavioral changes, similar treatment efficacy, and replicability of the model to other researchers (McKinney and Bunney, 1969). An animal model must therefore meet three types of validity as proposed by Naylor (1992): (i) face validity (similarity in symptoms and behaviors of the disorder, e.g., compulsive behavior in OCD), (ii) construct validity (similar underlying mechanisms, here, the cortical-basal ganglia circuitry), and (iii) predictive validity (similar pharmacological response, e.g., SSRI responsiveness). The heterogeneity of the disorder must be taken into account when applying these criteria in practice to potential animal models of OCD (Hoffman, 2011). In this context, this book chapter aims to provide a concise overview of various animal models of OCD research that offer insights for potential novel therapeutic options.

ANIMAL MODELS OF OCD

Animal models have proved their immense utility in understanding the disease and facilitating advancements to existing research. The animal models developed for OCD research can be divided into transgenic, pharmacological, behavioral, and neurodevelopmental models. The more types of validity criteria a model meets, the higher will be its utility and relevance for research.

Transgenic Models of OCD

Sapap3 Knockout Mice

SAP90/PSD95-associated protein 3 is a postsynaptic scaffolding protein that plays a role in the regulation of synaptic functions, particularly in the striatum. A study by Welch et al., 2007 demonstrated that the targeted deletion of *Sapap3* in mice resulted in behavioral phenotypes that closely align with the symptomatology of human OCD, consisting of increased anxiety, compulsive self-injurious grooming behavior, and treatment response to SSRIs. These mice also had immature cortico-striatal synapses characterized by issues in cortico-striatal synaptic transmission dependent on the glutamate receptors alpha-amino-3-hydroxy-5-methyl-4-isoxazole propionic acid receptor (AMPAR) and N-methyl-D-aspartate receptor (NMDAR). Lentivirus-mediated reintroduction of *Sapap3* specifically into the striatum rescued these behavioral and synaptic deficits (Welch et al., 2007). Additionally, the *Sapap3* mutant mice displayed impaired reversal learning (Manning et al., 2019), replicating the cognitive inflexibility in OCD patients reported in clinical studies (Dajani and Uddin, 2015). This supports the predictive validity of this model for OCD research. A recent study by Lamothe et al., 2023 discovered a previously undiscovered type of repetitive behavior in *Sapap3* mutant mice, redefining it as a mouse model of repetitive behaviors instead of compulsive-like behaviors. This consists of sudden and rapid repetitive single movements including head/body twitches and hind paw scratching, which they described as repeated hind paw beating against various body parts. All these symptoms, except long-term grooming behavior, were successfully rescued by pharmacological treatment using Aripiprazole, a first-line treatment for tic-like movements (Lamothe et al., 2023). Therefore, the presence of both tic- and compulsive-like traits in the *Sapap3-/-* model elevates its significance in investigating the neurobiological foundations of comorbid tic- and compulsive-like features in different disorders of repetitive behaviors.

Slitrk5 Null Mice

Soon after the initial identification of the *Sapap*3 knockout mice model of OCD, another synaptic protein, *Slitrk5*, emerged to be relevant to the neurobiology of OCD. *Slitrk5* is a member of the *Slitrk* family of transmembrane proteins, specifically enriched in striatal neurons, and is involved in the development and regulation of synaptic connections in the brain (Song et al., 2017). Mice with targeted inactivation of *Slitrk5* developed facial hair loss and severe skin lesions, which can be attributed to their overgrooming behavior (Shmelkov et al., 2010). Alongside, the mice developed enhanced marble burying and anxiety-like behaviors. Shmelkov et al. (2010) discovered that *Slitrk5-/-* mice showed upregulated neuronal activity in the orbitofrontal cortex (OFC) assessed by FosB expression, which is consistent with the findings of previous studies (Saxena and Rauch, 2000; Saxena, Bota, and Brody, 2001). The inference about the anatomical deficits in Slitrk5-/- mice is not, however, uniformly consistent across all the studies. Some studies have reported that striatal volumes are reduced in drug-naïve children with OCD (Rosenberg et al., 1997) while others have not observed a significant decrease in striatal volume (Stewart et al., 2013). Together these findings suggest that *Slitrk5* knockout mice model may help study CSTC circuit dysfunction in OCD pathogenesis.

5HT2C Knockout Mice

The hyperphagic behavior of the *5HT2C* receptor mutant mouse model provides evidence for the compulsive behavior of OCD. Chou-Green et al., 2003 observed persistent chewing behavior of kaolin clay, methodical neat chewing of the screen, and perseverative head dipping behavior in the serotonin receptor *5HT2C* knockout mice, which are similar to the compulsive phenotypes observed in OCD patients. The *5HT2C* mouse model provides a promising model to explore the role of serotonin in OCD pathophysiology.

HOXB8 Knockout Mice

HOXB8, a member of the mammalian homeobox-containing complex, is expressed in the orbitofrontal cortex, anterior cingulate cortex, and caudate nucleus; the regions comprising CSTC circuitry. *HOXB8* mutant mice are characterized by hair removal, skin lesions, excessive body licking and biting, and increased persistence of grooming behavior (Greer and Capecchi, 2002), which are some of the symptoms of OCD. This model was built on the basis of behavioral similarities of mutant mice with specific respects to OCD patients and does not fully replicate the complexity of OCD.

Slc1a1/EAAC1 Null Mice

Slc1a1, also known as *EAAC1* (excitatory amino acid carrier 1), is a functional candidate gene for OCD that encodes the glutamate transporter *EAAC1* (Wang et al., 2010). The link between OCD and the glutamatergic system, especially the glutamate transporter gene *Slc1a1*, has been reported in previous research (Stewart et al., 2013). Also, glutamatergic medications, including d-cycloserine (Wilhelm et al., 2008), riluzole (Grant et al., 2007) and memantine (Haghighi et al., 2013) seem to have an ameliorative effect on OCD symptoms. *Slc1a1* shows a notable association in a subset of OCD patients who exhibit sensitivity to physiological or environmental factors (Wang et al., 2010). A study by Aoyama et al., 2006 on *EAAC1-/-* mice showed impaired behavior in the Morris water maze and an increased susceptibility to oxidative injury as well as brain atrophy, both of which were supported by their hippocampal slices.

Dopamine Transporter Knockdown Model

Excessive stereotypic behavior in obsessive-compulsive related disorders is thought to be due to nigrostriatal dopaminergic dysfunction. The dopamine transporter (DAT) gene knockdown causes a hyperdopaminergic condition in mice due to impaired synaptic reuptake of dopamine, which is thought to be involved in their repetitive behavioral instincts. The dopamine knockdown mice were strongly stuck to rigid patterns of grooming behavior, like body licking, compared to the control mice. If the normal patterns of control mice are considered as control, this excessive stereotypic rigidity in DAT knockdown mice can be considered a compulsive-like behavior (Berridge et al., 2005). The DAT knockdown mice showed hyperactivity in the open field, which can be attributed to increased response to novelty (Zhuang et al., 2001).

D1CT-7 Transgenic Mice

In this model, a transgenic form of the neuropotentiating protein cholera toxin (CT) subunit A was expressed constitutively in the dopamine D1 receptor-expressing pyramidal neurons of the cortico-limbic circuit, thereby transgenically stimulating it. This caused a repetitive sibling-biting behavior in the mice. This further increased the cortical-striatal glutamate output as well as amygdalar glutamate output, leading to hyperactivation of these glutamatergic neurons (Campbell et al., 1999). In a similar study by McGrath et al., 2000, D1CT-7 mice displayed movements that resembled repetitive tics, as well as repetitive climbing and leaping behaviors; which under the administration of NMDA receptor antagonists were found to exacerbate and aggravate, thereby suggesting the increased glutamate output in the induction of these OCD behaviors.

Aromatase Knockout Mice

Aromatase knockout mice are estrogen-deficient because they lack a functional aromatase enzyme. They were developed initially to study the influence of estradiol on sexual development (Fisher et al., 1998). Male aromatase knockout mice displayed a heightened increase in wheel-running and grooming activities but a decrease in their overall behavioral performance. There was also decreased catechol-O-methyltransferase (COMT) activity within the hypothalamus (Hill et al., 2007). In addition to this, the male knockout mice exhibited reduced pre-pulse inhibition and increased amphetamine-induced hyperactivity (van den Buuse, Simpson, and Jones, 2003).

Leading Pharmacological Models of OCD

8-OH-DPAT-Induced Mice Model

The first pharmacologically induced mice model for perseverance (indicated by repetitive selection of the same arm in a T-maze) and indecisiveness seen in OCD patients came from the work of Yadin, Friedman, and Bridger, 1991. This was a simple model of OCD which showed reduced spontaneous alteration behavior in a T-maze upon acute injections of a selective 5-HT1A serotonergic agonist 8-hydroxy-2-(di-n-propylamino) tetralin hydrobromide (8-OH-DPAT), which was rescued by chronic treatment with SSRI fluoxetine (48 injections for 21 days). As the serotonergic system dysfunction is closely linked with OCD, enhancing the 5HT function in OCD patients worsens the OCD symptoms (Zohar and Insel, 1987). This symptom severity can be alleviated by the administration of selective serotonin reuptake inhibitors (Leonard et al., 1989) as clearly depicted in the aforementioned 8-OH-DPAT model of OCD. Further studies found that bilateral thalamic reticular nucleus (TRN) lesions, but not OFC lesions, were as effective as the tricyclic antidepressant clomipramine and SSRIs in reducing the perseverative behavior triggered by 8-OH-DPAT (Andrade et al., 2009). Findings by Agrati et al. (2005) that the 8-OH-DPAT-driven perseverant behavior varied with the reproductive cycles of female rats, offer a model for studying the correlation between ovarian hormones and the serotonergic system in 8-OH-DPAT mice.

Quinpirole-Induced Model

The quinpirole (QNP)-induced compulsive checking paradigm for OCD was first introduced by Szechtman, Sulis, and Eilam in 1998. Quinpirole hydrochloride (0.5 mg/kg) is a dopamine D2/D3 receptor agonist that induces compulsive checking behavior in rats, including ritual-like repetitive locomotory actions. These behavioral phenotypes were found to be partially rescued partially by the SSRI clomipramine. Repeated QNP administration (0.5 mg/kg subcutaneously) in the QNP model results in the expression of a hyperactive dopaminergic system. This eventually results in the obsessive behavior seen in the QNP mice model similar to human checking behavior, induced by ten injections of QNP given twice weekly (Szechtman, Sulis, and Eilam, 1998).

The compulsive checking behavior in QNP mice can be attenuated by inactivation of the subthalamic nucleus (STN) through high-frequency stimulation (HFS) or pharmacological inhibition using muscimol. However, the resulting anti-compulsive behavior is transient and does not affect the hyper-locomotory behavior of QNP-induced mice (Winter et al., 2008). A similar study by Djodari-Irani et al. (2011), found that bilateral HFS and pharmacological treatment using muscimol of the nucleus entopeduncularis (GPi in humans), and not globus pallidus (GPe in humans), partially reduced compulsive checking behavior in rats. The pattern of ritualistic motor behavior developed in QNP-induced rats changed with changes in environmental characteristics of location/objects. This indicates that QNP rats display spatio-temporal organization of compulsive behavior, restricted to chosen sites (Ben-Pazi, Szechtman, and Eilam, 2001). Also, the QNP-induced repetitive behaviors in mice were rescued by the adenosine A2A receptor antagonist Istradefylline (Asaoka et al., 2019).

Despite the fact that both 8-OH-DPAT and QNP are able to produce compulsive checking in animal models, they do so in different ways. QNP causes obsessive-compulsive checking behavior by stimulating the dopaminergic system, whereas 8-OH-DPAT maintains the motivational drive to check by blocking the serotonergic negative feedback signals, which would otherwise deactivate the obsessive-compulsive circuit (Alkhatib, Dvorkin-Gheva, and Szechtman, 2013).

mCPP-Induced Model of Reinforced Spatial Alteration

There are inconsistent findings on the pathophysiology of OCD and the usage of meta-chlorophenylpiperazine (mCPP), a nonspecific serotonin agonist of 5HT2C, 5HT1D, and 5HT1A serotonin receptors. While some studies indicate that mCPP worsens OCD symptoms (Broocks et al., 1998), other studies claim that mCPP has no impact on OCD (Ho Pian et al., 1998). The mCPP hypothesis is based on normal rats' side bias in T-maze reward alternation (referred to as "directional persistence") in the early days of training and the ability of mCPP to cause the bias to reappear even after extensive overtraining (Tsaltas et al., 2005). Treatment with mCPP dramatically reduced stereotypic behaviors in deer mice, which can be attributed to its stimulatory effect on 5HT2C receptors (Korff, Stein, and Harvey, 2008).

Behavioral Models of OCD

Signal Attenuation Model

The signal attenuation model postulates that deficits in feedback linked to normal goal-directed behaviors result in obsessions and compulsions (Alonso, López Solà, et al., 2015). In this model developed by Joel and Avisar (2001), the lever pressing for food pellets is the normal goal-directed behavior, and the associated feedback stimulus was a magazine light and tone, establishing it as a signal for food delivery. Later, the rats were subjected to a signal attenuation stage in which they were trained to unlearn the association between the stimulus and food delivery. In the last stage of the test, the behavior of the rat pressing the lever was observed under conditions where lever pressing resulted in the presentation of the stimulus but without food. Despite the extinction of food, the rats continued pressing the lever – a compulsive-like behavior observed in OCD. Furthermore, they discovered that the lever-pressing behavior was rescued by short-term administration of fluoxetine, a selective serotonin reuptake inhibitor.

Spontaneous Stereotypy in Deer Mice

Deer mice, a mouse model of spontaneous stereotypy, consistently perform typical stereotypic behaviors such as pattern running, jumping, and backward somersaulting (Powell et al., 1999) under standard laboratory conditions. Studies by Korff, Stein, and Harvey, 2008 categorized the deer mice into high, low, and non-stereotypic mice, based on how frequently these repetitive stereotypic behaviors occur. Among these, both high and low stereotypic deer mice have served as reliable models for OCD research. The study by Powell et al. (1999) aimed at discovering the influence of the environment on these repetitive behaviors and how it might be related to changes in brain chemistry. The finding that repeated treatment with fluoxetine decreased the stereotypic behaviors in deer mice makes them a relevant model for OCD (Korff, Stein, and Harvey, 2008). Additionally, the reduction of spontaneous repetitive behaviors upon stimulation of 5HT2C/D2 receptors suggests a hypoactivity of serotonergic and dopaminergic circuits in stereotypy.

Schedule-Induced Polydipsia

In this model developed by Woods et al. (1993), when the rats subjected to food deprivation were given food intermittently, they were observed to drink a large amount of water repeatedly within the test chamber. This behavior, called schedule-induced polydipsia (SIP), is an example of adjunctive behavior and is not caused by a lack of water in the body (Platt et al., 2008). Further, it was revealed that the administration of SSRIs like clomipramine, fluoxetine, and fluvoxamine reduced SIP. Since this behavior is an amplified version of normal drinking behavior, it can be considered a valid model for OCD. Moreover, it was further observed that high-frequency electrical stimulation of the NAc, medio dorsal thalamic nucleus (MD), or bed nuclei of stria terminalis (BNST) considerably reduced the repetitive drinking behavior in the polydipsic rats (van Kuyck et al., 2008).

Neurodevelopmental Animal Models

Neonatal Clomipramine Exposure

Exposure of mice specifically during the postnatal days 9–16 results in developmental reprogramming, and a novel mouse model with multiple OCD-like symptoms is born (Andersen, Greene-Colozzi, and Sonntag, 2010). Multiple symptoms of this model include increased anxiety, which was evident in the EPM and marble burying, perseverative behavior in the reversal task, impaired working memory, and increased hoarding behavior. Besides this, there was an upsurge in the RNA messenger expression for 5HT2C receptors in the OFC and D2 receptors in the striatum.

LIMITATIONS OF THE ANIMAL MODELS

Treatment outcomes vary depending on the subtype of OCD. To fully grasp the complexity of the disorder, multiple animal models are required. OCD encompasses multifaceted symptoms which cannot be fully replicated in animal models. Compulsive-like symptoms are often accompanied by non-OCD symptoms like obesity or hyperphagia in genetic animal models of OCD. It is not known how these non-specific symptoms interfere with the ability of medications to reduce obsessions and compulsions. The pharmacological QNP-induced and 8-OH-DPAT models concentrate specifically on the induction of compulsive behavior as a single facet of OCD. However, these compulsive and OCD-like behaviors are not confined to OCD; they can also occur in other neurodegenerative disorders like schizophrenia and Parkinson's disease. Another concern that questions the validity of animal models is the inconsistent results of lesional studies. Observed behaviors in the behavioral models may not fully capture the diversity and complexity of symptoms of the disorder. Sometimes, it is difficult to translate the drug responsiveness in animal models to human models, as what works for animals may not work for humans in the same way. The inability to directly test the cause and effect of multiple pharmacological treatments inherently limits these models. The heterogeneity of the disorder, as well as the overlapping set of symptoms with other neural disorders, makes it challenging to determine the precision and replicability of these animal models in OCD research.

CONCLUSION

Although it is unlikely for an animal model to replicate every aspect of OCD, research involving animal models is progressively illuminating significant neural processes underlying the disorder. Table 11.1 summarizes the phenotype expressed in each model, which evaluates the usefulness of the model for specific symptoms of OCD. It is important to select animal models for OCD investigation, taking into account the specific aspects they express. For example, the quinpirole-sensitized model will be more useful in studying complex aspects like cognitive flexibility and compulsive behaviors; compulsive lever pressing under signal attenuation is helpful for exploring the influence of cognitive and sensory feedback on repetitive behavior; genetic behaviors are excellent for studying the symptoms within the OCD spectrum. These models have demonstrated continuous advancement in OCD research using multiple animal models to unravel the neural basis underlying the disease.

TABLE 11.1
Current Rodent Models for OCD Research

Type of Animal Model	Animal Model	Phenotype	References
Transgenic models	SAPAP3/mouse	Self-injuring grooming behavior, anxiety-like behavior	Welch et al. (2007)
	SLITRK5/mouse	Increased self-grooming, marble burying, anxiety-like behavior	Shmelkov et al. (2010)
	5HT2C/mouse	Persistent chewing behavior	Chou-Green et al. (2003)
	HOXB8/mouse	Excessive body licking, biting, skin lesions	Greer and Capecchi (2002)
	Dopamine transporter knockdown (DAT)/mouse	Body licking, hyperlocomotion	Zhuang *et al.* (2001); Berridge *et al.* (2005)
	D1CT-7/mouse	Repetitive biting, tics, climbing, and leaping behaviors	Campbell et al. (1999); McGrath et al. (2000)
	Aromatase/mouse	Increased wheel-running activity and grooming	Fisher et al. (1998)
Pharmacological models	8-OH-DPAT/mouse	Motor perseveration	Yadin, Friedman, and Bridger (1991)
	QNP/mouse, rat	Increased locomotory, anxiety, decreased cognitive flexibility	Szechtman, Sulis and Eilam (1998); Asaoka et al. (2019)
	mCPP/mouse	Directional persistence in T-maze	Tsaltas *et al.* (2005)
Behavioral models	Signal attenuation model	Repetitive lever-pressing behavior	Joel and Avisar (2001)
	Spontaneous stereotype	Pattern running, jumping, and backward somersaulting	Powell et al. (1999)
	Schedule-induced polydipsia	Repetitive water drinking	Woods et al. (1993)
Neurodevelopmental models	Neonatal clomipramine exposure	Anxiety, hoarding, perseverance, impaired working memory	Andersen, Greene-Colozzi, and Sonntag (2010)

DECLARATIONS

Ethics Approval and Consent to Participate

Not Applicable

Consent for Publication

Not Applicable

Availability of Data and Materials

Not Applicable

Competing Interests

The authors declare that the research was conducted in the absence of any commercial or financial relationships that could be construed as potential conflicts of interest.

Funding

This work was supported by the Science and Engineering Research Board (SERB), Government of India (No. CRG/2020/004205) and the Ramalingaswami Re-entry Fellowship, Department of Biotechnology, Government of India (No. BT/RLF/Re-entry/49/2018) to GG. FJ and SFM were supported by the Dr. TMA Pai Ph.D. scholarship, Manipal Academy of Higher Education, Manipal, India.

Authors' Contributions

G.G. conceptualized the manuscript. All authors contributed to the literature search, writing, and editing of the manuscript. All authors read and approved this manuscript.

Acknowledgments

The authors thank Manipal School of Life Sciences, Manipal Academy of Higher Education (MAHE), Manipal, for the infrastructure and support.

REFERENCES

Agrati, D. *et al.* (2005) 'Compulsive-like behaviour according to the sex and the reproductive stage of female rats', *Behavioural Brain Research, 161*(2), pp. 313–319. doi: 10.1016/j.bbr.2005.02.017.

Ahmari, S. E. and Dougherty, D. D. (2015) 'Dissecting OCD circuits: From animal models to targeted treatments', *Depression and Anxiety, 32*(8), pp. 550–562. doi: 10.1002/DA.22367.

Alkhatib, A. H., Dvorkin-Gheva, A. and Szechtman, H. (2013) 'Quinpirole and 8-OH-DPAT induce compulsive checking behavior in male rats by acting on different functional parts of an OCD neurocircuit', *Behavioural Pharmacology, 24*(1), pp. 65–73. doi: 10.1097/FBP.0b013e32835d5b7a.

Alonso, P., Cuadras, D., *et al.* (2015) 'Deep brain stimulation for obsessive compulsive disorder: A meta-analysis of treatment outcome and predictors of response', *PLOS ONE, 10*(7), p. e0133591. doi: 10.1371/journal.pone.0133591.

Alonso, P., López Solà, C., *et al.* (2015) 'Animal models of obsessive–compulsive disorder: Utility and limitations', *Neuropsychiatric Disease and Treatment, 11*, pp. 1939–1955. doi: 10.2147/NDT.S62785.

Andersen, S. L., Greene-Colozzi, E. A. and Sonntag, K. C. (2010) 'A novel, multiple symptom model of obsessive-compulsive-like behaviors in animals', *Biological Psychiatry, 68*(8), pp. 741–747. doi: 10.1016/J.BIOPSYCH.2010.05.011.

Andrade, P. *et al.* (2009) 'Effects of bilateral lesions in thalamic reticular nucleus and orbitofrontal cortex in a T-maze perseverative model produced by 8-OH-DPAT in rats', *Behavioural Brain Research, 203*(1), pp. 108–112. doi: 10.1016/j.bbr.2009.04.026.

Aoyama, K. *et al.* (2006) 'Neuronal glutathione deficiency and age-dependent neurodegeneration in the EAAC1 deficient mouse', *Nature Neuroscience, 9*(1), pp. 119–126. doi: 10.1038/nn1609.

Asaoka, N. *et al.* (2019) 'An adenosine A2A receptor antagonist improves multiple symptoms of repeated quinpirole-induced psychosis', *eNeuro, 6*(1). doi: 10.1523/ENEURO.0366-18.2019.

Ben-Pazi, A., Szechtman, H. and Eilam, D. (2001) 'The morphogenesis of motor rituals in rats treated chronically with the dopamine agonist quinpirole', *Behavioral Neuroscience, 115*(6), pp. 1301–1317. doi: 10.1037//0735-7044.115.6.1301.

Berridge, K. *et al.* (2005) 'Sequential super-stereotypy of an instinctive fixed action pattern in hyper-dopaminergic mutant mice: A model of obsessive compulsive disorder and Tourette's', Wayne State University Associated BioMed Central Scholarship. Available at: https://digitalcommons.wayne.edu/biomedcentral/208.

Broocks, A. *et al.* (1998) 'Acute intravenous administration of ondansetron and m-CPP, alone and in combination, in patients with obsessive-compulsive disorder (OCD): Behavioral and biological results', *Psychiatry Research, 79*(1), pp. 11–20. doi: 10.1016/s0165-1781(98)00029-8.

van den Buuse, M., Simpson, E. R. and Jones, M. E. E. (2003) 'Prepulse inhibition of acoustic startle in aromatase knock-out mice: Effects of age and gender', *Genes, Brain, and Behavior, 2*(2), pp. 93–102. doi: 10.1034/j.1601-183x.2003.00014.x.

Campbell, K. M. *et al.* (1999) 'OCD-Like behaviors caused by a neuropotentiating transgene targeted to cortical and limbic D1+ neurons', *The Journal of Neuroscience: The Official Journal of the Society for Neuroscience, 19*(12), pp. 5044–5053. doi: 10.1523/JNEUROSCI.19-12-05044.1999.

Chou-Green, J. M. *et al.* (2003) 'Compulsive behavior in the 5-HT2C receptor knockout mouse', *Physiology & Behavior, 78*(4), pp. 641–649. doi: 10.1016/S0031-9384(03)00047-7.

Dajani, D. R. and Uddin, L. Q. (2015) 'Demystifying cognitive flexibility: Implications for clinical and developmental neuroscience', *Trends in Neurosciences, 38*(9), pp. 571–578. doi: 10.1016/j.tins.2015.07.003.

Djodari-Irani, A. *et al.* (2011) 'Activity modulation of the globus pallidus and the nucleus entopeduncularis affects compulsive checking in rats', *Behavioural Brain Research, 219*(1), pp. 149–158. doi: 10.1016/j.bbr.2010.12.036.

Fisher, C. R. *et al.* (1998) 'Characterization of mice deficient in aromatase (ArKO) because of targeted disruption of the cyp19 gene', *Proceedings of the National Academy of Sciences of the United States of America, 95*(12), pp. 6965–6970. doi: 10.1073/pnas.95.12.6965.

Gan, J. *et al.* (2017) 'Abnormal white matter structural connectivity in adults with obsessive-compulsive disorder', *Translational Psychiatry, 7*(3), p. e1062. doi: 10.1038/tp.2017.22.

Gershkovich, M., Wheaton, M. G. and Simpson, H. B. (2017) 'Management of treatment-resistant obsessive-compulsive disorder', *Current Treatment Options in Psychiatry, 4*(4), pp. 357–370. doi: 10.1007/S40501-017-0127-8/FIGURES/1.

Goodman, W. K., Storch, E. A. and Sheth, S. A. (2021) 'Harmonizing the neurobiology and treatment of obsessive-compulsive disorder', *American Journal of Psychiatry, 178*(1), pp. 17–29. doi: 10.1176/APPI.AJP.2020.20111601/ASSET/IMAGES/LARGE/APPI.AJP.2020.20111601F1.JPEG.

Grant, P. *et al.* (2007) 'An open-label trial of riluzole, a glutamate antagonist, in children with treatment-resistant obsessive-compulsive disorder', *Journal of Child and Adolescent Psychopharmacology, 17*(6), pp. 761–767. doi: 10.1089/cap.2007.0021.

Greer, J. M. and Capecchi, M. R. (2002) 'Hoxb8 is required for normal grooming behavior in mice', *Neuron, 33*(1), pp. 23–34. doi: 10.1016/S0896-6273(01)00564-5.

Haghighi, M. *et al.* (2013) 'In a double-blind, randomized and placebo-controlled trial, adjuvant memantine improved symptoms in inpatients suffering from refractory obsessive-compulsive disorders (OCD)', *Psychopharmacology, 228*(4), pp. 633–640. doi: 10.1007/s00213-013-3067-z.

Hill, R. A. *et al.* (2007) 'Estrogen deficient male mice develop compulsive behavior', *Biological Psychiatry, 61*(3), pp. 359–366. doi: 10.1016/j.biopsych.2006.01.012.

Ho Pian, K. L. *et al.* (1998) 'Effects of meta-chlorophenylpiperazine on cerebral blood flow in obsessive-compulsive disorder and controls', *Biological Psychiatry, 44*(5), pp. 367–370. doi: 10.1016/s0006-3223(97)00458-7.

Hoffman, K. L. (2011) 'Animal models of obsessive compulsive disorder: Recent findings and future directions', *Expert Opinion on Drug Discovery, 6*(7), pp. 725–737. doi: 10.1517/17460441.2011.577772.

Joel, D. and Avisar, A. (2001) 'Excessive lever pressing following post-training signal attenuation in rats: A possible animal model of obsessive compulsive disorder?', *Behavioural Brain Research, 123*(1), pp. 77–87. doi: 10.1016/S0166-4328(01)00201-7.

Korff, S., J. Stein, D. and H. Harvey, B. (2008) 'Stereotypic behaviour in the deer mouse: Pharmacological validation and relevance for obsessive compulsive disorder', *Progress in Neuro-Psychopharmacology and Biological Psychiatry, 32*(2), pp. 348–355. doi: 10.1016/j.pnpbp.2007.08.032.

van Kuyck, K. *et al.* (2008) 'Comparative study of the effects of electrical stimulation in the nucleus accumbens, the mediodorsal thalamic nucleus and the bed nucleus of the stria terminalis in rats with schedule-induced polydipsia', *Brain Research, 1201*, pp. 93–99. doi: 10.1016/j.brainres.2008.01.043.

Lamothe, H. *et al.* (2023) 'The Sapap3−/− mouse reconsidered as a comorbid model expressing a spectrum of pathological repetitive behaviours', *Translational Psychiatry, 13*(1), p. 26. doi: 10.1038/s41398-023-02323-7.

Lang, K. *et al.* (2009) 'Understanding obsessive compulsive disorder', *British Journal of School Nursing, 4*(8), pp. 390–394. doi: 10.12968/bjsn.2009.4.8.44656.

Leckman, J. *et al.* (1997) 'Symptoms of obsessive-compulsive disorder', *The American Journal of Psychiatry, 1547.* doi: 10.1176/ajp.154.7.911.

Leonard, H. L. *et al.* (1989) 'Treatment of obsessive-compulsive disorder with clomipramine and desipramine in children and adolescents. A double-blind crossover comparison', *Archives of General Psychiatry, 46*(12), pp. 1088–1092. doi: 10.1001/archpsyc.1989.01810120030006.

Manning, E. E. *et al.* (2019) 'Impaired instrumental reversal learning is associated with increased medial prefrontal cortex activity in Sapap3 knockout mouse model of compulsive behavior', *Neuropsychopharmacology, 44*(8), pp. 1494–1504. doi: 10.1038/s41386-018-0307-2.

McGrath, M. J. *et al.* (2000) 'Glutamatergic drugs exacerbate symptomatic behavior in a transgenic model of comorbid Tourette's syndrome and obsessive-compulsive disorder', *Brain Research, 877*(1), pp. 23–30. doi: 10.1016/s0006-8993(00)02646-9.

McKinney, W. T. and Bunney, W. E. (1969) 'Animal model of depression. I. Review of evidence: Implications for research', *Archives of General Psychiatry, 21*(2), pp. 240–248. doi: 10.1001/archpsyc.1969.01740200112015.

Naylor, R. J. (1992) 'Behavioural models in psychopharmacology: Theoretical, industrial and clinical perspectives: Edited by Paul Willner; Published 1991 Cambridge University Press, Cambridge; xiii + 540 pages; ISBN 0 521 39192 X £80.00 UK, $150.00 USA', *Journal of Pharmacy and Pharmacology, 44*(4), p. 370. doi: 10.1111/j.2042-7158.1992.tb03626.x.

Platt, B. *et al.* (2008) 'Schedule-induced polydipsia: A rat model of obsessive-compulsive disorder', *Current Protocols in Neuroscience, Chapter 9*, p. Unit 9.27. doi: 10.1002/0471142301.ns0927s43.

Powell, S. B. *et al.* (1999) 'A rodent model of spontaneous stereotypy: Initial characterization of developmental, environmental, and neurobiological factors', *Physiology & Behavior, 66*(2), pp. 355–363. doi: 10.1016/S0031-9384(98)00303-5.

Rosenberg, D. R. *et al.* (1997) 'Frontostriatal measurement in treatment-naive children with obsessive-compulsive disorder', *Archives of General Psychiatry, 54*(9), pp. 824–830. doi: 10.1001/archpsyc.1997.01830210068007.

Saxena, S., Bota, R. G. and Brody, A. L. (2001) 'Brain-behavior relationships in obsessive-compulsive disorder', *Seminars in Clinical Neuropsychiatry, 6*(2), pp. 82–101. doi: 10.1053/scnp.2001.21833.

Saxena, S. and Rauch, S. L. (2000) 'Functional neuroimaging and the neuroanatomy of obsessive-compulsive disorder', *Psychiatric Clinics of North America, 23*(3), pp. 563–586. doi: 10.1016/S0193-953X(05)70181-7.

Shmelkov, S. V. *et al.* (2010) 'Slitrk5 deficiency impairs corticostriatal circuitry and leads to obsessive-compulsive–like behaviors in mice', *Nature Medicine, 16*(5), pp. 598–602. doi: 10.1038/nm.2125.

Song, M. *et al.* (2017) 'Rare synaptogenesis-impairing mutations in SLITRK5 are associated with obsessive compulsive disorder', *PLoS ONE, 12*(1), p. e0169994. doi: 10.1371/journal.pone.0169994.

Stein, D. J. (2002) 'Obsessive-compulsive disorder', *The Lancet, 360*(9330), pp. 397–405. doi: 10.1016/S0140-6736(02)09620-4.

Stewart, S. E. *et al.* (2013) 'Meta-analysis of association between obsessive-compulsive disorder and the 3′ region of neuronal glutamate transporter gene SLC1A1', *American Journal of Medical Genetics Part B: Neuropsychiatric Genetics, 162*(4), pp. 367–379. doi: 10.1002/AJMG.B.32137.

Szechtman, H., Sulis, W. and Eilam, D. (1998) 'Quinpirole induces compulsive checking behavior in rats: A potential animal model of obsessive-compulsive disorder (OCD)', *Behavioral Neuroscience, 112*(6), pp. 1475–1485. doi: 10.1037/0735-7044.112.6.1475.

Tsaltas, E. *et al.* (2005) 'Reinforced spatial alternation as an animal model of obsessive-compulsive disorder (OCD): Investigation of 5-HT2C and 5-HT1D receptor involvement in OCD pathophysiology', *Biological Psychiatry, 57*(10), pp. 1176–1185. doi: 10.1016/j.biopsych.2005.02.020.

Wang, Y. *et al.* (2010) 'A screen of SLC1A1 for OCD-related alleles', *American Journal of Medical Genetics Part B: Neuropsychiatric Genetics, 153B*(2), pp. 675–679. doi: 10.1002/ajmg.b.31001.

Welch, J. M. *et al.* (2007) 'Cortico-striatal synaptic defects and OCD-like behaviours in Sapap3-mutant mice', *Nature, 448*(7156), pp. 894–900. doi: 10.1038/NATURE06104.
Wilhelm, S. *et al.* (2008) 'Augmentation of behavior therapy with d -cycloserine for obsessive-compulsive disorder', *American Journal of Psychiatry, 165*(3), pp. 335–341. doi: 10.1176/appi.ajp.2007.07050776.
Winter, C. *et al.* (2008) 'High frequency stimulation and temporary inactivation of the subthalamic nucleus reduce quinpirole-induced compulsive checking behavior in rats', *Experimental Neurology, 210*(1), pp. 217–228. doi: 10.1016/j.expneurol.2007.10.020.
Woods, A. *et al.* (1993) 'Selective serotonin re-uptake inhibitors decrease schedule-induced polydipsia in rats: A potential model for obsessive compulsive disorder', *Psychopharmacology, 112*(2–3), pp. 195–198. doi: 10.1007/BF02244910.
Yadin, E., Friedman, E. and Bridger, W. H. (1991) 'Spontaneous alternation behavior: An animal model for obsessive-compulsive disorder?', *Pharmacology Biochemistry and Behavior, 40*(2), pp. 311–315. doi: 10.1016/0091-3057(91)90559-K.
Zhuang, X. *et al.* (2001) 'Hyperactivity and impaired response habituation in hyperdopaminergic mice', *Proceedings of the National Academy of Sciences of the United States of America, 98*(4), p. 1982. doi: 10.1073/pnas.98.4.1982.
Zohar, J. and Insel, T. R. (1987) 'Obsessive-compulsive disorder: Psychobiological approaches to diagnosis, treatment, and pathophysiology', *Biological Psychiatry, 22*(6), pp. 667–687. doi: 10.1016/0006-3223(87)90199-5.

12 Multidimensional HPTLC

Instrumentation, Techniques, and Applications

Padmini Pai[#], Yashaswini Reddy[#], Manasa Gangadhar Shetty, Bipasa Dey, and Babitha Kampa Sundara[*]

Department of Biophysics, Manipal School of Life Sciences, Manipal Academy of Higher Education, Manipal, Karnataka, India
*Corresponding Author
# Equal Contribution

INTRODUCTION

High-performance thin-layer chromatography (HPTLC) is an improved form of thin-layer chromatography (TLC). It is sophisticated equipment with high separation efficiency. It adheres to a methodology that has been verified and is based on scientific data, ensuring that contemporary analytical labs meet quality standards and generate precise quantitative measurements with higher resolution. The application of samples, creation of standardized, reproducible chromatograms, and software assessment are all included in HPTLC [1].

The exceptional capability of HPTLC to separate and analyze a wide variety of chemicals is well recognized. This includes traditional Western medications, traditional Chinese medicines, Ayurvedic (Indian) medicines, herbal and botanical nutritional supplements, and nutraceuticals. Additionally, radio-labeled compounds in chemical, biochemical, pharmacological, and therapeutic samples can be found using this technique. The capacity to evaluate multi-component crude samples, a high number of samples, and standards that may be applied using the spray-on approach are crucial characteristics of this analytical method. Solvents of several types are available for HPTLC development. The standards and samples are treated evenly on the same plate as the mobile phases entirely evaporate before the detection procedure. This results in increased quantification accuracy and precision. There are several universally applicable selective detection techniques, and in situ spectra recording may be carried out sequentially to guarantee the accurate identification of fractions. Without any temporal restrictions, the entire sample can be kept on the layer. The application of the HPTLC technology might also reduce disposal issues and exposure to harmful organic effluents, lowering environmental contamination. HPTLC is a crucial analytical technique because of the use of contemporary equipment such as video scanners, densitometers, sophisticated chromatographic chambers, efficient elution procedures, high-resolution sorbents with chemically modified surfaces or selected particle sizes, the development of computer programs for method optimization, and the ability to combine with other experimental methods [2].

Only precoated layers and specific equipment may be used for HPTLC, which is mostly used to measure amounts. TLC is a method that is utilized all around the world to teach chromatography fundamentals. This procedure is simple to carry out, allows for sample observation during chromatography, and only requires inexpensive demonstration equipment. Resolution may be enhanced by several processes when capillary flow is regulated. Separations in one or two dimensions are possible using planar chromatography. Mobile-phase velocity can be regulated by outside forces, such as forced-flow development [3].

DOI: 10.1201/9781003545712-12

The application of HPTLC techniques has advanced significantly, including hyphenations with HPTLC-Mass Spectrometry (HPTLC-MS), HPTLC-Fourier Transform Infrared Spectroscopy (HPTLC-FTIR), and HPTLC-scanning diode laser. As a result of these developments, HPTLC is now a potent analytical tool. According to experts, HPTLC will become more crucial in the analysis of drug formulations, bulk drugs, and natural products in the future, with a focus on combinatorial approaches and the use of instrumental HPTLC [4].

PRINCIPLE

HPTLC is a separation method between a stationary phase and a mobile phase. On an inert substance like glass, aluminum, or polyester, the stationary phase is covered with a flat, thin coating. The mobile phase is a combination of aqueous and organic solvents. Samples are segregated based on their relative affinity for the stationary and mobile phases as well as the interaction between the mobile and stationary phases. After traversing various distances, separate bands or zones start to form [5]. The difference between TLC and HPTLC is given in Table 12.1.

Rf (retention or retardation factor):

Any solute spot's location on the TLC is identified by its retention/retardation factor (Rf).

$$Retention\ factor\,(Rf) = \frac{Distance\ travelled\ by\ the\ analyte\,(Zs)}{Distance\ travelled\ by\ the\ solvent\ font\,(zf)}$$

The range of Rf values varies from 1.0 for analytes that migrate to the solvent front to 0.0 for analytes that are strongly retained at the site of application. There are several factors that determine the reproducibility of Rf results, including sorbent quality, humidity, layer thickness, development distance, and temperature. Figure 12.1. represents the optimal method for placing the TLC plate in the chamber (Figure 12.1).

Selection of Stationary Phase

Smaller plates with a shorter development distance and analysis duration are commonly utilized for analysis. Pre-coated plates with aluminum oxide, cellulose, or silica gel are used as stationary phases. Over 90% of published analyses employ silica gel (fluorescence). Cellulose is utilized for

TABLE 12.1
Comparison between Parameters of HPTLC and TLC [6]

Parameters	HPTLC	TLC
Chromatographic plate used	Pre-coated	Handmade
Sorbent layer thickness	100–200 μm	250 μm
Pre-washing of plates	Compulsory	Not followed
Application of sample	Automatic	Manual
Sample volume	0.2 5 μL	1–10 μL
Analysis time	Greatly reduced	Slow
Shape	Spot/band	Spot
Spot size	0.5–1 mm	2–4 mm
Efficiency	High	Low
Mobile phase volume	A smaller quantity of solvent is needed	A greater quantity of solvent is needed
Scanning	Densitometer	Not possible

FIGURE 12.1 The optimal method for placing the TLC plate [7].

TABLE 12.2
Different Solvents Commonly Used in HPTLC [3]

Solvent	Molecular Weight (g/mol)	Elution Strength
N-Pentane	72.15	0.00
Hexane	86.17	0.01
Cyclohexane	84.16	0.04
Carbon tetrachloride	153.82	0.18
Toluene	92.14	0.29
Chloroform	199.38	0.40
Methylene chloride	84.93	0.42
Tetrahydrofuran	72.11	0.45
Acetone	58.08	0.56
Ethyl acetate	88.11	0.58
Aniline	93.13	0.62
Acetonitrile	41.05	0.65
Ethanol	46.06	0.88
Methanol	32.04	0.95

amino acids, dipeptides, sugars, and alkaloids. The most popular stationary phase is chemically bound octadecylsilane (ODS), whereas non-polar compounds including fatty acids, carotenoids, and cholesterol are handled by C8 and shorter alkyl chains [3].

Selection of Mobile Phase

The choice of mobile phase is influenced by the stationary phase and the characteristics of the analyte. Eluent strength, which is correlated with the polarity of the mobile phase components, determines the elution power. The mobile phase has to be 10–15 mL for twin trough chambers [3] (Table 12.2) (Figure 12.2).

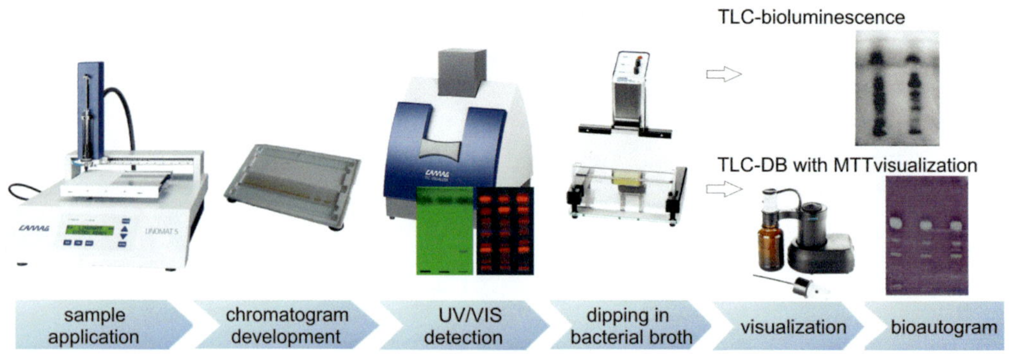

FIGURE 12.2 Workflow of the HPTLC sample preparation and analysis [8].

Steps Involved in HPTLC

Preparation of Plate

High molecular weight polymer is used to create precoated plates, which are abrasion-resistant, uniform, repeatable, preactivated, and ready to use. They are available with polyester, glass, or aluminum support. Plates made of aluminum foil are less expensive, more portable, and cuttable. High-quality results are best achieved using glass plates. UV cabinet visualization frequently uses fluorescent indicator F254 layers. TLC uses 20 × 20 cm plates, while HPTLC uses 20 × 10 cm or 10 × 10 cm plates [1].

Layer Prewashing

Unless chromatography creates impurity fronts, plates are typically used without preparation and handled at the upper edge to prevent contamination. Layers are prewashed using 20 mL of methanol per trough in a 20 × 10 cm twin-trough chamber (TTC) for repeatability testing and quantitative analysis. In each trough of the TTC, up to two 20 × 10 cm or four 10 × 10 cm plates can be developed back-to-back. The plate should be removed and dried for 20 minutes at 120°C in a clean drying oven. Equilibrate the plate in a suitable container that offers protection from dust and fumes with the laboratory environment (temperature, relative humidity) [9].

Activation of Precoated Plates

The newly opened package does not require activation. Plates that are exposed to high humidity for a long time require activation. Prior to spotting, the precoated plates can be activated by heating them to 110–120°C in an oven for 30 minutes. Between two glass plates, insert aluminum sheets, and bake for 15 minutes at 110–120°C [10].

Application of Sample

Simple analyses may be performed using TLC, which generally uses manual sample application with capillaries. This method permits the application of amounts of 0.5–5 μL without the need for intermediary drying. More complex analyses or separations require devices for bandwise application employing spray-on methods, which enhance separation power and repeatability. HPTLC layers may handle up to 1 L per spot [9]. The samples are sprayed on or transferred via contact transfer methods to the separating layer. Precision in volume, compact application zone size, and exact placement all contribute to the analysis's quality. Bandwise sample application delivers the best separation outcomes and flexibility in solvents and application quantities with Linomat 5 or Automatic TLC Sampler 4. Samples are sprayed onto the chromatographic layer using the Linomat in the shape of thin bands. The sample's solvent almost completely evaporates during the spraying, condensing the sample into a short, adjustable band [11].

The applicator makes it simple to apply precisely positioned samples in the form of spots onto TLC and HPTLC layers without causing any layer damage. A disposable capillary pipette that is precisely directed is used to administer the sample, enabling the chromatogram to be automatically scanned in accordance with a preprogrammed pattern. Dispensers accept capillary pipettes in dispenser magazines with quantities of 0.5, 1.0, 2.0, and 5.0 L, each needing a specific dispenser magazine. The sample solution is filled into universal capillary holder pipettes and set up against the applicator tip [12].

Chamber Separation

To achieve the highest level of consistency, HPTLC plates are produced within fully saturated TTCs. The development chamber must be lined with filter paper, an appropriate amount of mobile phase must be added, the chamber must be tilted to equalize the solvent volume in both troughs, the cover must be replaced, and the chamber must equilibrate for 20 minutes. The highest repeatability of the final samples is guaranteed by this technique [3].

Chromatogram Development

HPTLC plates are created in a variety of chambers, such as horizontal, twin-trough, and flat-bottom chambers. Low mobile phase volumes and simple pre-equilibrium with vapors or conditioning liquids are made possible by the TTC chamber architecture. A gas phase is present during development, and it can have a big impact on how well the separation works. The kind, dimension, and saturation state of the chamber all affect the gas phase. As a result of poor repeatability and bending solvent fronts, unsaturated chambers should be avoided. For difficult separations, two-dimensional development can enhance resolution. The first mobile phase is used to develop the plate, which is then dried before being developed with a second mobile phase in a perpendicular direction. For plate development, over-pressured layer chromatography with forced flow (OPLC) is also utilized [13].

In contrast to column chromatography, planar chromatography always begins in a non-equilibrium condition and never achieves equilibrium. When a liquid in the chamber comes into contact with the plate to which the sample was applied, it becomes the stationary phase. Chromatography cannot start until the sample has been dissolved by the mobile phase. Due to capillary action and rising flow resistance, the mobile phase's velocity decreases with migration distance [14].

Designed for TLC/HPTLC laboratories, the UV Lamp 4 is a functional device that offers straightforward, one-button operation for each UV tube. It has an integrated timer and tilt sensor to reduce the risk of UV radiation exposure. The lamp is a great option for laboratories since it also provides more consistent lighting and increased UV light intensity. The safety measures of the light protect the user's security and comfort. The UV Cabinet 4 was created for UV observation with little interference from ambient light. It consists of the UV Lamp 4 and Viewing Box 4. Its small size takes up little room, and the viewing port has an integrated UV screen for eye protection. Through a roller shutter, you may access the inside [15].

Derivatization

Derivatization can be done by dipping, spraying, or utilizing gas. It is necessary to transfer the reagent to the chromatogram uniformly. The chromatogram immersion device decreases fume formation, increases repeatability, and prevents tidemarks. To prevent tidemarks, dipping and withdrawal must be smooth. This method also removes fumes and reduces exposure to dangerous chemicals [14].

Immersing

The tank of the immersion device is filled with 200 mL of the reagent. The parameters are set in accordance with the technique, the plates are inserted into the immersion device's holder, and the process is initiated. Let any extra reagent drip off the plate and use paper towels to clean the back.

The plate should be taken out of the plate holder and dried with cold air (vertically, in the chromatographic direction).

Spraying

Fill the sprayer's bottle with a maximum of 50 mL of the reagent. Put the plate up against some filter paper in the spray cabinet. Spray the reagent on the plate in both horizontal and vertical motions until it is evenly distributed. Cold air should be used to dry the plate.

Detection

The quenching of fluorescence caused by UV radiation (which typically has a wavelength of 200–400 nm) improves the detection of separated chemicals on the sorbent layers. Fluorescence quenching is the term used to describe this phenomenon.

Visualisation at UV 254 nm

- A fluorescent indicator called F254 produces green fluorescence when stimulated by UV light with a wavelength of 254 nm. Shortly after excitation, this fluorescence persists for more than ten seconds. Compound zones appear as a dark violet patch on a green backdrop when compounds that absorb 254 nm light are present. Anthraglycosides, coumarins, flavonoids, propylphenol in essential oils, and various alkaloid types are examples of chemicals with conjugated double bonds that undergo quenching and should be identified under 254 nm.

Visualisation at UV 366 nm

- Hraglycosides, coumarins, flavonoids, phenol carboxylic acids, and several alkaloid kinds, such as rauwolfia and ipecacuanha family alkaloids, have been shown to exhibit the F366 phenomenon, in which fluorescence vanishes after stimulation.

Visualization at White Light

- By examining the natural color of the separated compounds in daylight, more especially white light, it is possible to identify zones that contain those compounds [10]

Quantification

Quantitative analysis of HPTLC is carried out using a fixed sample light beam and a chromatogram spectrophotometer, usually a densitometer or scanner. For quantitative evaluation, the TLC Scanner 3 makes use of WINCATS software, which may scan in the reflectance, transmittance, absorbance, or fluorescence modes. The scanner can calibrate single and multiple levels with linear or nonlinear regressions and quickly record spectra. For confirming target results, such as in stability tests and dissolution profiles, single-level calibration is appropriate. Based on the sample's composition and dilution parameters, the concentration of the analyte in the sample is computed. Two scanning methods have been employed: slit scanning and video scanning.

Slit Scanning and Documentation

- Scanning densitometry, which makes use of a fiber-optic bundle for sample zone illumination and reflected light collection, is an established technique for UV-visible absorption and fluorescence detection. The simultaneous detection of length and spectral recording is accomplished using the photodiode-array detector.

Video Scanning and Documentation

- Using a computer, video digitizer, light source, monochromator, and optics to illuminate a plate and focus pictures onto a charge-coupled device (CCD) video camera, video densitometry is a quick and simultaneous data-collecting procedure. It is helpful for TLC with image analysis and two-dimensional separations [10].

Documentation

Each developed plate is documented using a digital documentation system under UV light at 254 nm, UV light at 366 nm, and white light. If a type of light does not produce usable information, that fact must be documented. If a plate is derivatized, images are taken prior to and after derivatization [16].

HPTLC TROUBLESHOOTING

A TLC plate's solvent front may flow unevenly because of factors like a minor tilt, spilled eluent, or a jostled chamber during elution. When determining an Rf, it is advised to measure the front for each lane of the plate rather than only once in circumstances where it is noticeably different. Spotting the sample on the slide several times and letting the solvent dry between applications can fix the problem of no spots on the plate. If the solvent system level is too high, the development procedure needs to be modified because the chemical from the spot will dissolve in the solvent rather than moving up the plate. Due to sample overload, an inadequate solvent system polarity, or small changes in the polarity of several compounds in the sample, the substance runs as a streak, indicating the need for a new solvent system.

SAFETY MEASURES

Wearing the right personal protection equipment (PPE), such as lab coats, safety glasses, and disposable gloves, is crucial for safety to prevent chemical splashes and contact with dangerous substances. When handling chemicals, use caution and follow the safety data sheets. Compounds that are flammable or harmful should be used in fume hoods. Harmful chemicals should be handled carefully without contacting the skin or inhaling them. Solvents must be handled cautiously, kept away from open fires, stored in well-ventilated spaces, and disposed of in the proper waste containers. Make sure the HPTLC equipment is in excellent working condition and is calibrated appropriately by being familiar with its use and maintenance to prevent accidents and damage. Solvents must be handled cautiously, kept away from open fires, stored in well-ventilated spaces, and disposed of in the proper waste containers. Waste materials, such as spent TLC plates, solvents, and chemicals, must be disposed of correctly as part of the assignment in line with local regulations and lab protocols. Prior to each experiment, a risk assessment should be carried out to identify possible dangers and implement the required safety precautions. Reduce hazards associated with HPTLC and maintain a culture of safety in the lab to provide a secure working environment.

APPLICATION OF HPTLC IN DIFFERENT AREAS

HPTLC stands out as a highly advanced instrumental technique harnessing the full potential of TLC. The examination of pharmaceuticals, natural compounds, and emerging medications is a widespread practice throughout all phases of the drug discovery and development process. Its capabilities, such as automation, scanning, comprehensive optimization, selective detection principles, minimal sample preparation, and hyphenation, make it an exceptionally potent analytical method for obtaining chromatographic data from intricate mixtures encompassing pharmaceuticals, natural substances, herbal formulations, clinical specimens, food products, and more [17]. HPTLC continues to be an exceptionally versatile, dependable, precise, sensitive, and cost-effective separation method perfectly tailored for the examination of botanicals and herbal medicines. When employed in conjunction with standardized protocols, it ensures consistent outcomes, a crucial component in the quality control, verification, identification, and validation of intricate patterns found in different areas [10]. In contrast to alternative analytical methods, HPTLC generates chromatograms that provide comprehensive information about the entire sample in a visually interpretable manner. The advantage lies in its capacity to analyze multiple samples concurrently, facilitating the comparison of samples for accurate identification and validation. Comparing various plates allows for the easy identification of similarities and differences. Multiple chromatograms are simultaneously compared and analyzed. Analogous peaks are extracted from the chromatogram data. Evaluation is performed using image-based software such as a densitometer in conjunction with a TLC scanner or videoscan device. The absorption or fluorescence of the compound is measured and analyzed within the chromatogram [18]. The major applications of HPTLC in important areas of research are given in Figure 12.3.

Quality Control of Herbal Products

The pharmaceutical and herbal industries use HPTLC extensively to ensure the quality and consistency of herbal products such as medicinal plants, herbal medicines, and dietary supplements. It allows for the assessment of the presence and concentration of bioactive compounds, helping to verify if the product meets specific quality standards. HPTLC occupies a significant role in characterizing phytoconstituents present in herbal medicines, thereby contributing to the establishment of standardized criteria for various plant extracts. To underscore this, the following section offers illustrative instances featuring specific plant species from credible sources. Below are a few instances where HPTLC has been employed to analyze the components in herbal plants. *Curcuma longa Linn.* (turmeric), a member of the Zingiberaceae family, is extensively documented for its wide range of properties, including anti-inflammatory, cholagogue, hepatoprotective, blood-purifying, antioxidant, and detoxifying effects. HPTLC was carried out to determine the content of commercially available turmeric powder samples in reference to an in-house powder sample. The HPTLC separation was executed using HPTLC plates, employing a mobile phase consisting of a

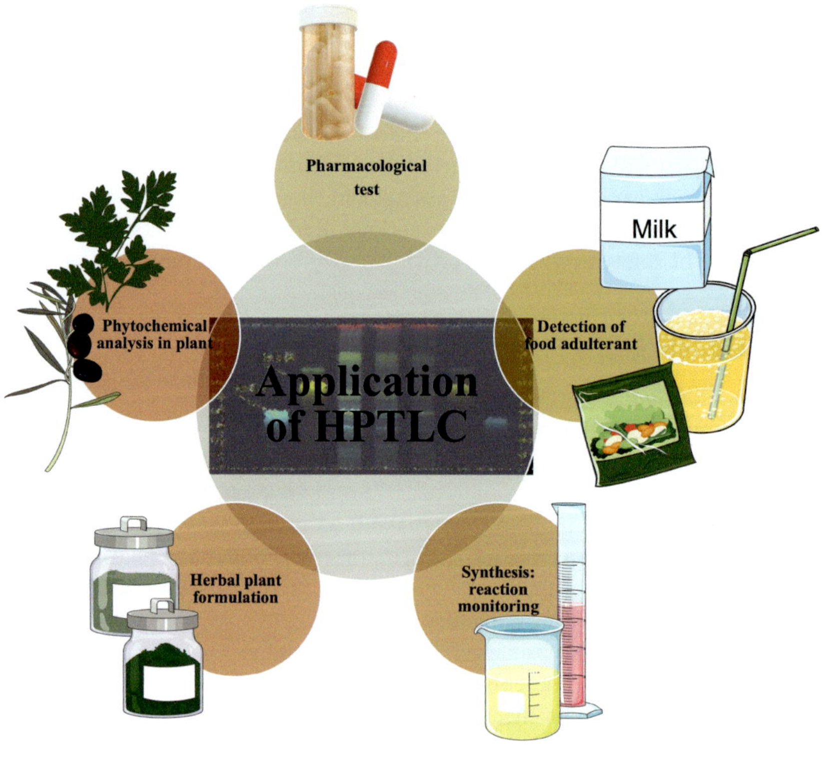

FIGURE 12.3 Major applications of HPTLC in different areas.

dichloromethane and methanol combination. The retention factor (Rf) for curcumin was measured at 0.43, and the compound's quantification was performed at its absorbance maximum (λmax) at 427 nm [19]. In traditional Indian systems of medicine like Ayurveda and Unani, dhatrinisha churna has a historical application for the treatment of diabetes, cancer, and hyperlipidemia. The HPTLC fingerprinting technique has been established to concurrently assess and quantify curcumin and ellagic acid in dhatrinisha churna. The chromatographic profile of dhatrinisha churna was quantitatively analyzed concerning both curcumin and ellagic acid. This developed HPTLC method has proven to be dependable, sensitive, precise, and accurate, making it suitable for quality control purposes concerning both raw materials and formulations [20].

Ginger contains various pungent and biologically active compounds, including 6-gingerol, 8-gingerol, zingerone, and paradol. Among these, 8-gingerol, as depicted in the provided structure, stands out as one of ginger's primary pungent constituents. The analysis was conducted on silica plates (E-Merck, Germany) and the mobile phase consisted of n-hexane:ethyl acetate. UV densitometric scanning at 569 nm was performed using a Camag TLC Scanner III [21]. The HPTLC technique has been formulated for the routine assessment of hesperidin content in both raw herbal materials and herbal or pharmaceutical formulations containing citrus fruits as a component. This method involved the use of silica plates and a mobile phase composed of ethyl acetate:methanol:water (15:3:2). The developed HPTLC method is characterized by its simplicity, reliability, sensitivity, and accuracy, making it suitable for quantifying hesperidin in various citrus fruit varieties [22]. β-sitosterol and gymnemagenin quantification in a commercially marketed herbal formulation was identified, validated, and quantified by the NP-HPTLC method. Gymnemic acid, derived from *Gymnema sylvestre* leaves, is credited with its antidiabetic properties. The NP-HPTLC method aimed to create and validate a novel, rapid, and accurate procedure for the quantitative measurement of β-sitosterol and gymnemagenin in a herbal formulation by densitometric detection. Detection was achieved by scanning at 423 nm [23].

Identification of Phytochemicals

HPTLC is instrumental in identifying and quantifying various phytochemicals present in plant samples. These phytochemicals include alkaloids, flavonoids, terpenoids, polyphenols, and many others. By separating the components in a plant extract or sample, HPTLC can provide detailed information about the presence and relative quantities of these compounds [24]. Below are a few instances where HPTLC has been employed to analyze the phytochemicals in plants (Figure 12.4).

The chromatographic profiles of the analyzed plant resin samples were evaluated using HPTLC settings optimized for achieving excellent resolution of phenolic acids and flavonoids. A normal-phase system is typically used to differentiate highly and moderately polar phenolics. Furthermore, formic acid is added to the mobile phase to reduce acidic group ionization and improve the morphology of chromatographic bands [25]. Another study was to compare the essential oil profiles of several *Annona muricata* plant organs using TLC and HPTLC. For fingerprint analysis, a toluene:ethyl acetate mobile phase was used, and essential oils were detected as blue, purple, or brownish spots after the vanillin-sulfuric acid reagent was applied. The HPTLC approach revealed a wide range of essential oil phytoconstituents in various plant sections, confirming the existence of nine different essential oils. Notably, three of these essential oil derivatives were found only in seeds, but an essential oil with an Rf value of 0.17 was consistently found in extracts from *A. muricata*'s rind, seed, leaf, and bark [26].

HPTLC was used in conjunction with microchemical derivatizations using chemicals such as DPPH, p-anisaldehyde, and ferric chloride to test antioxidant activity, phytosterols and terpenoids present, and polyphenolic content. The high concentration of aromatic chemicals, particularly polyphenolic acids, in oregano leaf extract accounts for its extraordinary antioxidant action [27]. HPTLC's advantages in terms of speed, cost-effectiveness, and versatility make it a valuable tool in the study of phytochemicals in plants. Researchers, pharmacists, and quality control professionals

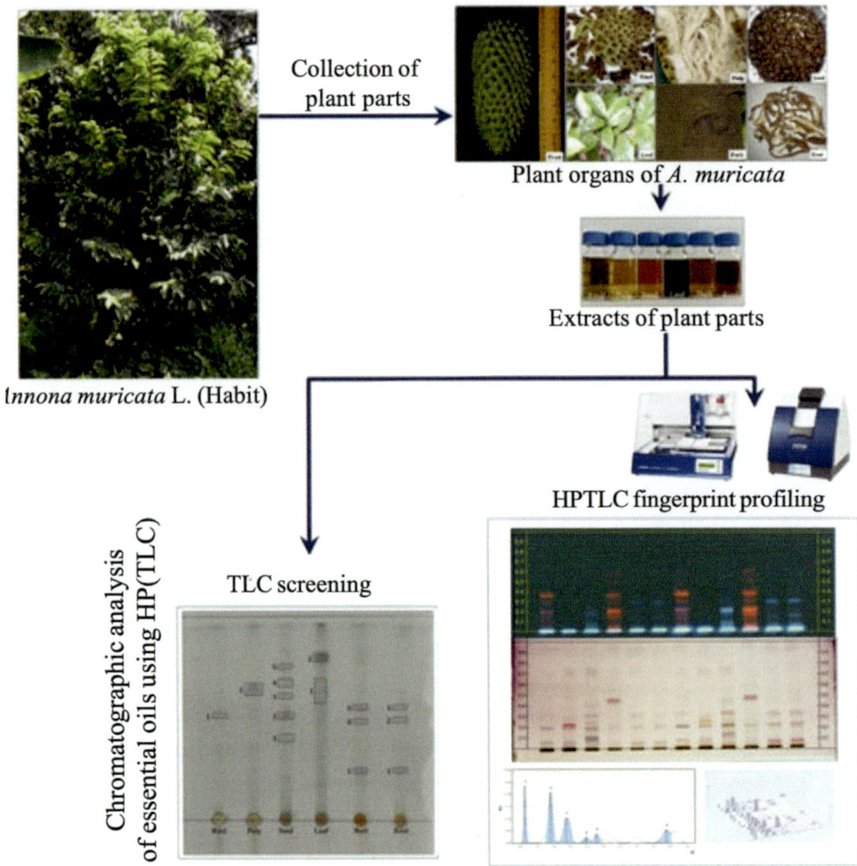

FIGURE 12.4 The workflow for phytochemicals detection in plant [26].

use HPTLC to evaluate the chemical composition of plant materials, contributing to the development of safe and effective plant-based products and medicines.

Quality Control and Authenticity Verification of Food Products

Detection of adulterants by HPTLC can identify the presence of adulterants or substitutes in food products. For example, it can be used to detect the addition of cheaper oils in premium olive oil or the presence of fillers in spices such as paprika. Pesticide residue analysis by HPTLC is employed to assess pesticide residues in fruits, vegetables, and grains. It helps ensure that food products meet safety standards by quantifying the levels of specific pesticides. Identification of food additives by HPTLC is used to identify and quantify food additives such as artificial colors, preservatives, and flavor enhancers. This is essential for regulatory compliance and consumer safety. Quality assurance in food processing carried out by HPTLC can be used at various stages of food processing to monitor the quality of raw materials, intermediates, and finished products, ensuring consistent quality. Determination of food colorants by HPTLC helps detect synthetic and natural colorants in food products, which is essential for regulatory compliance and consumer safety. HPTLC is utilized for the assessment of food product quality and authenticity. Mycotoxin contamination remains an ongoing concern across various food commodities, making it imperative to investigate the prevalence of this issue in baby food products. Factors such as technological advancements, urbanization, and hectic lifestyles have influenced dietary habits across all age groups [10]. Another study objective was to evaluate patulin (PAT) contamination in fruit-based baby food products in Turkey. The safety of these products was evaluated based on the provisional maximum tolerable daily intake (PMTDI) value established by regulatory authorities. This study encompassed the analysis of a total of 58 baby food products available in the Turkish market, employing an HPTLC method [28].

Furthermore, a newly developed HPTLC method for the analysis of seven steviol glycosides in six different food and Stevia formulation categories was verified. Following a shortened single-step sample preparation, 21 samples were prepared concurrently, allowing for quick food screening. This method proved suitable for the examination of food sample concentrations in the mg/kg range, depending on the volume of sample applied. Notably, this study established for the first time the recording of HPTLC-ESI-MS spectra using the TLC-MS Interface for *Rebaudioside A*. The study thoroughly investigated the operational costs and duration of analysis associated with the three separate approaches, specifically in the context of food product screening [29].

HPTLC's advantages, such as speed, cost-efficiency, and versatility, make it a valuable tool for food manufacturers, regulatory bodies, and quality control laboratories. It helps ensure that food products meet safety and quality standards and that consumers receive authentic and high-quality products.

Pharmacopoeial Testing

Identification of pharmaceuticals by HPTLC is used to confirm the identity of pharmaceutical substances by comparing their chromatographic profiles to established reference standards in pharmacopeias. This ensures that the correct active ingredients are present. Assay and quantitative analysis by HPTLC is employed to determine the quantity or concentration of active pharmaceutical ingredients in drug formulations. This quantitative analysis is crucial for ensuring that the drug product meets potency requirements. Stability testing carried out by HPTLC is employed in stability studies to monitor changes in the composition of pharmaceuticals over time. It helps identify degradation products and assess the shelf life of products. HPTLC is applied in pharmacokinetic studies to measure drug concentrations in biological fluids. It helps assess absorption, distribution, metabolism, and excretion (ADME) properties. Numerous combinations of synthetic drugs and naturally occurring bioactive compounds have been explored for their potential synergistic effects in cancer treatment. Among these combinations, sorafenib (SFN) and chrysin (CHR) have emerged as a promising duo with clear potential for synergistic cancer therapy. However, this poses analytical challenges as it becomes increasingly complex to simultaneously identify and quantify two or more active substances in the presence of one another. This report details the development and validation of an HPTLC densitometric method designed for the simultaneous analysis of SFN and CHR, followed by an examination of how biological matrices, specifically plasma, may impact the assay of both molecules [30].

In a separate context, the novel antitumor drug cabozantinib (CZN) was quantified using validated green RP-HPTLC and NP-HPTLC techniques in commercially available tablets and capsules. The green RP-HPTLC technique exhibited superior linearity, accuracy, precision, robustness, and sensitivity in comparison to the green NP-HPTLC technique. It is worth noting that the "green" aspect of the method pertains to the use of acetone/water as environmentally friendly solvents [31]. Additionally, a sensitive, quick, and environmentally friendly RP-HPTLC methodology was developed and compared to the NP-HPTLC method for the measurement of pterostilbene (PT) using traditional univariate calibration. For RP quantification of PT, green solvent solutions were utilized, whereas normal solvent systems were used for NP analysis of PT. Both techniques were used at 302 nm to detect PT. The robust RP-HPTLC test was a selective and stability-indicating method since it could detect PT even in the presence of its degradation products. Because of this, it was determined that the RP-HPTLC method with univariate calibration was better than the NP-HPTLC method for PT analysis [32]. The COVID virus causing the worldwide COVID-19 pandemic was then presented. Various containment measures were implemented in order to address this public health issue. The method utilized TLC silica gel aluminum plates as a stationary phase and a green mobile phase composed of ethyl acetate and ethanol with densitometric detection at 245 nm [33]. HPTLC is a valuable tool for ensuring the quality and safety of pharmaceutical products, as it allows for comprehensive testing of a wide range of parameters in a cost-effective and efficient manner. Pharmaceutical manufacturers and regulatory authorities rely on HPTLC to meet pharmacopeial requirements and maintain high standards of product quality.

Drug Discovery: Reaction Monitoring and Impurity Detection

During the synthesis of novel compounds, chemical reactions often involve multiple steps and intermediates. HPTLC allows researchers to monitor the progress of reactions by analyzing samples at different stages, ensuring that each step proceeds as intended. HPTLC is sensitive enough to detect impurities in synthesized compounds. Researchers can use it to identify and quantify impurities, ensuring the final product meets purity standards. Reaction optimization, carried out by monitoring the reaction, can make the necessary adjustments to optimize reaction conditions, such as temperature, time, or reagent concentrations, to enhance the yield of the desired compound. An HPTLC technique has been developed to separate and measure six specific non-polar, fluorescent pyrolytic heterocyclic aromatic amines from the carboline group. This contemporary planar chromatography method offers a speedy and cost-effective approach for assessing heterocyclic amines (HAs). It is applicable for both initial screening and precise quantification [34]. Intermediate detection in chemical reactions involves the formation of intermediate compounds before the final product is obtained. HPTLC can detect and identify these intermediates, providing valuable information about the reaction's mechanism. Early issue identification: by identifying by-products or side reactions that occur, HPTLC can detect them early in the synthesis process. This early identification allows issues promptly, preventing the production of unwanted compounds. Product confirmation: once the reaction is complete, HPTLC can be used to confirm the identity and purity of the final product. This is particularly crucial in industries where product consistency is essential. HPTLC is a versatile tool for monitoring chemical synthesis processes, offering real-time insights into reaction progress, purity assessment, and the identification of intermediates and impurities. Its applications are valuable across various scientific and industrial sectors, including pharmaceuticals, chemicals, and research laboratories, where precise control and optimization of chemical reactions are paramount.

Several companies that produce HPTLC apparatus and consumables which offer associated services, are listed below:

1) CAMAG: The Swiss company, CAMAG, is renowned for its HPTLC systems, which include hardware, software, and consumables. It has long held a key position in the industry.
2) Merck Millipore: High-performance TLC is supported by a variety of HPTLC plates, solvents, and other accessories from Merck Millipore, a branch of Merck KGaA.
3) Agilent Technologies: Leading provider of analytical and life sciences instruments, Agilent provides a variety of chromatography options, including HPTLC. Its goods are renowned for their excellence and accuracy.
4) Thermo Fisher Scientific: As a market leader in tools and equipment for scientific research, Thermo Fisher Scientific provides HPTLC devices and materials for labs.
5) Macherey–Nagel: HPTLC plates, accessories, and software are only a few of the chromatography items that are produced by the German business, Macherey–Nagel.
6) Analtech: The American company Analtech offers HPTLC plates and accessories for a range of analytical applications.
7) Sigma–Aldrich (now part of MilliporeSigma): Sigma–Aldrich was a well-known provider of lab supplies, including consumables for HPTLC. It is currently known as MilliporeSigma after merging with Millipore.

8) Sorbent Technologies: The HPTLC-focused business, Sorbent Technologies, focuses on offering HPTLC plates and equipment for chromatography applications.
9) Biosil: An Indian company called Biosil sells HPTLC plates, chambers, and associated chromatography items.
10) CILAS: For accurate analysis, CILAS, a French company, offers HPTLC equipment that includes densitometers and scanning tools.

CONFLICT OF INTEREST

The authors declare that they have no conflict of interest.

CREDIT AUTHORSHIP CONTRIBUTION STATEMENT

Padmini Pai: Writing – review & editing; **Yashaswini Reddy:** Writing – review & editing; **Manasa Gangadhar Shetty:** Review & editing; **Bipasa Dey:** Review & editing; **Babitha Kampa Sundara:** Supervision, review, & editing.

DECLARATION OF COMPETING INTEREST

The authors declare that they have no known competing financial interests or personal relationships that could have appeared to influence the work reported in this paper.

ACKNOWLEDGMENTS

The authors thank the Department of Biotechnology (DBT) BioCare Fellowship [Grant ID: BT/PR20046/BIC/101/683/2016], Government of India (GOI), and Manipal Academy of Higher Education (MAHE), Manipal, India for infrastructure.

REFERENCES

1. Sonia K, Lakshmi, KS (2017). HPTLC method development and validation: An overview. *Journal of Pharmaceutical Sciences and Research, 9*(5), 652.
2. Rashmin P, Mrunali P, Nitin D, Nidhi D, & Bharat P (2012). HPTLC method development and validation: Strategy to minimize methodological failures. *Journal of Food and Drug Analysis, 20*(4), 20.
3. Hussain SZ, Maqbool K, Naseer B (2019). High-performance thin layer chromatography: Principle, working and applications. *International Journal of Research in Pharmacy and Pharmaceutical Sciences, 4*(3), 83–88.
4. Ramu B, Chittela KB (2018). High-performance thin layer chromatography and its role pharmaceutical industry. *Open Science Journal of Bioscience and Bioengineering, 5*(3), 29–34.
5. Malik, A, Kumar, M, Kumar, P (2018). Basics of HPLC and HPTLC method developments, LAP Lambert Academic Publishing, 1, 1–41.
6. Ambati P, Bala Krishna P, Srinivasa Rao Y, Varaprasada Rao K, & Deepthi R (2021). Instrumentation and future prospects of HPTLC-A. *World Journal of Pharmaceutical Research, 10*, 650–661
7. Silver J (2020). Let us teach proper thin layer chromatography technique!. *Journal of Chemical Education, 97*, 4217–4219.
8. Choma IM & Jesionek W (2015). TLC-direct bioautography as a high throughput method for detection of antimicrobials in plants. *Chromatography, 2*, 225–238.
9. Hala A, Sethi GL, Bihani SD (2015). HPTLC instrumentation: An overview. Pharmatutor. Pharmacy Infopedia (www. pharmatutor.org), accessed on 20/09/2016.
10. Lalhriatpuii T (2020). HPTLC fingerprint in herbal drug formulations. In *Herbal Medicine in India: Indigenous Knowledge, Practice, Innovation, and its Value*, 337–362.DOI: 10.1007/978 981-13-7248-3_22, Editors: Saikat Sen, Raja Chakraborty, Springer Nature publisher.
11. https://www.hptlc-association.org/about/what_is_hptlc.cfm, accessed on 10/11/23.

12. Camag_Catalog_20191.pdf (kovalent.se), accessed on 10/11/23
13. Rashmin P, Mrunali P, Nitin D, Nidhi D, & Bharat P (2012). HPTLC method development and validation: Strategy to minimize methodological failures. *Journal of Food and Drug Analysis, 20*(4), 20.
14. https://www.lobov.com.py/downloads/CAMAG-basic-equipment-08.pdf, accessed on 10/11/23.
15. Camag_Catalog_2019.pdf (kovalent.se), accessed on 11/10/23.
16. Patel RB, Patel MR, Batel BG (2011). Experimental aspects and implementation of HPTLC. *High-Performance Thin-Layer Chromatography (HPTLC), 1*, 41–54.
17. Attimarad M, Ahmed KM, Aldhubaib BE, Harsha S (2011). High-performance thin layer chromatography: A powerful analytical technique in pharmaceutical drug discovery. *Pharmaceutical Methods, 2*, 71–75.
18. Affonso RS, Lima JA, Lessa BM, Caetano JV, Obara MT, Nóbrega AB, França, TC (2018). Quantification through TLC-densitometric analysis, repellency and anticholinesterase activity of the homemade extract of Indian cloves. *Biomedical Chromatography, 32*, 1–8.
19. Gantait A, Barman T, Mukherjee PK (2011). Validated method for estimation of curcumin in turmeric powder. *Indian Journal of Traditional Knowledge, 10*, 247–250.
20. Patel VR, Patel RK (2012). HPTLC method development and validation for quantification of markers of Dhatrinisha churna. *Pharmacognosy Journal, 4*, 26–29.
21. Alam P (2013). Densitometric HPTLC analysis of 8-gingerol in Zingiber Officinale extract and ginger- containing dietary supplements, teas, and commercial creams. *Asian Pacific Journal of Tropical Biomedicine, 3*, 634–638.
22. Alam P, Alam A, Anwer MK, Alqasoumi SI (2014). Quantitative estimation of hesperidin by HPTLC in different varieties of citrus peels. *Asian Pacific Journal of Tropical Biomedicine, 4*(4), 262–266
23. Potawale SE, Gabhe SY, & Mahadik KR (2014). Quantification of gymnemagenin and-sitosterol in marketed herbal formulation by validated normal phase HPTLC method. *Chromatography Research International, 1*, 1–6.
24. Altemimi A, Lakhssassi N, Baharlouei A, Watson DG, Lightfoot DA (2017). Phytochemicals: Extraction, isolation, and identification of bioactive compounds from plant extracts. *Plants, 6*, 1–23.
25. Ristivojević P, Trifković J, Vovk I, Milojković-Opsenica D (2017). Comparative study of different approaches for multivariate image analysis in HPTLC fingerprinting of natural products such as plant resin. *Talanta, 162*, 72–79.
26. Naik AV, Sellappan K (2020). Chromatographic fingerprint of essential oils in plant organs of Annona muricata L. (Annonaceae) using HPTLC. *Analytical Chemistry Letters, 10*, 214–226.
27. Rocamora CR, Ramasamy K, Lim SM, Majeed ABA, Agatonovic-Kustrin S (2020). HPTLC based approach for bioassay-guided evaluation of antidiabetic and neuroprotective effects of eight essential oils of the Lamiaceae family plants. *Journal of Pharmaceutical and Biomedical Analysis, 178*, 1–7.
28. Turkmen Z, Kurada O (2020). Rapid HPTLC determination of patulin in fruit-based baby food in Turkey. *JPC – Journal of Planar Chromatography – Modern TLC, 33*, 209–217.
29. Islam MK, Sostaric T, Lim LY, Hammer K, Locher C (2021). Development of an HPTLC-based dynamic reference standard for the analysis of complex natural products using Jarrah honey as test sample. *Plos One, 16*, 1–14.
30. Sharma T, Khurana RK, Borges B, Kaur R, Katare OP, Singh B (2021). An HPLC densitometric method for simultaneous quantification of sorafenib tosylate and chrysin: Analytical method development, validation and applications. *Microchemical Journal, 162*, 2–8.
31. Alam P, Shakeel F, Alqarni MH, Foudah AI, Faiyazuddin MD, Alshehri S (2021). Rapid, sensitive, and sustainable reversed-phase HPTLC method in comparison to the normal-phase HPTLC for the determination of pterostilbene in capsule dosage form. *Processes, 9*, 1–15.
32. Alam P, Salem-Bekhit M M, Al-Joufi FA, Alqarni MH, Shakeel F (2021). Quantitative analysis of cabozantinib in pharmaceutical dosage forms using green RP-HPTLC and green NP-HPTLC methods: A comparative evaluation. *Sustainable Chemistry and Pharmacy, 21*, 1–8
33. Abo-Gharam AH, El-Kafrawy DS (2022). Eco-friendly stability-indicating HPTLC micro-determination of the first FDA approved SARS-CoV-2 antiviral prodrug Remdesivir: Study of degradation kinetics and structural elucidation of the degradants using HPTLC-MS. *Sustainable Chemistry and Pharmacy, 29*, 1–14.
34. Morlock G (2004). New HPTLC method, with systematic mobile-phase optimization, for determination of six apolar heterocyclic aromatic amines. *JPC – Journal of Planar Chromatography – Modern TLC, 17*, 431–434.

13 High-Performance Liquid Chromatography

Analytical Technique for Component Identification and Purification Across Diverse Industries

Bipasa Dey[#], Manasa Gangadhar Shetty[#], Padmini Pai, Yashaswini Reddy, and Kampa Sundara Babitha[*]

[1]Department of Biophysics, Manipal School of Life Sciences, Manipal Academy of Higher Education, Manipal, Karnataka, India
*Corresponding Author
#Equal Contribution

INTRODUCTION TO HIGH-PERFORMANCE LIQUID CHROMATOGRAPHY

The term "chromatography" refers to a laboratory technique used for separation [1]. Furthermore, it aids in the identification and purification of components within a mixture for both qualitative and quantitative analyses. The underlying principle of chromatography lies in the separation of mixtures of molecules as they traverse solid or fluid stationary phases while being conveyed by a mobile phase. Depending on the state of matter of the stationary phase, there are three different types of chromatographic procedures: paper, thin layer, and column chromatography. Among the three types, column chromatography stands out as a widely employed technique for isolating proteins from a mixture. Liquid chromatography (LC) and gas chromatography (GC) are two classifications of chromatography based on the state of matter of the mobile phase employed [2]. When a liquid or gaseous mobile phase is used, the terms LC and GC are used accordingly. Mikhail Semyonovich Tsvet is credited with pioneering adsorption chromatography in 1901, utilizing a technique to separate plant pigments by employing petroleum ether and a small glass tube containing calcium carbonate. Tsvet extensively investigated 126 different powdered adsorbents, as detailed in his 1903 publication. The term "chromatography" was introduced by Tsvet in his articles published in 1906. In 1941, R.L.M. Synge and A.J.P. Martin presented a groundbreaking method of partition chromatography, using silica gel in columns to maintain a stationary state for water while chloroform descended the column to separate amino acids [3]. This innovation laid the foundation for high-performance liquid chromatography (HPLC), which evolved from column chromatography in the 1960s. HPLC, a highly modified version of column liquid chromatography, involves pushing a solvent through a column at pressures of up to 400 atmospheres, as opposed to allowing it to flow through by gravity.

The inception of HPLC research can be traced back to the experiment by A.J.P. Martin and R.L.M. Synge, marking the beginning of HPLC advancements. However, it took an additional 30 years before pumps were employed to compel a liquid phase along the packed column [4]. HPLC

DOI: 10.1201/9781003545712-13

plays a crucial role in separating components within a chemical mixture, employing the pressure-driven flow of a mobile phase through a column filled with a stationary phase [5].

EVOLUTION OF HPLC

In 1964, J.C. Moore from the Dow Chemical Company commenced research on gel permeation chromatography, a method discovered in the 1960s. Waters Corporation developed the first commercially used HPLC, the ALC100, in 1969. Dionex Corp. introduced the initial industrial ion-exchange chromatography device for charge-based separation in 1971, featuring enhanced ion detection through groundbreaking suppression technology. The first variable wavelength monitor for HPLC that was commercially available was developed by Dionex, founded by Cecil Tarbet. By the late 1970s, improvements in the separation of chemically similar molecules were made possible by techniques like reversed-phase liquid chromatography. In 1979, Agilent Technologies, Inc. introduced a cutting-edge diode array detector that provided a quick optical method for chemical analysis.

The 1980s witnessed widespread use of HPLC for chemical substance separation, aided by computerization and automation, leading to enhanced reproducibility. ESA Biosciences, Inc. submitted a patent for the Coulochem electrochemical detector in 1982. Pharmacia developed fast protein liquid chromatography (FPLC), a chromatographic technology suitable for sensitive protein separation, using lower pressures than HPLC. Waters introduced the Alliance HPLC system in 1996, setting a new standard for performance in pharmaceutical research. In 2004, Waters introduced ultra performance LC (UPLC), a liquid chromatography system offering significantly faster run times, greater peak capacity, and improved routine sensitivity. This innovation, utilizing smaller particles, increased laboratory productivity. Whatman, Inc. introduced the Mini-UniPrep Syringeless Filter series, streamlining sample preparation for HPLC. Pickering Laboratories advanced analyte detection with a pulse-free syringe pump. In 2005, ESA Biosciences introduced the Corona Charged Aerosol Detector, the first new HPLC detection technique. JASCO introduced the X-LC Xtreme HPLC system in 2006, and Agilent developed technology to eliminate common proteins in human plasma.

In 2008, IDEX Health & Science introduced ultra high-performance (UHP) fittings and connectors, capable of handling higher temperatures and larger column pressures. Agilent launched the 1290 Infinity Liquid Chromatography System in 2009, enhancing UHPLC performance. In 2010, Phenomenex introduced Kinetex core-shell HPLC columns, allowing UHPLC performance on existing HPLC systems. The future may see a demand for more precise chiral separation columns in pharmaceutical manufacturing and increased use of HPLC in environmental toxin monitoring. Advancements in computer technology may lead to the norm of multiple detectors per system, and robots could be employed for handling hazardous materials in HPLC processes. Despite future developments, HPLC is poised to remain a significant laboratory separation technique for analytical and preparative purposes [6] (Figure 13.1).

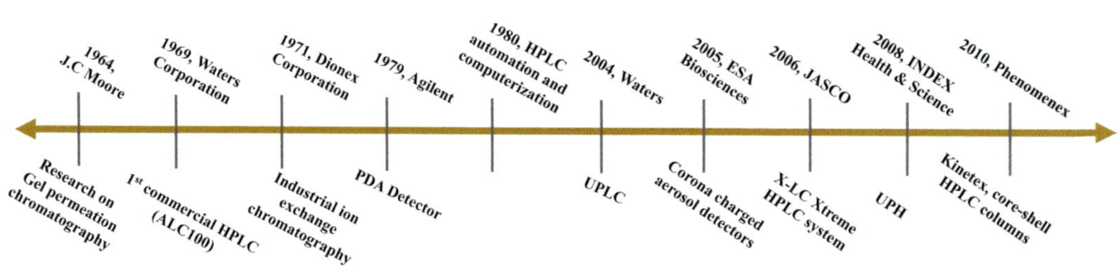

FIGURE 13.1 Timeline representing the major discoveries related to HPLC development.

TYPES OF HPLC

Size exclusion, ion exchange, reverse, and normal phase are the four types of HPLC. The mobile phase in normal phase HPLC is non-polar, but the column packing, such as silica, is polar. It is used for water-sensitive chemicals, geometric isomers, cis-trans isomers, and chiral compounds. Reverse phase is a type of phase where the mobile phase is polar, such as water with a miscible solvent (methanol/ethanol, etc.), and the packing of the column is non-polar, like C18. It may be used for samples that are polar, non-polar, ionizable, and ionic. When the mobile phase is a buffer, and the column packing contains ionic groups and is used to separate ionic compounds, the chromatography technique is known as ion exchange chromatography. In a kind of HPLC known as size exclusion, molecules diffuse into the pores of a porous media and are then separated based on how big they are in comparison to the pores. Larger molecules elute first, followed by smaller ones [7].

MODES OF HPLC

There are two primary modes in HPLC based on the mobile phase composition. In an isocratic elution system, the mobile phase remains constant throughout the separation process. Conversely, a gradient elution system involves changing the mobile phase composition during separation. Two techniques within gradient elution are low-pressure gradient (LPG), where solvent mixing occurs before the pump (suction side), and high-pressure gradient (HPG), where individual pumps supply solvents, and mixing takes place after the pumps (discharge side) [8]. Table 13.1 demonstrates all the properties of various solvents used as mobile phases in HPLC.

SEMI-PREPARATIVE AND PREPARATIVE HPLC

Semi-preparative HPLC facilitates the purification of specific quantities of substances, ranging from micrograms to lower-milligram amounts, after analytical identification. It serves as a bridge for the seamless transition and upscaling from analytical HPLC to the preparative scale. On the other hand, preparative HPLC is the preferred method for swiftly and effectively isolating high-purity substances, from milligrams to multi-gram quantities. Its primary applications include the purification of biochemicals for research or clinical trials, the generation of standards, and the isolation of commercial products like active pharmaceutical compounds or chemical substances. While

TABLE 13.1
Properties of HPLC Mobile Phase

Mobile Phase	Polarity Index	UV Cutoff (nm)
Cyclohexane	0.04	210
n-Hexane	0.1	210
Carbon tetrachloride	1.6	265
i-Propyl ether	2.4	220
Toluene	2.4	286
Diethyl ether	2.8	218
Tetrahydrofuran	4.0	220
Ethanol	4.3	210
Ethyl acetate	4.4	255
Dioxane	4.8	215
Methanol	5.1	210
Acetonitrile	5.8	190
Water	10.2	—

preparative HPLC employs methods and instrumentation similar to analytical HPLC, the pumps and detectors are designed for larger flow rates. Moreover, the ultimate objective differs between the two; analytical HPLC aims for high-quality data, whereas preparative HPLC focuses on the successful purification of substances [9].

PRINCIPLE OF HPLC

Based on the variation in the degree of interaction between the sample components and the column, each mixture's individual components are separated using a column. The first to be eluted are substances that interact with the column infrequently. Adsorption, polarity, hydrophilic and hydrophobic contacts, electron affinity, penetration, and exclusion are some of these interactions [10].

INSTRUMENTATION

The mobile phase/solvent reservoir, the solvent delivery system, the sample introduction device, the column, sample vials, the post-column apparatus, the detector, the data gathering and output system, the post-detector eluent processing, and the connective tubing and fittings are the fundamental components that make up the majority of HPLC instruments (Figure 13.2).

Solvent Reservoir

A clean, inert container that can store 0.5–2 L of solvent and has a top that enables a tubing intake line to supply the mobile phase to the solvent delivery system as the mobile phase reservoir. The cover also prevents dust from entering and reduces solvent evaporation. It is advisable to conduct fresh filtration and, ideally, degassing for all mobile phases. To get rid of any particles or precipitates that might form while using the mobile phase, an extra filter is frequently added at the end of the input line for the mobile phase. Depending on the application, the material selection for the mobile phase reservoir is frequently a crucial factor. For small molecule analysis, glass is frequently employed to stop plastic compounds from leaking and potentially jeopardizing the results. However, for studies involving RNA, proteins, electrochemistry, and similar applications, plastic mobile phase reservoirs are frequently preferred to avoid the leaching of trace ions from glass containers that could interfere with the study.

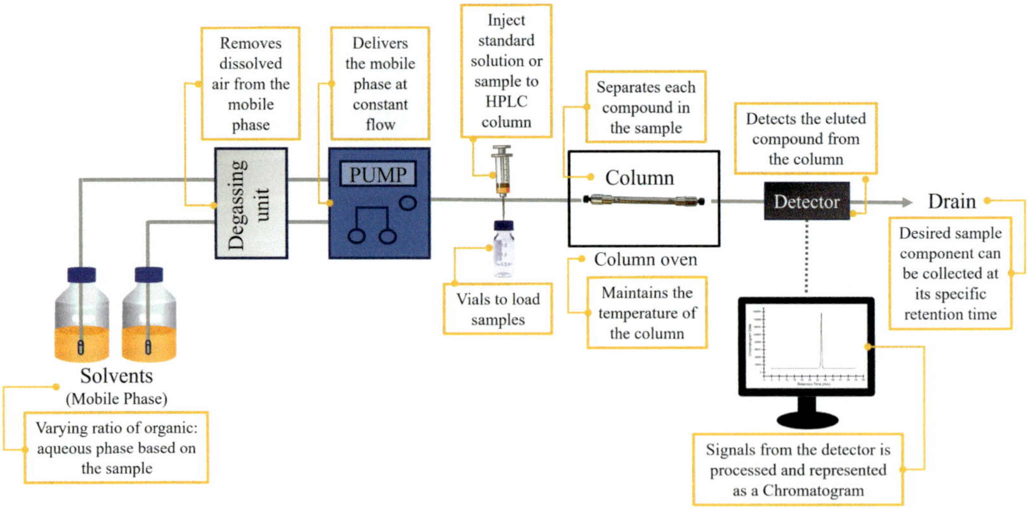

FIGURE 13.2 Schematic representation of HPLC instrumentation.

Solvent Delivery System

The pump provides the mobile phase to the column in a precise, accurate, repeatable, continuous, and pulse-free flow. The high-pressure pump may operate at pressures between 500 and 20,000 psi. The three most common types of HPLC pumps used today are constant pressure pumps, syringe-based or displacement pumps, and constant flow pumps. Since neither syringe-based pumps nor constant pressure pumps can be easily adjusted for the delivery of gradient liquids, constant flow pumps are commonly used. A simple reciprocating piston pump serves as the foundation for most commercial high-pressure pumps. It is possible to do several separations in an isocratic manner, in which case the solvent supplied to the column doesn't vary in composition as the separation advances. For more challenging separations, gradient elution is required. Throughout the separation, the eluent ratio is changed, which is the most common method of gradient elution. As a result, the early-eluting molecules retain their sharpness while the more strongly bound compounds elute more quickly. Gradient elution is merely a separation method in which the solvent strength is programmed to change. To achieve the necessary separation, a gradient can take on various forms such as linear, convex, concave, stepped, or a combination of these. The generation of gradient flow necessitates computer-controlled pumping. Recently, there have been introductions of systems designed for ultra-high-performance liquid chromatography (UHPLC), featuring pumps with maximum operating pressures ranging from 15,000 to 20,000 psi. The UHPLC pump operates based on the same fundamental principles previously discussed, although the management of significantly higher pressures requires additional considerations in system design. Exploring UHPLC is worthwhile for certain applications due to its advantages, including accelerated analytical rates, reduced eluent consumption, and other benefits attributed to the elevated system pressure.

Sample Introduction Device

The repeatability of the sample introduction method is frequently what prevents an HPLC system from operating with the accuracy it is capable of. The sample is introduced into the HPLC system using the sample introduction device, sometimes referred to as a sample injector, without depressurizing it first. The prevalent method for sample injection involves a sample loop that can be readily inserted and removed from the mobile phase flow pathway by valve rotation. In the load position, the sample loop is filled with air at atmospheric pressure. Typically, sample loop capacities range from 5 to 500 μL. Simple robotic technology can quickly automate the valve-based sample introduction mechanism. In addition to increasing injection repeatability, using autoinjectors – also known as autosamplers – allows for the continuous processing of several (i.e., tens to thousands of) samples at once. Autoinjectors work by guiding a syringe-pump-driven needle to various sample places using an XYZ positioning mechanism. Samples may be kept on well plates with designated places or in vials kept in sample racks. The selected location is used to remove a predetermined volume of sample. It is occasionally possible to utilize temperature-controlled autoinjectors to keep thermally delicate samples at a consistent temperature before, during, and after examination. Some commercially marketed autoinjectors provide a temperature control range of 4–60°C for sample trays. Pre-column derivatization techniques have also been employed with autoinjectors, particularly for amino acid studies. Using pre-column derivatization, analytes' chromatographic and/or detection characteristics can be enhanced.

Column

The HPLC column is where the separation takes place. Due to the high pressure involved in HPLC, smoothbore stainless-steel tubing is typically used in HPLC columns. To give the analyst superior performance for analytes, current column technology is occasionally built from heavy-walled glass, titanium, and plastic (for example, PEEK). Analytical scale columns typically have lengths of 10–30 cm and a thickness of 4–10 mm. The most typical packing particle sizes are 3, 5, and 10 μm. The employment of columns with better performance and higher speeds, which have smaller dimensions than those mentioned above, has become the current trend. The irreversible adsorption of contaminants from samples and solvents causes columns to readily deteriorate. To safeguard the integrity of the analytical column, which is significantly more expensive, a guard column is frequently utilized. Many applications do not require precise control of column temperature and HPLC separations are carried out in ambient conditions. However, temperature management offers chances to increase separation effectiveness and can improve chromatographic repeatability. Column heating is frequently necessary for UHPLC procedures. The column temperature of contemporary instruments may be controlled to within a few tenths of a degree between near ambient and 150°C by the use of column heaters/ovens (Figure 13.3(A) and (B)).

Sample Vials

HPLC vials ensure the accuracy and integrity of the analysis. They are constructed from unreactive materials to inhibit any reactivity between the samples and the vial walls. Common materials for these vials encompass borosilicate glass, polypropylene, and polyethylene. Vial capacity typically ranges from 1.5 to 4 mL, although larger volumes are utilized for specific applications. The vials are securely sealed using caps or septa to prevent contamination, evaporation, and inadvertent sample leakage. These caps incorporate septa, usually made of materials like polytetrafluoroethylene or silicone, allowing the injection needle to puncture and collect the sample. The vial designs include crimp top, screw thread, snap cap, and shell vials. The vials need to be clean, dry, and should be compatible with the autosampler to avoid any interference during sample analysis (Figure 13.3(C)).

Post-Column Apparatus

Post-column equipment will be in the flow path of the system if post-column alteration of the mobile phase or analyte is necessary. While post-column derivation of the analyte may be necessary to enhance the analyte's detection capabilities following their separation, altering the mobile phase (e.g., by adding buffers, adjusting the pH, or increasing the solvent strength) may be necessary to enhance

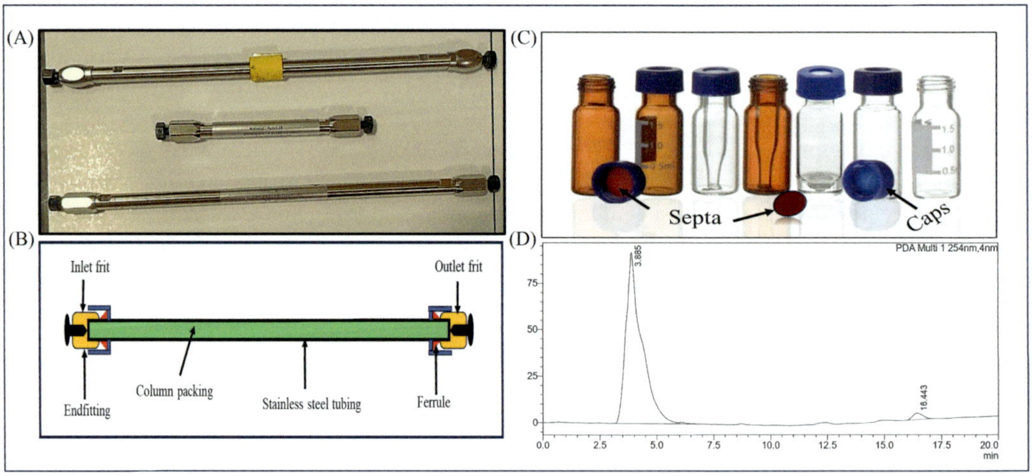

FIGURE 13.3 The images represent (A) HPLC columns with varying column lengths and packing, (B) the various parts of an HPLC column, (C) the different types of HPLC vials with their cap features, and (D) the chromatograph displaying the components present in the eluted sample at different retention times.

the mobile phase's compatibility with the detector. A post-column addition system is made up of just three components in each scenario: a reagent supply pump, a mixing tee, and a mixing coil.

Detector(s)

The solutes are removed from the column as soon as an analyte reaches the detector, which converts an analyte's chemical or physical characteristics into an electrical signal. The amount of a certain characteristic of the mobile phase or solutes is proportional to the electrical signal, which may be amplified and controlled by the appropriate electronics. The two main types of HPLC detectors are bulk property detectors, which react to an eluent's bulk characteristics like conductivity or refractive index (RI), and solute property detectors, which react to analyte characteristics like UV absorbance or fluorescence detection. The presence and concentration of the analyte affect the detector's response in both scenarios. Solute property detectors are usually 1000 times or more sensitive than those for bulk properties. High sensitivity, good stability, linearity, quick reaction times, dependability, non-destructiveness, simplicity of use, and minimum dead volume are desired qualities in an HPLC detector.

Data Collection and Output

The electronic output of the detector is linked to a device for data collection and output, such as a computer, integrator, or recorder. The data collection tool generates a response versus time plot using the electrical signal generated by the detector. The resulting chromatogram can then be assessed for both qualitative and quantitative data. Nowadays, recorders are seldom used by themselves. The peaks of a chromatogram may be integrated by both integrators and computers, and both have the advantage of being able to store chromatograms for post-collection processing. It is also possible to export data from computer-based collection systems to several other programs and apps. The computer can often interface with and operate the complete HPLC apparatus. As a result, most contemporary HPLC systems have computer-based output, data gathering, and control systems [11].

RESULTS AND DATA INTERPRETATIONS

The results are recorded in the form of peaks, each representing a compound in the mixture as it passes through the detector. This recorded output is known as a chromatogram, featuring x- and y-axes as shown in Figure 13.3(D). The x-axis represents time in minutes, while the y-axis represents absorbance units. The retention time of a compound is the time point at which the display exhibits the maximum peak height for that specific compound. This time corresponds to the duration taken by a particular compound to traverse the column and reach the detector, measured from the moment the sample is injected into the column. The area beneath the peak is proportionate to the quantity of the component that has passed the detector, and this area can be automatically calculated by the computer linked to the display. The shaded region in green in the diagram (albeit simplified) illustrates the area that the computer would measure [12].

SAFETY

It is important to avoid sudden variations in temperature or pressure. The filling state in HPLC columns may be affected by temperature changes or an accidentally dropped column. Similarly, abrupt changes in column pressure can affect the packing inside the column. Therefore, it is advisable to regulate the flow rate cautiously when injecting the valve to prevent abrupt pressure changes. Additionally, the rotation of the valve should not be too slow. When using reversed-phase chromatography, the solvent's composition should be altered gradually rather than going from water to an organic solvent or vice versa. HPLC-grade solvents and reagents should be used and prepared following the recommended methods and parameters, such as temperature, pH, and composition. Filtering the samples is necessary to get rid of any debris that can clog or affect the column. Additionally, it's best to refrain from injecting samples that include a lot of salts or other contaminants because doing so could harm the function of the column. To guarantee good solubility and stability, the sample should be thoroughly prepared to prevent contamination or deterioration. The right solvent and pH conditions should also be used. The HPLC columns should not be backflushed. The contaminants, if present in the column head, can only be backflushed when the manufacturer specifies that it is safe to do so. Otherwise, the column's effectiveness will be reduced by the recoil. To prevent damage to the column, use an appropriate mobile phase (particularly pH). Pre-columns can occasionally be linked in front of injectors. To prevent the silica matrix from dissolving inside the analytical column, the pre-column used when the analytical column is bonded to silica gel is also silica gel. This allows the mobile phase to be "saturated" by silica gel before entering the analytical column. It is not recommended to directly inject biological samples or samples with complex matrices into the HPLC columns. Pretreatment of the sample is necessary, or the injector and column must be connected to a guard column. A brief column filled with a comparable stationary phase serves as the HPLC guard column in most cases. The guard column can and ought to be changed on a regular basis. To get rid of any contaminants inside the column, it should be rinsed frequently with a powerful solvent. When cleaning, a miscible solvent should be used to progressively transition the mobile phase out of the flow route system. For regular analysis, each mobile phase's volume should be 50 to 75 mL, or around 20 times the column volume. When not in use, HPLC columns need to be stored properly. To avoid degradation, the column should be stored in a tidy, dry location away from direct sunlight, heat, and moisture and should not be exposed to corrosive compounds or chemicals. To ensure the column will operate well over the long run, abide by the manufacturer's storage instructions. The HPLC column should be correctly installed in the device with proper orientation, connections, and fitting torque requirements as per the manufacturer's instructions. To avoid leaks, which can contaminate the system and reduce resolution, the column should be placed accurately and tightly but should not be overtightened. Configure the HPLC column before use in accordance with the manufacturer's instructions. This often entails washing the column with the mobile phase to get rid of any storage solvents, air, or remaining impurities. The performance of the chromatographic process is stable and repeatable with proper conditioning. To avoid the accumulation of particulates or deposits that could impair performance, the column should be cleaned often utilizing compatible solvents and cleaning conditions while adhering to the manufacturer's cleaning instructions. In order to identify any issues, such as clogging or packing anomalies, which influence the separation quality and column lifetime, the column backpressure should be monitored while it is being used. Following the manufacturer's guidelines, routine maintenance should be performed, such as replacing frits/end fittings, flushing with solvent, or repacking the column. The lifespan of HPLC columns is finite, and eventually, they need to be replaced, so a column replacement should be scheduled based on routine maintenance and performance review. HPLC columns should be handled carefully since they are delicate and easily get damaged. Dropping the columns or putting too much pressure on them should be avoided. When handling and moving columns, use column storage and transport equipment, such as column guards or protective containers. To prevent increased backpressure and poor separation, the HPLC column flow rate should be in accordance with the manufacturer's guidelines. The column may degrade, have less clarity, or even fail if the flow rate specifications are exceeded. To correctly manage the flow rate, use flow rate control hardware or software. A thorough record should be kept regarding the usage of HPLC columns, including the type of column used, its serial number, the date of installation, and its history. This documentation supports column traceability, performance monitoring, and troubleshooting [13].

THE ADVANTAGES OF HPLC

The great degree of accuracy and precision across individual sample runs is one of HPLC's main advantages. The capacity of HPLC to separate and even detect minute quantities of contaminants and degradation products in raw materials and finished items makes it the perfect tool for tracking variations between various manufacturing batches. HPLC is sensitive and capable of detecting trace amounts of substances in products, ranging from nanogram to picogram amounts. This factor is crucial in the production of pharmaceuticals since even trace levels of contaminants can affect the efficacy and safety of the final product. HPLC can also be used to measure degradation of products, which can aid in determining a product's stability and shelf life. Since HPLC is a selective method, it can be used to separate and purify chemically similar substances, such as isomers, from a complex mixture. When you need to extract and purify enantiomers from products before packing, this capability is helpful. A variety of compounds, including polar or nonpolar, acidic or basic, large or small molecules, can be analyzed using HPLC due to various column chemistries. The adaptability of HPLC makes it a perfect instrument for quality control in contract manufacturing, which frequently produces a variety of products including small molecules, proteins, and antibodies. In manufacturing, HPLC is an extensively used method for quality control, and regulatory bodies like the Food and Drug Administration (FDA) frequently demand its use. It ensures the safety of the product as per the regulations. Although the initial cost of HPLC equipment and training may be high, a company will save a lot of money in the long run. HPLC can assist in preventing pricey product recalls and preserving the integrity of a company by precisely identifying product quality issues. Additionally, HPLC can reduce the time and money required for manual testing, generating sizable savings over time. HPLC enables trace-level analysis since it has excellent sensitivity in detecting substances at low concentrations. High selectivity is another feature, enabling the separation and quantification of chemically similar substances. A wide range of chemicals, including pharmaceuticals, environmental samples, food and drink, biological samples, polymers, and more, can be processed using HPLC. It can isolate and analyze a variety of chemical compounds. The components of a combination can be precisely and accurately quantified using HPLC. It offers trustworthy and repeatable results, making it a crucial tool for quality control and academic study. HPLC allows for quick analysis and high throughput, enabling the examination of several samples in a small amount of time. In the pharmaceutical industry, where prompt product release and quality control are critical, this efficiency is crucial. HPLC frequently involves less sample preparation than some other analytical methods. This minimizes the time and effort required for pre-analysis preparation while also lowering the possibility of sample degradation. UV-Vis spectroscopy, fluorescence, mass spectrometry (MS), and diode array detection (DAD) are only a few of the detection techniques supported by HPLC. This adaptability enables the detection and measurement of many compound types according to their characteristics. HPLC offers versatility in method creation and optimization for a variety of applications and target chemicals by allowing the use of a large range of mobile phases and stationary phases. Due to their high levels of automation and computerization, HPLC systems enable remote monitoring, accurate parameter control, and simple data analysis software integration. As a result, analyses are more effective, accurate, and reproducible. In general, HPLC utilizes less solvent than conventional chromatographic methods, making it ultimately more cost- and environmentally friendly. Through calibration curves, HPLC enables quantitative analysis by establishing a linear relationship between a compound's concentration and detector response. This makes precise quantification easier [14].

LIMITATIONS OF HPLC

The amount of sample required for HPLC analysis usually ranges from 10 to 100 μL, and the injection volume is taken to be 1–2% of the column volume. This loading volume can be a challenge for expensive or limited samples, which can be resolved by using detectors with higher sensitivity and sample concentrators. For samples that aim at separating impurities and extracting pure compounds, the maximum sample loading size of 100 μL can also seem inadequate compared to other techniques such as gas chromatography or preparative chromatography. To elute the samples in HPLC, the samples must be in liquid form or dissolved in a suitable solvent. Ideally, the sample should be thermally stable and non-volatile; for volatile samples, gas chromatography is mostly preferred. Only liquids that can dissolve the analytes and are compatible with the stationary phase may be used as the mobile phase in HPLC. In contrast to gas chromatography (GC), which offers a wider selection of suitable mobile phases, this restricts the variety of substances that can be analyzed. To achieve separation in HPLC, the selection of the stationary phase is essential. However, occasionally it can be difficult to find a suitable stationary phase for analytes, which reduces the analysis's efficacy. Due to the limitations in stationary phases and interactions between the mobile and stationary phases, HPLC may have trouble providing sufficient resolution for isomeric substances. For isomers, gas chromatography (GC) frequently offers greater resolution. HPLC has a detection limit ranging from 50 to 250 nM, with most of the compounds lying within the limit of 50–100 nM. In comparison to HPLC, several alternative chromatographic methods, such as mass spectrometry combined with gas chromatography (GC-MS), may be more sensitive and have lower detection limits. Pre-concentration or derivatization may be necessary for HPLC to analyze some chemicals at the trace level. Other chromatographic methods, such as gas chromatography (GC) or solid-phase extraction (SPE), may require less pre-concentration or none. Although the UV-vis detector detects at fixed wavelengths and photodiode array (PDA) detectors have addressed this with their large detection range spanning from 200 to 700 nM, the lack of a universal detector is frequently emphasized. The solution is refractive index detection, but it has poor sensitivity and is incompatible with gradient elution. Charged aerosol detection (CAD) won the competition, defeating evaporative light scattering detection (ELSD). Compared to ELSD, CAD offers better sensitivity (low ng) and usability due to the use of a nebulizer with corona discharge detection. For ionic or ionizable chemicals, mass spectrometry is evolving into a technology that is incredibly quick, sensitive, and selective. When utilizing columns with 20,000 plates under gradient conditions, conventional HPLC has a practical peak capacity (Pc) of about 200, making it ineffective for very complex samples. The introduction of UHPLC has increased Pc's range from 400 to 1000 in a matter of about 60 minutes. For thorough analyses of extremely complex samples in proteomics and metabolomics, 2D-LC can further boost Pc. The columns used as the stationary phase in HPLC are consumable and, hence, need to be altered periodically [15].

APPLICATIONS

Pharmaceutical Industry

In the pharmaceutical industry, HPLC is used for both quantitative and qualitative analysis of bulk-manufactured drugs. In particular, Reversed-phase high-performance liquid chromatography (RP-HPLC) is extensively utilized within the pharmaceutical sector for drug evaluation due to its high specificity, sensitivity, and reproducibility [16]. RP-HPLC is primarily used for the assessment of drug purity. The active pharmaceutical ingredients (API) are first separated from the impurities, and then the amount of API is quantified. These impurities can originate from different sources like starting materials, intermediates, and degradation products [17]. Using RP-HPLC these impurities can be separated completely from the API. Additionally, RP-HPLC can also be used to monitor the rate of degradation of the drug and precisely measure the formation of byproducts during the process of manufacturing or storage [18]. RP-HPLC finds applications in analyzing drug–drug interactions as well. This encompasses separating and measuring both the drug and its metabolites within biological samples like plasma or urine. Evaluating drug–drug interactions is crucial for assessing drug safety and efficacy, as well as determining the pharmacokinetic parameters of the drugs [19]. Beyond these functions, RP-HPLC is instrumental in analyzing drug stability, bioavailability, and pharmacokinetics. Assessing drug stability involves subjecting the drug substance to various stress conditions and analyzing the resulting degradation products. RP-HPLC is utilized for separating and quantifying these degradation products that emerge during stability studies [20, 21]

Food Sciences

To identify various components and quantify their presence in food, HPLC also has application in food analysis. When residual pesticides from crops are ingested, they are known to affect our health negatively. So, HPLC is used to monitor the levels of pesticides and ensure the levels are within government regulations [22]. HPLC also identifies the presence of additives, antibiotics, veterinary drugs, and allergens in various food items derived from both plants and animals [23, 24]. Food contaminated with molds and fungi contains mycotoxins such as aflatoxins and fumonisins, which is accurately determined and quantified by HPLC to ensure the safety of the food [25]. Apart from toxins, HPLC is also used to determine the composition of fatty acids, vitamins, and nutrients for labeling and quality control [26].

Textile Industry

The analysis of dyes, pigments, and other chemical elements used in the production and coloring of textiles is one of the many uses of HPLC in the textile industry. To ensure uniform color matching and formulation, HPLC aids in establishing the composition, purity, and stability of dyes [27]. HPLC is used to analyze and verify the quality of textile auxiliaries used in the processing of textiles, such as surfactants, dispersion agents, wetting agents, and softeners. It aids in maintaining the desired chemical characteristics and regulatory compliance [28]. To ensure optimal dye uptake and color fastness, HPLC is used to monitor the concentration and distribution of dyes during the dyeing process. It aids in process optimization and enhances dyeing productivity. After processing, leftover chemicals, including dyes and auxiliary agents found in textile goods, are detected and quantified using HPLC. This is necessary to guarantee adherence to safety and environmental laws. The environmental impact of textile processes, particularly the presence of contaminants and dangerous substances in wastewater produced during dyeing and finishing operations, is measured and analyzed using HPLC. A variety of additives, including UV stabilizers, flame retardants, and antimicrobial agents that are added to textile materials for particular functionality, can be identified and quantified with the aid of HPLC [29]. In order to make sure that textile dyes adhere to industry requirements for color fastness in terms of washability, lightfastness, sweat, and other factors, HPLC is used to test the color fastness of the dyes. To ensure that the inks used in textile printing are compatible with various textile materials and printing methods, HPLC is used to characterize the composition and properties of the inks. The chemical makeup of fibers and yarns may be determined by HPLC, which helps with material selection, quality assurance, and process improvement in the textile industry. HPLC is used in R&D efforts to create and refine new dye formulas, investigate environmentally friendly and sustainable dyeing procedures, and enhance the general quality of textile products. The examination of chemicals and compounds used in various phases of textile production is precise and accurate due to the use of HPLC [30]. It helps the textile sector with quality assurance, process improvement, and the creation of sustainable and environmentally friendly practices.

Drug Delivery

HPLC holds a pivotal position in drug delivery, guaranteeing pharmaceutical products' quality, effectiveness, and safety. HPLC has utility in formulating and assuring the quality of medicinal products, allowing precise quantification of active pharmaceutical ingredients (APIs) and excipients. This precision ensures uniformity in formulation and compliance with established regulatory standards. Additionally, HPLC is instrumental in monitoring and assessing drug release from diverse drug delivery systems like tablets, capsules, patches, and nanoparticles. It aids in the assessment of drug release dynamics and formulation enhancement for controlled release. Following administration, medication concentrations in biological samples (such as blood, plasma, and urine) are assessed using HPLC. In order to create bioavailability and pharmacokinetic profiles, it is essential to understand how drugs are absorbed, distributed, metabolized, and eliminated [31]. Utilizing HPLC ensures that products have a long shelf life and are effective by assessing the stability of medications and formulations under various storage circumstances, and providing information on degradation products [32]. HPLC is used to characterize nanoparticles and assess drug loading, release kinetics, and efficacy of drug encapsulation in nanoparticle-based drug delivery systems. It helps in formulating nanoparticles more effectively for regulated and targeted medication delivery [33]. Impurities and degradation products in pharmaceutical formulations are identified and measured using HPLC. This is essential for guaranteeing the pharmaceutical product's safety and purity [34]. HPLC is used in dissolution testing to assess the rate at which a medication is released from a dosage form, providing vital details regarding the efficacy and bioavailability of the formulation [35]. In order to ensure compliance with regulatory requirements like the International Conference on Harmonisation (ICH) guidelines, HPLC is used to validate analytical procedures used in drug development. Excipients and other formulation elements are analyzed with the aid of HPLC to make sure of their quality, compatibility, and suitability for use in drug delivery systems [36]. During clinical trials, drug concentrations in biological samples are measured using bioanalytical assays that are developed and validated using HPLC. The safety and effectiveness of a medicine must be determined by accurate quantification [37]. HPLC is a vital tool in drug delivery because of its accuracy, sensitivity, and capacity to separate and quantify compounds. This allows researchers and pharmaceutical specialists to optimize formulations, evaluate drug release, and uphold high-quality standards throughout the drug development process [38].

Tissue Engineering

HPLC provides precise and accurate analytical capabilities essential for characterizing biomolecules, polymers, growth factors, and cellular components in tissue engineering. It contributes to advancing the design and development of functional tissue-engineered constructs for regenerative medicine applications. Proteins, peptides, nucleic acids, and amino acids – as well as other biomolecules – are analyzed and quantified using HPLC. This helps in understanding the biological processes and developing tissue scaffolds with the right bioactive chemicals. The polymer composition and purity used in scaffold fabrication are also analyzed by HPLC. HPLC ensures the desired molecular weight distribution and structural integrity of the scaffold and monitors its rate of degradation [39–41]. Growth factors like transforming growth factor-beta (TGF-β), vascular endothelial growth factor (VEGF), and others play a vital role in tissue regeneration and are analyzed and quantified by HPLC to ensure their purity and potency in tissue engineering applications [42–45]. Components of cell culture media, including glucose, lactate, and other metabolites, are analyzed using HPLC [46, 47]. Monitoring these elements helps to improve culture conditions for cell development and tissue generation while also revealing information about cell metabolism. Drug release characteristics from various drug delivery systems integrated into tissue engineering scaffolds are evaluated using HPLC. In cell culture systems, HPLC is used to measure cellular metabolites such as ATP, ADP, AMP, NADH, NAD+, and other energy-related compounds, which provides insight into cellular energy metabolism and how it affects tissue growth. To create biomimetic scaffolds that closely resemble the original tissue, an understanding of extracellular matrix (ECM) composition is essential [48]. The composition of ECM elements such as collagen, elastin, glycosaminoglycans (GAGs), and proteoglycans is analyzed using HPLC [49]. Cyclic adenosine monophosphate (cAMP), cyclic guanosine monophosphate (cGMP), and other signaling pathways essential for cell behavior and tissue regeneration are also examined using HPLC [50]. To evaluate cell viability and tissue health in tissue engineering projects, it is crucial to comprehend oxidative stress. Reactive oxygen species (ROS) and other antioxidants are utilized as oxidative stress markers, and HPLC is used to detect and quantify these markers [51].

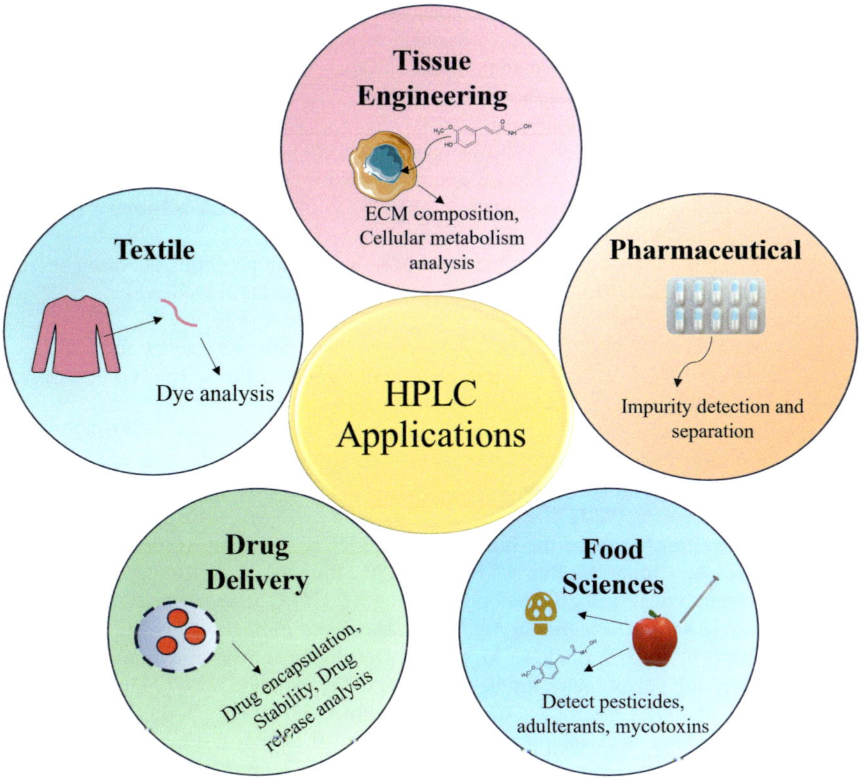

FIGURE 13.4 Schematic representation of various applications of HPLC.

All the applications of HPLC in diverse fields are shown in Figure 13.4.

DISCLOSURE OF POTENTIAL CONFLICTS OF INTEREST

None

AUTHOR CONTRIBUTIONS

Bipasa Dey and Manasa Gangadhar Shetty: Original draft, writing, and editing; Padmini Pai and Yashaswini Reddy: Review and editing; Dr. Babitha KS: Supervision, review, and editing.

ACKNOWLEDGMENTS

The authors thank the Department of Biotechnology (DBT) BioCare grant and Fellowship [Grant ID: BT/PR20046/BIC/101/683/2016], Government of India (GOI), and Manipal Academy of Higher Education (MAHE), Manipal, India.

REFERENCES

1. https://www.shimadzu.com/an/service-support/technical-support/analysis basics/basic/what_is_hplc.html, accessed on 01/10/2023.
2. Coskun, O. (2016). Separation techniques: chromatography. *Northern Clinics of Istanbul, 3*(2), 156.

3. .t https://www.thermofisher.com/in/en/home/industrial/chromatography/chromatography-learning-center/liquid-chromatography-information/hplc-basics.html accessed on 05/10/2023.
4. https://microbenotes.com/high-performance-liquid-chromatography-hplc/, accessed on 16/10/2023.
5. https://www.thermofisher.com/in/en/home/technical-resources/research-tools/image-gallery/image-gallery-detail.2492.html, accessed on 16/10/2023.
6. https://www.labmanager.com/evolution-of-hplc-systems-19579 , accessed on 02/10/2023.
7. https://microbenotes.com/high-performance-liquid-chromatography-hplc/ , accessed on 14/10/2023.
8. https://www.knauer.net/Application/application_notes/VSP0019_HPLC%20Basics%20-%20principles%20and%20parameters_final%20-web-.pdf, accessed on 11/10/2023.
9. https://www.hta-it.com/blog/preparative-and-semi-prep-lc-autosampler.html, accessed on 11/10/2023.
10. https://www.jasco-global.com/principle/principles-of-hplc-3-separation-modes/, accessed on 05/10/2023.
11. LaCourse, W. R., & LaCourse, M. E. (2017). Chapter 17-General instrumentation in HPLC. In *Liquid Chromatography: Fundamentals and Instrumentation, Second Edition* (pp. 61–73). Elsevier. DOI: 10.1016/B978-0-12-805393-5.00017-8
12. https://www.chemguide.co.uk/analysis/chromatography/hplc.html, accessed on 16/10/2023.
13. https://chromtech.com/how-does-column-temperature-affect-hplc-resolution/, accessed on 16/10/2023.
14. https://www.thermofisher.com/blog/analyteguru/top-7-benefits-of-using-hplc-for-manufacturing-quality-control/, accessed on 16/10/2023.
15. https://labioanalytics.com/what-are-the-limitations-of-hplc/, accessed on 16/10/2023.
16. Kumar, D., Kumar, A., Kumar, V., Raj, A., Rai, R. R. M., Baliyan, V., & Kumar, N. (2023). A comprehensive review on analytical method development using RP-HPLC and recent advances in pharmaceutical applications. *Journal for Research in Applied Sciences and Biotechnology*, *2*(2), 53–60.
17. Dutta, A. K., Avery, B. A., & Wyandt, C. M. (2006). Development and validation of a stability-indicating reversed-phase high performance liquid chromatography method for NPC 1161C, a novel 8-aminoquinoline anti-malarial drug. *Journal of Chromatography A*, *1110*(1–2), 35–45.
18. Deore, K. R. (2017). Method development and validation for Apremilast by RP-HPLC. *International Journal of Advanced Engineering and Management Research*, *2*, 68–72.
19. Annadi, A. M., El Zahar, N. M., Abdel-Sattar, N. E. D. A., Mohamed, E. H., Mahmoud, S. A., & Attia, M. S. (2022). Development and validation of molnupiravir assessment in bulk powder and pharmaceutical formulation by the RP-HPLC-UV method. *RSC Advances*, *12*(53), 34512–34519.
20. Mangla, B., Beg, S., Alam, O., Ahsan, W., Haque, A., Patel, K. S., ... & Kohli, K. (2020). Systematic development and validation of RP-HPLC method for simultaneous estimation of tamoxifen and sulphoraphane with specific application for nanolipidic formulations. *Arabian Journal of Chemistry*, *13*(11), 7909–7920.
21. Hicks, M. B., Oriana, S., & Liu, Y. (2022). Assessment of analytical testing: The impact of metrics for the sustainable measurement of pharmaceuticals. *Current Opinion in Green and Sustainable Chemistry, 38*, 100689.
22. Watanabe, E., Kobara, Y., Baba, K., & Eun, H. (2014). Aqueous acetonitrile extraction for pesticide residue analysis in agricultural products with HPLC– DAD. *Food Chemistry*, *154*, 7–12.
23. Won, S. Y., Lee, C. H., Chang, H. S., Kim, S. O., Lee, S. H., & Kim, D. S. (2011). Monitoring of 14 sulfonamide antibiotic residues in marine products using HPLC-PDA and LC-MS/MS. *Food Control*, *22*(7), 1101–1107.
24. Monaci, L., De Angelis, E., Montemurro, N., & Pilolli, R. (2018). Comprehensive overview and recent advances in proteomics MS based methods for food allergens analysis. *TrAC Trends in Analytical Chemistry*, *106*, 21–36.
25. Hickert, S., Gerding, J., Ncube, E., Hübner, F., Flett, B., Cramer, B., & Humpf, H. U. (2015). A new approach using micro HPLC-MS/MS for multi-mycotoxin analysis in maize samples. *Mycotoxin Research*, *31*, 109–115.
26. Fontannaz, P., Kilinç, T., & Heudi, O. (2006). HPLC-UV determination of total vitamin C in a wide range of fortified food products. *Food Chemistry*, *94*(4), 626–631.
27. Balakina, G. G., Vasiliev, V. G., Karpova, E. V., & Mamatyuk, V. I. (2006). HPLC and molecular spectroscopic investigations of the red dye obtained from an ancient Pazyryk textile. *Dyes and Pigments*, *71*(1), 54–60.
28. Mouri, C., & Laursen, R. (2012). Identification of anthraquinone markers for distinguishing Rubia species in madder-dyed textiles by HPLC. *Microchimica Acta*, *179*, 105–113.
29. Koren, Z. C. (1994). HPLC analysis of the natural scale insect, madder and indigoid dyes. *Journal of the Society of Dyers and Colourists*, *110*(8), 273–277.

30. Hu, C., Zhu, J., Mei, H., Shi, H., Guo, H., Zhang, G., ... & Zheng, X. (2018). A sensitive HPLC-MS/MS method for the analysis of fiber dyes. *Forensic Chemistry*, *11*, 1–6.
31. Md, S., Haque, S., Fazil, M., Kumar, M., Baboota, S., Sahni, J. K., & Ali, J. (2014). Optimised nanoformulation of bromocriptine for direct nose-to-brain delivery: biodistribution, pharmacokinetic and dopamine estimation by ultra-HPLC/mass spectrometry method. *Expert Opinion on Drug Delivery*, *11*(6), 827–842.
32. Shen, H. R., Li, Z. D., & Zhong, M. K. (2006). HPLC assay and pharmacokinetic study of atorvastatin in beagle dogs after oral administration of atorvastatin self-microemulsifying drug delivery system. *Die Pharmazie-An International Journal of Pharmaceutical Sciences*, *61*(1), 18–20.
33. Corrie, L., Gulati, M., Kaur, J., Awasthi, A., Vishwas, S., Ramanunny, A. K., ... & Singh, S. K. (2023). Quality by Design-based RP-HPLC method for estimation of curcumin in rat plasma and fecal microbiota extract-based solid self-nano emulsifying drug delivery system. *Current Drug Research Reviews Formerly: Current Drug Abuse Reviews*, *15*(3), 272–285.
34. Rao, R. N., & Nagaraju, V. (2003). An overview of the recent trends in development of HPLC methods for determination of impurities in drugs. *Journal of Pharmaceutical and Biomedical Analysis*, *33*(3), 335–377.
35. Tucker, R. M., Parcher, B. W., Jones, E. F., & Desai, T. A. (2012). Single-injection HPLC method for rapid analysis of a combination drug delivery system. *AAPS PharmSciTech*, *13*, 605–610.
36. Karmarkar, S., Garber, R., Genchanok, Y., George, S., Yang, X., & Hammond, R. (2011). Quality by design (QbD) based development of a stability indicating HPLC method for drug and impurities. *Journal of Chromatographic Science*, *49*(6), 439–446.
37. Wu, W., Tang, M. H., Tang, H., Chen, K., Fu, J., Wang, L., ... & Chen, L. J. (2018). Identification, characterization and HPLC quantification of formulation-related impurities of honokiol, an antitumor natural drug candidate in clinical trials. *Journal of Pharmaceutical and Biomedical Analysis*, *153*, 186–192.
38. Evrim Kepekci Tekkeli, S., & Volkan Kiziltas, M. (2017). Current HPLC methods for assay of nano drug delivery systems. *Current Topics in Medicinal Chemistry*, *17*(13), 1588–1594.
39. Zhang, J., Deng, A., Yang, Y., Gao, L., Xu, N., Liu, X., ... & Yang, S. (2015). HPLC detection of loss rate and cell migration of HUVECs in a proanthocyanidin cross-linked recombinant human collagen-peptide (RHC)–chitosan scaffold. *Materials Science and Engineering: C*, *56*, 555–563.
40. Wang, H. J., Zhang, Y., Kato, S., Nakagawa, K., Kimura, F., Miyazawa, T., & Wang, J. Y. (2018). HPLC-MS/MS: A potential method to track the in vivo degradation of zein-based biomaterial. *Journal of Biomedical Materials Research Part A*, *106*(2), 606–613.
41. Balmayor, E. R., Tuzlakoglu, K., Marques, A. P., Azevedo, H. S., & Reis, R. L. (2008). A novel enzymatically-mediated drug delivery carrier for bone tissue engineering applications: Combining biodegradable starch-based microparticles and differentiation agents. *Journal of Materials Science: Materials in Medicine*, *19*, 1617–1623.
42. Aguilar, L. M. C., Silva, S. M., & Moulton, S. E. (2019). Growth factor delivery: Defining the next generation platforms for tissue engineering. *Journal of Controlled Release*, *306*, 40–58.
43. Wenk, E., Meinel, A. J., Wildy, S., Merkle, H. P., & Meinel, L. (2009). Microporous silk fibroin scaffolds embedding PLGA microparticles for controlled growth factor delivery in tissue engineering. *Biomaterials*, *30*(13), 2571–2581.
44. Sommer, N., Sattler, M., Weise, J. M., Wenck, H., Gallinat, S., & Fischer, F. (2013). A tissue-engineered human dermal construct utilizing fibroblasts and transforming growth factor β1 to promote elastogenesis. *Biotechnology Journal*, *8*(3), 317–326.
45. Seyednejad, H., Ji, W., Yang, F., van Nostrum, C. F., Vermonden, T., van den Beucken, J. J., ... & Jansen, J. A. (2012). Coaxially electrospun scaffolds based on hydroxyl-functionalized poly (ε-caprolactone) and loaded with VEGF for tissue engineering applications. *Biomacromolecules*, *13*(11), 3650–3660.
46. Ju, Y. M., Yu, B., West, L., Moussy, Y., & Moussy, F. (2010). A dexamethasone-loaded PLGA microspheres/collagen scaffold composite for implantable glucose sensors. *Journal of Biomedical Materials Research Part A: An Official Journal of The Society for Biomaterials, The Japanese Society for Biomaterials, and The Australian Society for Biomaterials and the Korean Society for Biomaterials*, *93*(1), 200–210.
47. Cebotari, S., Tudorache, I., Jackel, T., Hilfiker, A., Dorfman, S., Ternes, W., ... & Lichtenberg, A. (2010). Detergent decellularization of heart valves for tissue engineering: Toxicological effects of residual detergents on human endothelial cells. *Artificial Organs*, *34*(3), 206–210.
48. Loebel, C., Saleh, A. M., Jacobson, K. R., Daniels, R., Mauck, R. L., Calve, S., & Burdick, J. A. (2022). Metabolic labeling of secreted matrix to investigate cell–material interactions in tissue engineering and mechanobiology. *Nature Protocols*, *17*(3), 618–648.

49. Braun, M., Rýglová, Š., & Suchý, T. (2021). Determination of glycosaminoglycans in biological matrices using a simple and sensitive reversed-phase HPLC method with fluorescent detection. *Journal of Chromatography B*, *1173*, 122626.
50. Bähre, H., Danker, K. Y., Stasch, J. P., Kaever, V., & Seifert, R. (2014). Nucleotidyl cyclase activity of soluble guanylyl cyclase in intact cells. *Biochemical and Biophysical Research Communications*, *443*(4), 1195–1199.
51. Martins, R. M., Alves, G. D. A. D., Martins, S. D. S., de Freitas, L. A. P., Rochette, P. J., Moulin, V. J., & Fonseca, M. J. V. (2020). Apple Extract (Malus sp.) and rutin as photochemopreventive agents: Evaluation of ultraviolet b-induced alterations on skin biopsies and tissue-engineered skin. *Rejuvenation Research*, *23*(6), 465–475.

14 Starch-Based Analytical Schemes for Various Applications

*Nikhil Kumar Daimari, Upama Das, Pohar Borah, and Rajib Biswas**

[1]Applied Optics and Photonics Research Laboratory, Department of Physics, Tezpur University, Assam
*Corresponding Author

INTRODUCTION

Starch, a homopolysaccharide and a natural constituent of food, is produced by plants through glucose polymerization, which occurs via photosynthesis of carbon dioxide. It is mainly found in fruits, tubers, cereals, roots, and legumes. Basically, it is a polymer which has a semi-crystalline structure and is composed of nearly 1000–2,000,000 anhydrous glucose monomers [1, 2]. Two polymers, viz., amylose and amylopectin, along with two types of glycosidic linkages mainly constitute the overall structure of starch [3, 4]. Amylose contains about 99% of (1-4)-α-linkage and 1% of (1-6)-α-linkages, whereas amylopectin is composed of about 95% of (1-4)-α-linkage and 5% of (1-6)-α-linkages [5]. Amylose is an amorphous polymer possessing a linear chained structure with an average molecular weight (MW) of 1×10^6 g.mol^{-1}, whereas amylopectin is crystalline in nature with a highly branched double helix structure having a higher MW of 1×10^8 g.mol^{-1} [6–8]. Starch obtained from various botanical sources has different ratios of the two polysaccharides, which affects its chemical properties and functionality. Additionally, some other factors such as geographic origin, plant source, climate conditions, and soil, etc., determine its structure [6]. In addition to these two constituents, some starch from mutant plants contains intermediate materials that are different from these two polysaccharides [2, 8].

On the grounds of structural category, they are mainly classified into three types: A, B, and C based on their dimensions. A-type structure possesses closed packing with water molecules between two double helices, whereas B-type structure contains more water molecules. But the third type possesses an intermediate structure between A- and B-type. Cereal starches are mainly of A-type while tubers are mainly composed of B-type starch. C-type is more common in starches obtained from roots, legumes, and fruits. There is one more kind of starch, known as V-type, which is formed when amylose is compounded with lipids and other substances [9, 10].

Non-carbohydrate compounds such as proteins and lipids contribute a small proporation of the total weight of the starch granules. Some tightly bound proteins take part in the synthesis of amylose. Lipids occur abundantly as fatty acids in cereals, which form complexes with amylose. Some trace elements, such as phosphorus and potassium, can also be found in starch [2, 8].

Due to the low cost, abundance, ease of dissolution, water retention, biocompatibility, binding ability, degradation and some other chemical and physical properties, starch has drawn the attention of researchers in recent times. However, retrogradation, low thermal stability, hydrophilicity, brittleness, etc. limit its use in industrial applications [1, 6]. In order to mitigate these pitfalls, starches are modified by various protocols to attain the necessary properties. The modifications

DOI: 10.1201/9781003545712-14

could be physical, thermal, or chemical modifications. The hydroxyl groups present in starch could be replaced by selective other functional groups to enhance desirable qualities [1, 11, 12]. The modified starch has better thermal stability, solubility, biodegradability, lower swelling power etc. [5]. In recent times, starch has found applications in drug delivery, water treatment, the pharmaceutical industry, self-healing polymeric materials, porous foam structures, sensing etc. [5, 6, 13–15]. Starches obtained from various sources are used as substrates, templates, thin films, or stabilizing agents in a wide range of sensing applications owing to their extraordinary properties such as biocompatibility, ability to form films, good adhesion characteristics, easy structural modification, high mechanical resistance, gel-forming capacity, and excellent elasticity. It is a new generation green replacement for the fabrication of sensing architectures with high sensitivity and stability along with a lower limit of detection. Figure 14.1 provides schematic representation of starch.

This chapter gives insight into the fundamental distinctive properties of different kinds of starch, along with a detailed emphasis on the various sensing abilities and mechanisms of starch in various domains such as chemical, biological, and optical.

PROPERTIES OF STARCH

Starch, also referred to as amylum, acts as an essential source of energy. Photosynthetic green plants store carbohydrates in the form of small compact granules known as starch, where it is stored as amyloplasts in the case of grains and tubers [16, 17]. Smaller starch granules are less than 1 μm in size, whereas larger granules are more than 100 μm in various shapes such as oval, spherical, polygonal, etc. [16, 18, 19]. They are primarily composed of two different kinds of polysaccharide units, one of which is amorphous known as amylose, containing 500–2000 glucose units, and another possessing a crystalline structure known as amylopectin, which contains more than 1,000,000 glucose units [20]. Its properties and structure depend significantly on the amount and composition of both homopolysaccharides. Generally, starch contains only about 20–30% of amylose, but its concentration may vary depending on its type, where starch containing less than 15% of amylose is termed as waxy starch [16, 21]. Amylose is a single-chained linear-structured polysaccharide where the polymeric groups are connected via α-d-(1-4) glycosidic linkages. In contrast, amylopectin possesses a higher molecular weight with a branched structure where the linear polymeric groups are connected via α-d-(1-4) glycosidic linkages, constituting 95% of the bonding, whereas the branched chains in the polysaccharide are connected via α-d-(1-6) glycosidic linkages, which is less than 5% of the total linkages [16, 18, 22]. Amylopectin forms a double helix structure, with the amylose remaining scattered between its helix chain. These polysaccharides contribute to about 98–99% of the dry weight of starch granules. The shape and size of each starch granule are determined by the shape of the amylopectin polysaccharide, which is also a controlling factor determining its functional properties along with other conditions such as the concentration of phospholipids and phosphate-monoester, etc. Starches with a higher content of amylopectin are highly stable to retrogradation, whereas as starches with high amylose content act as highly resistant starches [23–26].

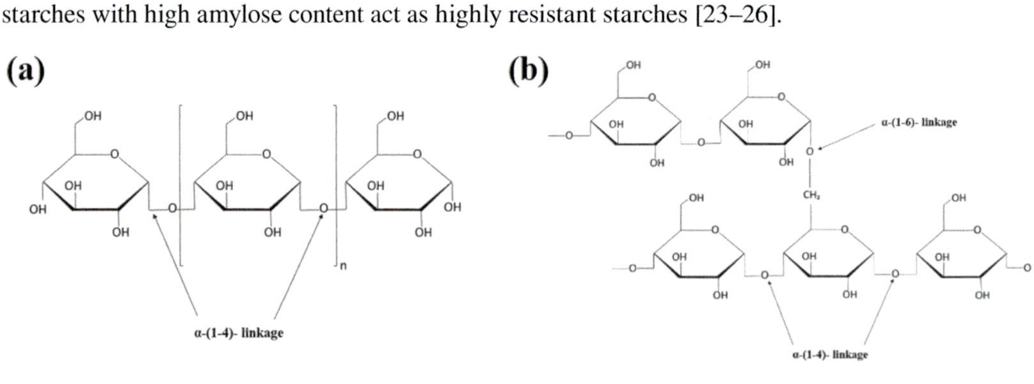

FIGURE 14.1 Schematic representation of the structure of (a) amylose and (b) amylopectin [5].

Depending on the functional properties and applications of starch, it is classified into three categories: "native starch," "modified starch," and "hydrolyzed starch." The first one is a raw form of starch, which is insoluble in water, highly inert, and also retrogrades in different pH and temperature conditions. To enhance their functionalities and properties, these native starches are modified, which forms another class of starch. Another category of starch is "hydrolyzed starch," where the long polymeric chains are broken into smaller units, such as maltose [16, 22, 27].

The structure, chemical compositions, and morphological parameters significantly determine the properties and characteristics of the starch for its varied application across a wide range of fields.

Physical Properties of Starch

The physical properties of starch depend on its physical modifications of starch granules without involving any change in the chemical bonding. These properties can be altered by changing the length of polymeric chains, the ratio of the two polysaccharides, and the degree of branching in amylopectin, where the concentration of phosphate groups also has a crucial impact on determining the structural aspects of starch [22].

Gelatinization and Retrogradation Property

At ambient room temperature conditions, starch is insoluble in water, but on raising the temperature to 52°C, gelatinization of starch granules of starch occurs in the presence of solvents, where they act as "plasticizers." It is an endothermic process in which the starch becomes amorphous due to the breakage of the double helix structure as it loses the Maltese cross linkages. Therefore, a significant increase in randomness in the starch granules is obtained. Without the involvement of any plasticizer, starch cannot be broken down and they can thermally decompose at higher temperatures, more than 250°C [16, 28].

The amount of gelatinization in starch is monitored via different sophisticated routes such as differential scanning calorimetry (DSC), nuclear magnetic resonance (NMR) spectroscopy, thermomechanical analysis, Fourier transform infrared spectroscopy, and X-ray scattering [29–34], etc. Out of all of these methods, DSC is a highly utilized mode, where a pan is used to heat starch in the presence of water, and the gelatinization temperature and the energy required for the dissociation of bonds are measured [34, 35]. The gelatinization temperature depends on the internal characteristics of starch such as structure, source of origin of starch, and some external parameters such as pH, concentration and type of plasticizer, sugars, salts, and lipids. The length of branched chain significantly determines the gelatinization temperature; with an increase in the later also showing an increasing trend. But B-type polymorph potato starch shows an opposite trend where in spite of having a longer branched chain of amylopectin, it displays a lower gelatinization temperature compared to A-type starches of potato [34, 36–38].

Sugars such as fructose, sucrose, and glucose, etc., increase the temperature at which gelatinization of starch occurs as these sugars bind with the plasticizers rather than decreasing gelatinization in starch [34]. Some types of plasticizers with higher viscosity, lower H-bonding capability, and higher molecular weight also increase the temperature of gelatinization due to the reduction of interaction with the starch granules [39].

In the case of modified starch, this property also depends upon the amount of oxidation, cross-linking, chemical substitutes used for modification, and different treatments such as hydrothermal, high hydro-pressure, and enzymatic treatments, etc., used for modification of starch granules [34, 40, 41].

Rheological Property

Rheological properties are determined by chemical structure, concentration, temperature, heating rate, and viscosity, flow characteristics of starch granules under different stress conditions, including the storage conditions under which the starch granules are kept. These properties are studied with the help of a rapid visco analyzer (RVA) or by using a rheometer [16, 34, 42, 43]. When gelatinization of starch is performed under shear stress, the starch granules swell, eventually rupture, and the amylose comes out of the starch granule, resulting in formation of a highly viscous paste. This process is called starch pasting [44]. This property determines the thickening ability of starch for toward wide range of applications, especially in the food industry [34, 45]. These starches paste at the appropriate concentration after cooling, and can form gels, as the molecules present in the granules form a certain network that displays viscoelastic property [34].

In the case of modified starch, rheological properties also depend on amount of oxidation, the level of modification, chemical substitution, and various treatments made on starch for its modification [34, 40, 46].

A. **Pasting Property**

Brabender amylograph is commonly used to determine the pasting properties in starch [47]. Starch containing a higher quantity of amylopectin displays a lower temperature of pasting but a higher value of peak viscosity, as it enhances the swelling power of starch while heating. On the contrary, amylose, in the presence of phosphate-monoester derivatives and lipids, prevents its swelling and thereby degrades this property [34, 48, 49].

B. **Viscosity**

This parameter can easily be measured by using a viscometer [50]. Starch paste is a non-Newtonian fluid that possesses a varied viscosity depending on the stress applied. It is thixotropic, displaying a decreasing trend in accordance with shear time and rate and also increasing with amylose concentration, but increases with an increase in concentration of starch paste [51, 52].

Morphological Property

An atomic force microscope (AFM) and a scanning electron microscope (SEM) are the most commonly used techniques to determine the morphological aspects of starch granules [44]. It depends on the geographic origin, physiology of the plant, biochemistry of the chloroplast, its metabolic and developmental changes, and their type [53, 54], along with the ratio of amylose to amylopectin. In the case of potato starch, each granule ranges of size between 1 and 110 μM, with the small granules ranging from 1 to 20 μM, whereas larger granules range of 20–110 μM, possessing an oval, irregular, or cuboidal shape [44, 55]. The corn starch smaller granules range in size from 1 to 7 mm whereas larger granules range of 15–20 mm and have a singular shape [44, 56]. In the case of rice starch, the granules range in size between 3 and 5 mm and are angular or pentagonal shape [57]. When the wheat endosperm matures, it possesses two kinds of granules known as large A granules and small B granules; the former are disk-shaped and range in size between 10 and 35 mm, whereas the latter are polygonal, roughly spherical and are of size ranging between 1 and 10 mm. Corn, wheat, and rice starch granules display a rougher surface structure compared to potato starch granules [44, 56]. Some "pin holes" were observed in corn starch granules, which are larger in size [57]. AFM images depicted the presence of large "protuberances" ranging between 200 to 500 mm [44, 58]. Larger starch granules are more prone to enzymatic degradation. Compared to modified starch granules, native starch has a smoother outer surface with a slight depression at the center of each granule. Large granules of starch display more pore formation and surface erosion compared to the smaller granules due to their high content of amylose [44, 59].

Structural Property

The X-ray diffraction (XRD) analysis method is a popularly utilized method to determine the structural orientation of starch [60]. Depending on botanical origin and crystalline pattern, the native variety of starch is classified into various types, namely A, B, and C [34, 61]. *Type A:* Composed of amylopectin chains having 23–29 glucose units, such as normal cereal starch; *Type B:* Composed of amylopectin chains having 30–44 glucose units, such as amylose rice starch and tubers; *Type C:* Composed of amylopectin chains having 26–29 glucose units, such as legume starch [16]. Though the first two types are crystalline starches, the first one has less binding ability with water due to high density, as in them the glucose helix structure is tightly packed compared to type B where the glucose helix structure is loosely packed, leaving sites for the binding of water molecules. On the contrary, type C starch is an amalgamation of both crystalline and amorphous starch particles [16, 62, 63].

After modification of native starch by physical, chemical, or enzymatic methods, significant changes are observed in the XRD spectra. In fact, thermal and heat treatment may also alter the crystallinity of starch granules. It was observed that the amorphous portion of the starch granules is accessible to enzymatic degradation, contrary to the crystalline one [16]. Effective enzyme hydrolysis was found in A-type polymorph compared to the B-type one. Acetylated modified starch displayed more crystallinity compared to native starch but the diffraction pattern remained the same [62]. Under ultrasonic treatments displayed weaker crystalline structure than the non-modified ones [61–64].

Chemical Properties of Starch

Modification of starch can enhance its chemical properties. The different chemical reactions that may occur in starch include oxidation, hydrolysis, etherification, and esterification which may result from the breakage of certain bonds or the creation of new bonds due to substitution. This alteration or modification can be performed without damaging its size and shape, but it may alter its physicochemical properties [20].

Starch molecules chemically react to transform into modified starch, where its chemical properties are dependent on the amount of reactivity, which is, in turn, determined by the hydroxyl group of the long polysaccharide chains. The free hydroxyl groups contain some secondary and primary alcoholic groups, which are ready for substitution due to the absence of any glycosidic linkages in these sites. These are the sites where any kind of reaction or substitution may occur, but the reaction requires prior activation of these sites in basic or acidic conditions [20, 65].

Hydrolysis

This is an enzymatic or chemical process in which a bond breaks down upon addition of water or hydrochloric acid under constant heating, resulting in formation of products containing hydroxyl groups. When hydrolysis is performed in an acidic medium it is known as roasting, and the native starch gets modified by acid, whereas hydrolysis in basic media results in the formation of alkaline modified starch [20, 66].

Esterification

Esterification is a process of condensation of carboxylic acid and alcohol in acidic media [67]. In the case of starch, the reaction occurs between the alcoholic group and the carboxylic group of the glucose unit of starch, which adds a lipophilic group to the starch granules. However, this process weakens the intermolecular forces, thereby changing its shape and size [20, 68].

Succinylation

Esterification of starch granules with succinic anhydride at high temperatures, under aqueous pyridine, is known as a succinylation reaction, which generates succinyl starch. This process facilitates an increase in solubility and decreases the gelatinization temperature by weakening the intermolecular bonding in starch granule [20].

Etherification

Epoxide reagents are used for this kind of reaction, where they are first reduced to diols. They are utilized for modification of starch molecules by introducing lipophilic alkyl groups, which alters the intramolecular hydrogen bonding capacity and reduce the hydrophilicity [20, 69].

Acetylation

An acetylation reaction occurs in starch when it is modified by some acetic anhydride, which yields the formation of acetylated starch. The hydroxyl group gets replaced by an acetyl group in starch after the reaction. The heavy acetyl group causes steric hindrance, which allows easy percolation of water between the chains, increasing its solubility and thereby lowering the temperature of gelatinization. This modification of starch helps in the improvement of freeze–thaw stability of starch along with its paste clarity [20, 70].

Hydroxypropylation

Modifying starch with propylene epoxide in alkaline conditions is called a hydroxypropylation reaction, which increases its viscosity and freeze–thaw stability by modifying the inter- and intramolecular hydrogen bonds [20, 69].

Oxidation

This type of reaction in starch occurs in the presence of some oxidizing agents such as sodium hypochlorite. These oxidants interact with the glycosidic bonds of the starch molecules and hydrolyze them, where a higher pH favors the formation of carboxylate groups. Modification of starch by these groups increases the steric hindrance and electrostatic repulsion in them, thereby decreasing chances of retrogradation. Oxidized starches offer good adhesive properties and are thereby used in meat products or the baked food industry. Oxidation may also lead to the formation of bleached starch when weaker oxidizing agents such as hydrogen peroxide, sodium hypochlorite, etc., are used it improves whiteness in the starch and eliminates microbial contamination [20, 71].

Cross-Linking of Starch

This type of linking in starch molecules may occur by employing any one of the above-mentioned reactions. These reactions alter the intra- and intermolecular bonding in the starch granules. These reactions are utilized to modify the starch according to need by altering its gelatinization, viscosity, retrogradation, and textural properties towards a wide range of applications. In fact, they display high stability towards extreme temperature and pH conditions [20, 72].

TYPES OF ANALYTICAL SCHEMES

Chemical Sensing

Starch-based chemical sensing refers to the incorporation of starch on a material substrate for the detection of specific chemicals or analytes. During the last two decades, researchers have made significant progress in harnessing starch-based materials for various chemical sensing mechanisms. As an alternative to conventional methods, sensing based on electrochemical techniques offers advantages in various properties such as accurate sensitivity, stable performance, and reproducibility. Additionally, devices based on electrochemical techniques have proven to be reliable with rapid response time, cost-effectiveness, and potential for seamless integration into portable devices [73, 74].

A fascinating approach to the synthesis of carbon microspheres (CM) from potato starch (PS) was developed by Lin et al. They employed PS to make glassy carbon electrodes (GCEs) to fabricate an electrochemical sensor for detecting Hg(II) in aqueous samples. This method involved a gentle hydrothermal process, where PS was first made soluble and kept stirring for 5 hours, followed by autoclaving at 180°C for 16 hours and then subsequent cooling to room temperature. To complete the process, the solution was heated at 4°C/min under a nitrogen atmosphere. It was found that the resulting sensor exhibited remarkable reproducibility, minimal susceptibility to interference from metal ions such as Zn^{2+}, Cu^{2+}, and Fe^{3+}, and showed stable performance for a week [75].

A modified GCE as an electrochemical sensor was reported by Zambianco et al., using carbon-based nanodiamonds (ND) and manioc starch (MS). MS was made soluble in 100 mL of acetic acid by stirring for 5 minutes at 95°C followed by 1 hour of stirring at 75°C. In this work, the authors employed cyclic voltammetry and square wave voltammetry to study the electrochemical response of herbicide diquat (DQ) in water samples. The ND-based film proved to enhance the electron transfer of DQ in the ND-MS glassy carbon electrode system. The DQ^{2+} ions underwent irreversible reduction to DQ^{+} ions, which was considered to be the possible sensing mechanism of the analyte. This novel modified electrode system also exhibited a good linear response to DQ with a LOD of 1.1×10^{-7} mol L^{-1} demonstrating a simple and efficient sensor [76].

Utilization of starch as renewable and biodegradable films has been broadly studied due to its abundance in nature, low cost, and wide functional capabilities [77, 78]. Use of such a thin starch film for the development of a chemodosimeter was reported by Isaad et al., where cyanide anions were detected in an aqueous medium with a low detection limit. The starch film was first synthesized via the sol-gel method under acidic catalysis and then dipped in quinolinium merocyanine dye to produce a yellow-colored film. The film was then immersed in various aqueous solutions of anions, which changed color from yellow to purple in the presence of cyanide anions only. This suggested a selective detection of cyanide anions attributed to nucleophilic addition between the anions and dye film [79].

Another electrochemical non-enzymatic sensor based on a starch-modified carbon paste electrode was developed by Gautam et al. for reliable cholesterol sensing. Electrocatalytic analysis of the carbon paste electrode modified with Polyaniline/MWCNTs/starch (PCS) nanocomposite revealed several –OH groups from the starch helped in the inhibition of the positive charge of the polymer. However, conducting MWCNTs provided a charge transfer process to obtain equilibrium between the hydrophilic and hydrophobic nature of the nanocomposite system. The sensor also exhibited highly selective detection of cholesterol among other different analytes with a sharp response time within 4–6 s and a low detection limit of 0.01 mM owing to the stability of the nanocomposite material for electrochemical application [80].

An example of the role of PS in gas sensitivity was reported by Peregrino et al., where graphene oxide (GO) films and PS were assembled layer by layer on quartz substrates. Here, starch played various roles in the functioning of the sensing unit. To begin with, starch formed hydrogen bonds with –C–O–C– and =C=O groups of GO, leading to superior film assembly. Secondly, during the photoreduction of GO to reduced GO, starch provided extra electron density, enhancing the process. Additionally, the sensing performance of the composite in response to humidity also improved due to the hydrophilic and hygroscopic nature of starch [81].

Tapioca (TP) starch extracted from manioc root also proved to be an excellent modifier of electrochemical sensors. This thin biofilm combined with nitrogen-doped titanium dioxide nanoparticles (N-TiO_2) as a nanocomposite material on a GCE was presented Maurúto de Oliveira et al., as a new sensing architecture for the detection of 17-β estradiol. The preparation of the starch film with N-TiO_2 nanoparticle dispersion was done until complete homogenization. Morphological characterization also showed the TP film to be uniform, and when included with N-TiO_2, granules were evident on the film, indicating a strong interaction between the film and the nanoparticles. The proposed modified electrode showed excellent repeatability and reproducibility with a detection limit of 1.7×10^{-7} mol L^{-1} [82].

Maghemite nanoparticles γ-Fe_2O_3 chemically synthesized using biopolymer starch were proposed by Ramu et al. for electrochemical sensing of folic acid (FA) in pharmaceutical samples. A starch solution of 1% was used in the preparation of the nanoparticles, which produced particles with a mean size of 24.20 ± 7.43 nm. The nanoparticles were immobilized on the electrode to fabricate the sensing platform, which showed enhanced electrocatalytic activity towards the oxidation of FA into dehydrofolic acid. Based on differential pulse stripping voltammetry analysis, the sensor exhibited dual linear concentration ranges of FA from 0.05 to 1.0 μM and 1.0 to 80 μM, with detection limits of 2.8 nM and 48.2 nM, respectively. The sensor demonstrated good reproducibility, high stability, and also remarkable selectivity to ascorbic acid and other common interferents, posing as a suitable candidate in the field of sensors [83].

Optical Sensing

Higher stability, excellent biocompatibility, and unique optical properties of starch have enabled it to be used in optical sensors in a diverse array of applications, where starch granules are modified into films or converted into quantum dots to facilitate the detection of various analytes. Its ability to withstand extreme conditions has enabled it to be used as a stabilizing factor in various sensing applications.

A work reported by Basu et al. utilized starch from sabudana "tapioca sago" to prepare carbon dots (C-dots). These C-dots are surface functionalized with Fe^{3+} ions, for quenching the photoluminescence intensity of the C-dots through a charge transfer method. But in the presence of F^- ions, the Fe^{3+} ions get removed from the surface of the C-dots, and the PL intensity returns to its original form, which enables optical detection of the harmful F^- ions in water. The synthesis protocol that this group used was also quite simple, as they first treated starch with hot water then stirred it with potassium hydroxide (KOH) for 12 hr at pH 7, which resulted in the formation of C-dots. This sensor was found to be highly spontaneous and selective towards detection of F^- ions [84].

Vasileva et al. utilized greener synthesized starch-stabilized silver nanoparticles (AgNPs) for the detection of hydrogen peroxide (H_2O_2). AgNPs display significant absorbance at around 420 nm, but in presence of H_2O_2 there was a significant decrease in absorbance. The presence of hydrogen peroxide results in the catalytic degradation of the starch stabilizer, causing significant decrease in absorbance. The presence of the starch coating enables proper encapsulation of the NPs, leading to highly efficient detection of H_2O_2 up to a limit of detection (LOD) of 0.9 μM [85]. In another study, Miao et al. utilized starch-stabilized AgNPs for the detection Cu^{2+} ions based on a redox reaction between the starch-stabilized AgNPs and Cu^{2+}, which displayed a colorimetric change [86].

Another fascinating work was reported by Boonkanon et al., where they prepared a colorimetric sensor by using tapioca to prepare a starch film, which was then doped with curcumin nanoparticles. The fabricated starch-based sensor was utilized for colorimetric-based sensing boron in various water samples. The sensor was fabricated on a plastic spoon, which was encapsulated with a yellow-colored biodegradable film that turned reddish brown in color in the presence of boron. The relative standard deviation of their fabricated device was found to be 1.52% [87].

Starch was also utilized as a substrate for various categories of sensing comprising fluorescence-based sensing technology. Qiang et al., fabricated a water-soluble carbon quantum dot (CQD) by utilizing PS as the raw material. The synthesized nanoparticles displayed a strong green luminescence, which was independent of excitation wavelength but highly sensitive to pH variation. Thus, this kind of starch-based compound can be utilized as pH sensors [88].

In a work reported by Hooshyar et al., they embedded glutathione-capped CdTe/ZnS QDs into cationic starch biopolymer (CS-GSH-CdTe/ZnS QDs) to enhance their fluorescence intensity which could further be utilized towards the detection of rifampicin. When these QDs interacted with this analyte, their florescence intensity was highly quenched compared to that of bare QDs without the embedment of the copolymer. The technique developed by them was found to be highly sensitive, simple, and selective towards detection of rifampicin. The limit of detection of their fabricated device was found to be 0.06×10^{-6} mol/L [89].

In another paper drafted by Yazid et al., they demonstrated the use of Sago starch carbon nanodots (C-dots) as an optical probe for detecting the presence of Sn(II) ions. They were synthesized by carbonization followed by surface oxidation of the starch nanoparticles. The fluorescence emitted by the C-dots was significantly quenched in the presence of Sn(II) ions. The fabricated sensor was found to be highly selective, with a limit of detection of 0.36 μM [90].

Starch granules were also utilized for a wide range of photoluminescent-based sensing applications. In a work reported by Zhang et al., they fabricated boron-doped carbon spheres (B-CSs) via a solvothermal route using maize starch as a source of carbon and boric acid as the dopant. This fabricated sensor was then utilized for PL-based sensing of Cr(VI) ions and pH in an aqueous solution. The sensor was found to be highly selective and sensitive, with a limit of detection of ~0.34 μmol L^{-1} [91].

Sign et al. reported the use of water-soluble starch-capped CdSe quantum dots functionalized with thiourea (QDs) for PL-based sensing of certain heavy metals such as Cu^{2+}, Cr^{6+}, and Hg^{2+}. Encapsulation with starch and functionalization with thiourea increased their PL intensity which was quenched in the presence of Cu^{2+}, Cr^{6+}, and Hg^{2+} metal ions. The limits of detection of the fabricated device were found to be 27.6, 196.0, and 279.2 μM, respectively [92].

Xie et al. utilized cationic-etherified starch as the raw material for the synthesis of nitrogen-doped graphene quantum dots (N-GQDs). The synthesized GQDs exhibited excellent blue-green photoluminescence, with high fluorescence stability, along with water solubility. These PL probes could detect the presence of clenbuterol in a wide range of samples with high selectivity and recoveries. The nanoprobe worked on the phenomenon of quenching between the analyte and the N-GQDs. The fabricated probe was found to be linear with an LOD of 2.083×10^{-13} M [93].

Another work, reported by Vantasin et al., demonstrated use of starch-stabilized gold nanoparticles for the colorimetric estimation of amount of protein in milk samples. The phenomenon of sensing is based on an aggregation-based mechanism, which brings about a change in color of the nanoparticle solution, estimating the amount of protein in milk. Here, the starch acts as a stabilizing agent, inhibiting the aggregation of nanoparticles, but in the presence of milk protein, the nanoparticles become destabilized, and a red shift in the absorbance plot is observed. This nano-sensing technique could detect protein content in milk ranging from 2.93×10^{-1} to 2.93×10^{-3} mg/mL by displaying visual colorimetric change [94].

Naghdi et al. developed a PS-based thin film modified with betacyanin, which helped to increase the hydrophobicity and water vapor barrier capacity for the detection of ammonia and pH sensing in a wide category of samples. The film displayed a change in color in response to a pH change in pH from 2 to 13. The film could detect the presence of ammonia between 0.1 and 0.01 mg/mL in water. They used this film in the packaging of fish samples, where the label changed its color, suggesting an increase in total volatile nitrogen content [95].

Biological Sensing

Due to the unique texture and binding properties of starch, it acts as an excellent coating material used for the fabrication of biological sensors. Biological sensors are mostly used in a diverse range of applications, and starch-based biological sensors have gained attention due to their clean, green nature, that could be utilized as an agent to enhance the quality of sensing in a variety of biosensing applications.

In line with this, colorimetric-based biological sensors using starch as the sensing element have gained a lot of importance in today's research. Supporting this, a group of researchers, Suea-Ngam et al., fabricated a sandwich-shaped colorimetric biosensor that utilizes starch-iodide for the targeted detection of the mycotoxin ochratoxin A (OTA). The fabricated device was based on an antibody-aptamer capture mechanism to perform the detection. The limit of detection of the fabricated device was found to be 20 pg/mL. The sandwich-shaped sensor displayed excellent sensing ability even in complex matrices such as human serum, artificial urine, and beer, along with higher cost-effectiveness and eco-friendliness. The sensing scheme developed by them was highly robust, accurate, selective, and sensitive [96].

In a work reported by Chin et al., they used starch for fabrication of an electrochemical immunosensor that could quantitively sense the presence of the Japanese encephalitis virus (JEV). The fabricated sensor was developed by a screen-printed setup which was then modified with carbon nanoparticles procured directly from sago starch nanoparticles. It was prepared by the dropwise addition of a starch solution to ethanol, followed by the dehydration of the starch nanoparticles in the presence of sulfuric acid to convert them to carbon nanoparticles. The addition of starch increased the current intensity and electron transfer kinetics by 63%. The fabricated device was found to be cost-effective compared to the traditionally available devices [97, 98].

In line with biological sensing, another fascinating work was reported by Zhang et al., where they used nano sensing probes based on colorable zeolitic imidazolate framework termed as ZIF for the detection of various biomolecules. For the fabrication of the framework, they encapsulated starches such as amylopectin, dextrin, or amylose in ZIF-90 frameworks assembled with polyethylenimine (PEI) and aptamers of microRNA/DNA, followed by the addition of chromogenic reagents such as I_2/KI or $FeCl_3$. The aptamers separate from the surface of the framework after reacting with the targeted microRNA/DNA, and when the chromogenic reagents are added to this framework, different colorimetric changes can be observed. The biosensor displayed excellent linearity between (1–180 nM) concentration of microRNA with a low limit of detection, which can be tuned by altering the amount of PEI and aptamers that are absorbed on the surface of framework nanoparticles [99].

An antibody-assisted biological sensor utilizing starch as a substrate or coating was developed by Chang et al. They developed a magnetic force-assisted electrochemical aptamer-antibody sandwich assay called MESA for the detection of the protein thrombin in various serum samples. Electrochemical detection of the protein thrombin is performed by determining the cathodic currents of toluidine blue O (TBO), which is attached to a magnetic nanoparticle (MNP) whose surface is modified with a thrombin antibody. For the fabrication of this sensor, an aptamer was bound to a functionalized conducting polymer layer (poly-(2,2′:5′,5″-terthiophene-3′-p-benzoic acid) (pTBA)), and streptavidin and starch-coated MNP was conjugated with biotinylated thrombin antibodies (Ab) and TBO as the bioconjugate (MNP@Ab-TBO). This sensing setup enabled the amount of thrombin even in tiny droplets of samples. The device was handheld, simple, and cost effective with an LOD of 0.49 ($\pm$ 0.06) nM [100].

Chen et al. used starch to develop a functional DNA biosensor to detect adenosine, interferon-γ, and Pb^{2+} ion. They used modified super-paramagnetic microbeads with biotinylated DNA to selectively detect the analyte. The process involved capturing a streptavidin-labeled SiO_2 microbead with glucoamylase tags. Glucoamylase catalyzed the hydrolysis of starch polymer, producing glucose molecules. The reaction mixture was then applied to a strip-like μPAD, altering the ink's flow distance based on the target concentration. This method is simple, cost-effective, and suitable for detecting analytes in various samples like human serum, tap water, and pond water without the need for specialized equipment. It is the first report of applying this starch-hydrolysis-based wettability change to design equipment-free functional DNA biosensors. The fabricated device was highly efficient with a LOD of 1.6 μM for adenosine, 0.3 nM for interferon-γ and 0.5 nM for Pb2+ [101].

You et al. conducted a study utilizing the surface-enhanced Raman scattering (SERS) phenomenon. In their research, they coated the surface of gold nanoparticles (AuNPs) by using starch magnetic beads (SMBs), followed by specific surface functionalization with a specific antibody, which enabled successful separation and concentration of the targeted pathogenic bacteria, *Escherichia coli* O157:H7, from the aqueous sample. On the surface of the nanoparticles, the antibody was immobilized using a bifunctional linker protein known as 4× gold-binding peptide-tagged Streptococcal protein G (4GS), which is a Raman reporter, thus giving a strong signal for SERS measurement. Here, the antibody conjugated-AuNP@SMBs acted as a hotspot for SERS-based detection. The amplitude of the SERS signal exhibited an increase with the concentration of target bacteria. Remarkably, single-cell detection was possible with excellent specificity and colloidal stability of the immuno-AuNP@SMBs and SERS tags. The starch-enabled SERS-based detection system demonstrated high sensitivity, offering a promising approach for detecting various micro-organisms [102].

FUTURE PROSPECTS

In the recent era of scientific modernization, research focuses more on eco-friendly and sustainable methods, where starch has emerged as an attractive candidate because of its excellent biodegradability properties. Its cost effectiveness and abundant availability make it even more appealing for use in a wide range of sensing applications. Due to its intriguing flexibility, it can be used in the fabrication of green electronic sensing devices. This property of starch also enables its use in the modification of electrodes fabricated via 3D printing mechanisms. Starch, being a natural, and renewable polymer, can be used as a sustainable alternative for designing easy-to-use biodegradable sensing architectures, thereby minimizing the harmful and toxic impacts on nature caused by present-day sensors.

CONCLUDING REMARKS

The growing pace of research is inclined towards the development of starch for a wide variety of uses. The geographic origin, along with the polymeric chain conformation of starch, plays a crucial role in determining its specific physical properties such as gelatinization, rheology, morphology, and structure. Though some native starch possesses highly qualified functional properties, these can be enhanced by various routes of starch modification. The high chemical reactivity of starch monomers has paved the way for a wide range of chemical reactions of starch monomers. Their use in the modification of materials and composites has been significantly successful in the development of sensors in various fields. The fascinating physical and chemical properties of starch allow for its exploitation in the preparation of thin films and membranes, which substantially enhance sensing performance. Furthermore, several sensing schemes incorporated with starch-based materials show different interaction mechanisms with different modifiers, leading to more sensitive, selective, and reproducible sensors.

ACKNOWLEDGEMENT

Author N.K.D. would like to acknowledge CSIR for the financial support received as a fellowship. Author U.D. would like to acknowledge DST Inspire for the financial assistance received as an INSPIRE fellowship [IF190914]. Author P.B. would like to acknowledge Tezpur University for the financial help received as an institutional fellowship.

REFERENCES

1. E. Ojogbo, E.O. Ogunsona, T.H. Mekonnen, Chemical and physical modifications of starch for renewable polymeric materials. *Mater. Today Sustain.* **7**, 100028 (2020).
2. E. Bertoft, Understanding starch structure: recent progress. *Agronomy* **7**(3), 56 (2017).
3. A. Bahaji, J. Li, Á.M. Sánchez-López, E. Baroja-Fernández, F.J. Muñoz, M. Ovecka, ... J. Pozueta-Romero, Starch biosynthesis, its regulation and biotechnological approaches to improve crop yields. *Biotechnol. Adv.* **32**(1), 87–106 (2014).
4. B.R. Hamaker, Current and future challenges in starch research. *Curr. Opin. Food Sci.* **40**, 46–50 (2021).
5. Q. Chen, H. Yu, L. Wang, Z. ul Abdin, Y. Chen, J. Wang, ... X. Chen, Recent progress in chemical modification of starch and its applications. *RSC Adv.* **5**(83), 67459–67474 (2015).
6. E. Ogunsona, E. Ojogbo, T. Mekonnen, Advanced material applications of starch and its derivatives. *Eur. Polym. J.* **108**, 570–581 (2018).

7. R.K. Mishra, S.K. Ha, K. Verma, S.K. Tiwari, Recent progress in selected bio-nanomaterials and their engineering applications: an overview. *J. Sci-Adv. Mater. Dev.* **3**(3), 263–288 (2018).
8. S.M. Chisenga, T.S. Workneh, G. Bultosa, B.A. Alimi, Progress in research and applications of cassava flour and starch: a review. *J. Food Sci. Technol.* **56**, 2799–2813 (2019).
9. Y.I. Cornejo-Ramírez, O. Martínez-Cruz, C.L. Del Toro-Sánchez, F.J. Wong-Corral, J. Borboa-Flores, F.J. Cinco-Moroyoqui, The structural characteristics of starches and their functional properties. *CYTA J. Food.* **16**(1), 1003–1017 (2018).
10. N.W. Cheetham, L. Tao, Variation in crystalline type with amylose content in maize starch granules: an X-ray powder diffraction study. *Carbohydr. Polym.* **36**(4), 277–284 (1998).
11. F. Haq, H. Yu, L.I. Wang, L. Teng, M. Haroon, R.U. Khan, ... A. Nazir, Advances in chemical modifications of starches and their applications. *Carbohydr. Res.* **476**, 12–35 (2019).
12. U. Shah, F. Naqash, A. Gani, F.A. Masoodi, Art and science behind modified starch edible films and coatings: a review. *Compr. Rev. Food Sci. Food Saf.* **15**(3), 568–580 (2016).
13. X. Tang, S. Alavi, Recent advances in starch, polyvinyl alcohol based polymer blends, nanocomposites and their biodegradability. *Carbohydr. Polym.* **85**(1), 7–16 (2011).
14. S.M. Asharuddin, N. Othman, W.A.H. Altowayti, N.A. Bakar, A. Hassan, Recent advancement in starch modification and its application as water treatment agent. *Environ. Technol. Innov.* **23**, 101637 (2021).
15. R.V. Gadhave, A. Das, P.A. Mahanwar, P.T. Gadekar, Starch based bio-plastics: the future of sustainable packaging. *Open J. Polym. Chem.* **8**, 21–33 (2018).
16. K. Bashir, M. Aggarwal, Physicochemical, structural and functional properties of native and irradiated starch: a review. *J. Food Sci. Technol.* **56**, 513–523 (2019).
17. D.J. Gallant, B. Bouchet, A. Buleon, S. Perez, Physical characteristics of starch granules and susceptibility to enzymatic degradation. *Eur. J. Clin. Nutr.* **46**(2), 3–16 (1992).
18. O.T. Fatokun, Micrometrics and Morphological Properties of Starch. *Chem. Prop. Starch* **143** (2019), doi: 10.5772/intechopen.90286.
19. A. Apriyanto, J. Compart, J. Fettke, A review of starch, a unique biopolymer – structure, metabolism and in planta modifications. *Plant Sci.* **318**, 111223 (2022).
20. H.O. Egharevba, Chemical properties of starch and its application in the food industry. *Chem. Prop. Starch* **9**, 1–17 (2019).
21. N. Singh, N. Isono, S. Srichuwong, T. Noda, K. Nishinari, Structural, thermal and viscoelastic properties of potato starches. *Food Hydrocoll.* **22**(6), 979–988 (2008).
22. C. Takeda, Y. Takeda, S. Hizukuri, Structure of amylomaize amylose. *Cereal Chem.* **66**(1), 22–25 (1989).
23. R.F. Tester, J. Karkalas, The effects of environmental conditions on the structural features and physicochemical properties of starches. *Starch-Stärke* **53**(10), 513–519 (2001).
24. D. French, Chemical and physical properties of starch. *J. Anim. Sci.* **37**(4), 1048–1061 (1973).
25. A.L. Santana, M.A.A. Meireles, New starches are the trend for industry applications: a review. *Public Health Nutr.* **4**(5), 229–241 (2014).
26. J.J.M.S. Jane, Starch properties, modifications, and applications. *J. Macromol. Sci. A* **32**(4), 751–757 (1995).
27. F. Eltaboni, A. Alabidi, Physical and chemical modifications of starches, in *Proceeding of 2nd Libya Conference on Chemistry and its Applications* (LCCA-2), 2017, pp. 9–11. https://www.researchgate.net/publication/319876735_Physical_and_Chemical_Modifications_of_Starches
28. H.J. Chung, Q. Liu, Effect of gamma irradiation on molecular structure and physicochemical properties of corn starch. *J. Food Sci.* 74(5), C353–C361 (2009).
29. Q. Li, Q. Xie, S.J. Yu, Q.Y. Gao, New approach to study starch gelatinization applying a combination of hot-stage light microscopy and differential scanning calorimetry. *J. Agric. Food Chem.* **61**, 1212–1218 (2013).
30. C.G. Biliaderis, C.M. Page, T.J. Maurice, B.O. Juliano, Thermal characterization of rice starches – a polymeric approach to phase transitions of granular starch. *J. Agric. Food Chem.* **34**, 6–14 (1986).
31. A. Gonera, P. Cornillon, Gelatinization of starch/gum/sugar systems studied by using DSC, NMR, and CSLM. *Starch-Stärke* **54**, 508–516 (2002).
32. P.J. Jenkins, R.E. Cameron, A.M. Donald, W. Bras, et al., In-situ simultaneous small and wide-angle x-ray-scattering – a new technique to study starch gelatinization. *J. Polym. Sci. B Polym. Phys.* **32**, 1579–1583 (1994).
33. P. Rubens, K. Heremans, Pressure-temperature gelatinization phase diagram of starch: an in situ fourier transform infrared study. *Biopolymers* **54**, 524–530 (2000).

34. Y. Ai, J. L. Jane, Gelatinization and rheological properties of starch. *Starch-Stärke* **67**(3–4), 213–224 (2015).
35. J.W. Donovan, Phase-transitions of the starch-water system. *Biopolymers* **18**, 263–275 (1979).
36. J. Jane, Y.Y. Chen, L.F. Lee, A.E. McPherson, et al., Effects of amylopectin branch chain length and amylose content on the gelatinization and pasting properties of starch. *Cereal Chem.* **76**, 629–637 (1999).
37. A. Beleia, R.A. Miller, R.C. Hoseney, Starch gelatinization in sugar solutions. *Starch-Stärke* **48**, 259–262 (1996).
38. K. Kohyama, K. Nishinari, Effect of soluble sugars on gelatinization and retrogradation of sweet-potato starch. *J. Agric. Food Chem.* **39**, 1406–1410 (1991).
39. P.A. Perry, A.M. Donald, The role of plasticization in starch granule assembly. *Biomacromolecules* **1**, 424–432 (2000).
40. Y.J. Wang, L.F. Wang, Physicochemical properties of common and waxy corn starches oxidized by different levels of sodium hypochlorite. *Carbohydr. Polym.* **52**, 207–217 (2003).
41. K.S. Woo, P.A. Seib, Cross-linked resistant starch: preparation and properties. *Cereal Chem.* **79**, 819–825 (2002).
42. H. Nawaz, R. Waheed, M. Nawaz, D. Shahwar, Physical and chemical modifications in starch structure and reactivity. *Chem. Prop. Starch* **9**, 13–35 (2020).
43. G. Tabilo-Munizaga, G.V. Barbosa-Canovas, Rheology for the food industry. *J. Food Eng.* **67**, 147–156 (2005).
44. N. Singh, J. Singh, L. Kaur, N.S. Sodhi, B.S. Gill, Morphological, thermal and rheological properties of starches from different botanical sources. *Food Chem.* **81**(2), 219–231 (2003).
45. T. Yoneya, K. Ishibashi, K. Hironaka, K. Yamamoto, Influence of cross-linked potato starch treated with POCl3 on DSC, rheological properties and granule size. *Carbohydr. Polym.* **53**, 447–457 (2003).
46. L. Kaur, J. Singh, N. Singh, Effect of cross-linking on some properties of potato (Solanum tuberosum L.) starches. *J. Sci. Food Agric.* **86**, 1945–1954 (2006).
47. D.S. Suh, J.L. Jane, Comparison of starch pasting properties at various cooking conditions using the micro visco-amylo-graph and the rapid visco analyser. *Cereal Chem.* **80**, 745–749 (2003).
48. H.J. Liu, L. Ramsden, H. Corke, Physical properties and enzymatic digestibility of phosphorylated ae, wx, and normal maize starch prepared at different pH levels. *Cereal Chem.* **76**, 938–943 (1999).
49. Z.A.N. Fadzlina, A.A. Karim, T.T. Teng, Physicochemical properties of carboxymethylated sago (Metroxylon sagu) starch. *J. Food Sci.* **70**, C560–C567 (2005).
50. L.M. Che, D. Li, L.J. Wang, N. Ozkan, et al., Rheological properties of dilute aqueous solutions of cassava starch. *Carbohydr. Polym.* **74**, 385–389 (2008).
51. S. Sahin, S.G. Sumnu, in *Rheological properties of foods*, ed. by S. Sahin, S.G. Sumnu (Springer, New York, 2006), pp. 39–105.
52. Q.D. Nguyen, C.T.B. Jensen, P.G. Kristensen, Experimental and modelling studies of the flow properties of maize and waxy maize starch pastes. *Chem. Eng. J.* **70**, 165–171 (1998).
53. K. Svegmark, A.M. Hermansson, Microstructure and rheological properties of composites of potato starch granules and amylose: a comparison of observed and predicted structures. *Food Struct.* **12**(2), 6 (1993).
54. R. Sheldrake, *The Biogenesis of Starch Granules in Higher Plants* (1969). Appleton-Century-Crofts. https://books.google.co.in/books/about/The_Biogenesis_of_Starch_Granules_in_Hig.html?id=fdw5AAAAMAAJ&redir_esc=y
55. R. Hoover, Composition, molecular structure, and physico-chemical properties of tuber and root starches: a review. *Carbohydr. Polym.* **45**, 253–267 (2001).
56. B.R. Baum, L.G. Bailey, A survey of endosperm starch granules in the genus Hordeum: a study using image analytic and numerical taxonomic techniques. *Can. J. Plant Sci.* **65**(8), 1563–1569 (1987).
57. J.H. Li, T. Vasanthan, B. Rossnagel, R. Hoover, Starch from hull-less barley: I. Granule morphology, composition and amylopectin structure. *Food Chem.* **74**(4), 395–405 (2001).
58. P. Baldwin, Studies on the surface chemistry, minor component composition and structure of granular starches, University of Nottingham, Nottingham, UK, 1995.
59. C.M. Franco, S.J. do Rio Preto, C.F. Ciacco, Factors that affect the enzymatic degradation of natural starch granules-effect of the size of the granules. *Starch-Stärke* **44**(11), 422–426 (1992).
60. R. Hoover, Composition, molecular structure, and physicochemical properties of tuber and root starches: a review. *Carbohydr. Polym.* **45**(3), 253–267 (2001).
61. S. Hizukuri, Polymodal distribution of the chain lengths of amylopectins, and its significance. *Carbohydr. Res.* **147**, 342–347 (1986).

62. P. J. White, Properties of corn starch, in *Specialty corns* Second edition (2000), pp. 29–54. CRC Press DOI:https://doi.org/10.1201/9781420038569
63. J. Bao, Z. Ao, J.L. Jane, Characterization of physical properties of flour and starch obtained from gamma-irradiated white rice. *Starch-Stärke* **57**(10), 480–487 (2005).
64. A.O. Ashogbon, E.T. Akintayo, Recent trend in the physical and chemical modification of starches from different botanical sources: a review. *Starch-Stärke* **66**(1–2), 41–57 (2014).
65. J.N. BeMiller, K.C. Huber, Physical modification of food starch functionalities. *Annu. Rev. Food Sci. Technol.* **6**, 19–69 (2015).
66. A.J. Palacios-Fonseca, J. Castro-Rosas, C.A. Gómez-Aldapa, T. Tovar-Benítez, B. M. Millán-Malo, A. del Real, et al., Effect of the alkaline and acid treatments on the physicochemical properties of corn starch. *CYTA J. Food.* **11**(1), 67–74 (2013).
67. F.A. Carey, *Organic chemistry*, 5th edn. (McGraw-Hill, New York, 2003), pp. 638–684.
68. Đ. Ačkar, J. Babić, A. Jozinović, B. Miličević, S. Jokić, R. Miličević, ... D. Šubarić, Starch modification by organic acids and their derivatives: a review. *Molecules* **20**(10), 19554–19570 (2015).
69. K. Neelam, S. Vijay, S. Lalit, Various techniques for the modification of starch and the applications of its derivatives. *Int. Res. J. Pharm.* **3**(5), 25–31 (2012).
70. V. Karma, A.D. Gupta, D.K. Yadav, A.A. Singh, M. Verma, H. Singh, H. Recent developments in starch modification by organic acids: a review. *Starch-Stärke* **74**(9–10), 2200025 (2022).
71. J. Singh, L. Kaur, O.J. McCarthy, Factors influencing the physico-chemical, morphological, thermal and rheological properties of some chemically modified starches for food applications—a review. *Food Hydrocoll.* **21**(1), 1–22 (2007).
72. S.A. Korma, S.N. Kamal-Alahmad, A.F. Ammar, F. Zaaboul, T. Zhang, Chemically modified starch and utilization in food stuffs. *Int. J. Food Sci. Nutr.* **5**(4), 264 (2016).
73. T. Kant, K. Shrivas, K. Dewangan, A. Kumar, N.K. Jaiswal, M. K. Deb, S. Pervez, Design and development of conductive nanomaterials for electrochemical sensors: a modern approach. *Mater. Today Chem.* **24**, 100769 (2022).
74. F.C. Vicentini, A.E. Ravanini, L.C. Figueiredo-Filho, J. Iniesta, C.E. Banks, O. Fatibello-Filho, Imparting improvements in electrochemical sensors: evaluation of different carbon blacks that give rise to significant improvement in the performance of electroanalytical sensing platforms. *Electrochim. Acta* **157**, 125–133 (2015).
75. Q. Lin, X. Peng, Z. Zhang, Electrochemical determination of Hg (II) ions based on biosynthesized spherical activated carbon from potato starch. *Int. J. Electrochem. Sci.* **12**(3), 2232–2241 (2017).
76. N.A. Zambianco, T.A. Silva, H. Zanin, O. Fatibello-Filho, B.C. Janegitz, Novel electrochemical sensor based on nanodiamonds and manioc starch for detection of diquat in environmental samples. *Diam. Relat. Mater.* **98**, 107512 (2019).
77. R. Thakur, P. Pristijono, C.J. Scarlett, M. Bowyer, S.P. Singh, Q.V. Vuong, Starch-based films: major factors affecting their properties. *Int. J. Biol. Macromol.* **132**, 1079–1089 (2019).
78. D. Lourdin, G. Della Valle, P. Colonna, Influence of amylose content on starch films and foams. *Carbohydr. Polym.* **27**(4), 261–270 (1995).
79. J. Isaad, A. El Achari, F. Malek, Bio-polymer starch thin film sensors for low concentration detection of cyanide anions in water. *Dyes Pigm.* **97**(1), 134–140 (2013).
80. V. Gautam, K.P. Singh, V.L. Yadav, Polyaniline/MWCNTs/starch modified carbon paste electrode for non-enzymatic detection of cholesterol: application to real sample (cow milk). *Anal. Bioanal. Chem.* **410**, 2173–2181 (2018).
81. P.P. Peregrino, M.R. Cavallari, F.J. Fonseca, S.G. Moreira, M.J.A. Sales, L.G. Paterno, Starch-mediated immobilization, photochemical reduction, and gas sensitivity of graphene oxide films. *ACS Omega* **5**(10), 5001–5012 (2020).
82. G.C. Mauruto de Oliveira, E. P. de Palma, M.H. Kunita, R. Antigo Medeiros, R. de Matos, K.R. Francisco, B.C. Janegitz, Tapioca biofilm containing nitrogen-doped titanium dioxide nanoparticles for electrochemical detection of 17- β estradiol. *Electroanalysis* **29**(11), 2638–2645 (2017).
83. J. Ramu, M. Mahanthappa, S. Yellappa, K.B. Chandrasekhar, γ-Fe_2O_3 nanoparticles modified glassy carbon electrode for the sensitive detection of folic acid. *Mater. Res. Express* **6**(10), 105070 (2019).
84. A. Basu, A. Suryawanshi, B. Kumawat, A. Dandia, D. Guin, S.B. Ogale, Starch (Tapioca) to carbon dots: an efficient green approach to an on–off–on photoluminescence probe for fluoride ion sensing. *Analyst* **140**(6), 1837–1841 (2015).
85. P. Vasileva, B. Donkova, I. Karadjova, C. Dushkin, Synthesis of starch-stabilized silver nanoparticles and their application as a surface plasmon resonance-based sensor of hydrogen peroxide. *Colloids Surf. A: Physicochem. Eng* **382**(1–3), 203–210 (2011).

86. L.J. Miao, J.W. Xin, Z.Y. Shen, Y.J. Zhang, H.Y. Wang, A.G. Wu, Exploring a new rapid colorimetric detection method of Cu^{2+} with high sensitivity and selectivity. *Sens. Actuators B: Chem.* **176**, 906–912 (2013).
87. C. Boonkanon, K. Phatthanawiwat, W. Wongniramaikul, A. Choodum, Curcumin nanoparticle doped starch thin film as a green colorimetric sensor for detection of boron. *Spectrochim. Acta A Mol. Biomol. Spectrosc.* **224**, 117351 (2020).
88. R. Qiang, S. Yang, K. Hou, J. Wang, Synthesis of carbon quantum dots with green luminescence from potato starch. *New J. Chem.* **43**(27), 10826–10833 (2019).
89. Z. Hooshyar, G.R. Bardajee, Fluorescence enhancement of glutathione capped CdTe/ZnS quantum dots by embedding into cationic starch for sensitive detection of rifampicin. *Spectrochim. Acta A Mol. Biomol. Spectrosc.* **173**, 144–150 (2017).
90. S.N.A. Mohd Yazid, S.F. Chin, S. C. Pang, S.M. Ng, Detection of Sn (II) ions via quenching of the fluorescence of carbon nanodots. *Microchim. Acta* **180**, 137–143 (2013).
91. W. Zhang, L. Hao, L. Shang, D.F. Chai, Y. Gao, J. Li, ... G. Dong, Maize starch derived boron doped carbon spheres via facile solvothermal route as the photoluminescence sensor for determination of pH and Cr (VI). *Nanotechnology* **33**(27), 275707 (2022).
92. A. Singh, A. Guleria, S. Neogy, M.C. Rath, UV induced synthesis of starch capped CdSe quantum dots: functionalization with thiourea and application in sensing heavy metals ions in aqueous solution. *Arab. J. Chem.* **13**(1), 3149–3158 (2020).
93. H. Xie, C. Chen, J. Lie, R. You, W. Qian, S. Lin, Y. Lu, Sensitive and selective detection of clenbuterol in meat samples by a graphene quantum dot fluorescent probe based on cationic-etherified starch. *Nanomaterials* **12**(4), 691 (2022).
94. S. Vantasin, P. Pienpinijtham, K. Wongravee, C. Thammacharoen, S. Ekgasit, Naked eye colorimetric quantification of protein content in milk using starch-stabilized gold nanoparticles. *Sens. Actuators B: Chem.* **177**, 131–137 (2013).
95. S, Naghdi, M. Rezaei, M. Abdollahi, A starch-based pH-sensing and ammonia detector film containing betacyanin of paperflower for application in intelligent packaging of fish. *Int. J. Biol. Macromol.* **191**, 161–170 (2021).
96. A. Suea-Ngam, L.T. Deck, P.D. Howes, A.J. deMello, An ultrasensitive non-noble metal colorimetric assay using starch-iodide complexation for Ochratoxin A detection. *Anal. Chim. Acta* **1135**, 29–37 (2020).
97. S.F. Chin, L.S. Lim, S.C. Pang, M.S.H. Sum, D. Perera, Carbon nanoparticle modified screen printed carbon electrode as a disposable electrochemical immunosensor strip for the detection of Japanese encephalitis virus. *Microchim. Acta* **184**, 491–497 (2017).
98. F.C. Vicentini, L.R. Silva, J.S. Stefano, A.R. Lima, J. Prakash, J.A. Bonacin, B.C. Janegitz, Starch-based electrochemical sensors and biosensors: a review. *Biomed Mater.* **1**, 1–20 (2022).
99. J. Zhang, X.Y. Wang, Y.H. Wang, D.D. Wang, Z. Song, C.D. Zhang, H.S. Wang, Colorable zeolitic imidazolate frameworks for colorimetric detection of biomolecules. *Anal. Chem.* **92**(18), 12670–12677 (2020).
100. S. Chung, J.M. Moon, H. Choi, Y.B. Hwang, Y.B. Shim, Magnetic force assisted electrochemical sensor for the detection of thrombin with aptamer-antibody sandwich formation. *Biosens. Bioelectron.* **117**, 480–486 (2018).
101. Y. Chen, L. Zhang, J. Huang, Z. Deng, Y. Yuan, J. Zou, ... Y. Zhang, Enhanced functional DNA biosensor for distance-based read-by-eye quantification of various analytes based on starch-hydrolysis-adjusted wettability change in paper devices. *RSC Advances* **10**(47), 28121–28127 (2020).
102. S.M. You, K. Luo, J.Y. Jung, K.B. Jeong, E.S. Lee, M.H. Oh, Y.R. Kim, Gold nanoparticle-coated starch magnetic beads for the separation, concentration, and SERS-based detection of E. coli O157: H7. *ACS Appl. Mater. Interfaces* **12**(16), 18292–18300 (2020).

Index

K

L

M

N

O

P

Q

R

S

T

U

V

W

X

For Product Safety Concerns and Information please contact our
EU representative GPSR@taylorandfrancis.com Taylor & Francis
Verlag GmbH, Kaufingerstraße 24, 80331 München, Germany